Introduction to

Number Theory

DISCRETE MATHEMATICS AND ITS APPLICATIONS

Series Editor
Kenneth H. Rosen, Ph.D.

Juergen Bierbrauer, Introduction to Coding Theory

Kun-Mao Chao and Bang Ye Wu, Spanning Trees and Optimization Problems

Charalambos A. Charalambides, Enumerative Combinatorics

Henri Cohen, Gerhard Frey, et al., Handbook of Elliptic and Hyperelliptic Curve Cryptography

Charles J. Colbourn and Jeffrey H. Dinitz, Handbook of Combinatorial Designs, Second Edition

Martin Erickson and Anthony Vazzana, Introduction to Number Theory

Steven Furino, Ying Miao, and Jianxing Yin, Frames and Resolvable Designs: Uses, Constructions, and Existence

Randy Goldberg and Lance Riek, A Practical Handbook of Speech Coders

Jacob E. Goodman and Joseph O'Rourke, Handbook of Discrete and Computational Geometry, Second Edition

Jonathan L. Gross and Jay Yellen, Graph Theory and Its Applications, Second Edition

Jonathan L. Gross and Jay Yellen, Handbook of Graph Theory

Darrel R. Hankerson, Greg A. Harris, and Peter D. Johnson, Introduction to Information Theory and Data Compression, Second Edition

Daryl D. Harms, Miroslav Kraetzl, Charles J. Colbourn, and John S. Devitt, Network Reliability: Experiments with a Symbolic Algebra Environment

Leslie Hogben, Handbook of Linear Algebra

Derek F. Holt with Bettina Eick and Eamonn A. O'Brien, Handbook of Computational Group Theory

David M. Jackson and Terry I. Visentin, An Atlas of Smaller Maps in Orientable and Nonorientable Surfaces

Richard E. Klima, Neil P. Sigmon, and Ernest L. Stitzinger, Applications of Abstract Algebra with Maple™ and MATLAB®, Second Edition

Patrick Knupp and Kambiz Salari, Verification of Computer Codes in Computational Science and Engineering

William Kocay and Donald L. Kreher, Graphs, Algorithms, and Optimization

Donald L. Kreher and Douglas R. Stinson, Combinatorial Algorithms: Generation Enumeration and Search

Continued Titles

Charles C. Lindner and Christopher A. Rodgers, Design Theory

Hang T. Lau, A Java Library of Graph Algorithms and Optimization

Alfred J. Menezes, Paul C. van Oorschot, and Scott A. Vanstone, Handbook of Applied Cryptography

Richard A. Mollin, Algebraic Number Theory

Richard A. Mollin, Codes: The Guide to Secrecy from Ancient to Modern Times

Richard A. Mollin, Fundamental Number Theory with Applications

Richard A. Mollin, An Introduction to Cryptography, Second Edition

Richard A. Mollin, Quadratics

Richard A. Mollin, RSA and Public-Key Cryptography

Carlos J. Moreno and Samuel S. Wagstaff, Jr., Sums of Squares of Integers

Dingyi Pei, Authentication Codes and Combinatorial Designs

Kenneth H. Rosen, Handbook of Discrete and Combinatorial Mathematics

Douglas R. Shier and K.T. Wallenius, Applied Mathematical Modeling: A Multidisciplinary Approach

Jörn Steuding, Diophantine Analysis

Douglas R. Stinson, Cryptography: Theory and Practice, Third Edition

Roberto Togneri and Christopher J. deSilva, Fundamentals of Information Theory and Coding Design

W. D. Wallis, Introduction to Combinatorial Designs, Second Edition

Lawrence C. Washington, Elliptic Curves: Number Theory and Cryptography

Jonathan L. Gross, Combinatorial Methods with Computer Applications

Francine Blanchet-Sadri, Algorithmic Combinatorics on Partial Words

DISCRETE MATHEMATICS AND ITS APPLICATIONS
Series Editor KENNETH H. ROSEN

INTRODUCTION TO NUMBER THEORY

MARTIN ERICKSON
ANTHONY VAZZANA

Chapman & Hall/CRC is an imprint of the
Taylor & Francis Group, an **informa** business

Maple is a trademark of Waterloo Maple Inc.

Chapman & Hall/CRC
Taylor & Francis Group
6000 Broken Sound Parkway NW, Suite 300
Boca Raton, FL 33487-2742

Printed in the United States of America on acid-free paper
10 9 8 7 6 5 4 3 2 1

International Standard Book Number-13: 978-1-58488-937-3 (Hardcover)

Library of Congress Cataloging-in-Publication Data

Erickson, Martin J., 1963-
Introduction to number theory / Martin Erickson, Anthony Vazzana.
p. cm.
ISBN 978-1-58488-937-3 (hardback : alk. paper)
1. Number theory--History. I. Vazzana, Anthony. II. Title.

QA241.E775 2006
512.7--dc22 2007038689

Visit the Taylor & Francis Web site at
http://www.taylorandfrancis.com

and the CRC Press Web site at
http://www.crcpress.com

For Christine and Dana

Contents

Preface

This book is an introduction to the main concepts of number theory at the undergraduate level. We hope that our treatment of number theory reads almost like a story, with each new topic leading naturally to the next. We begin with the ancient Euclidean algorithm for finding the greatest common divisor of two integers, and we end with some modern developments, including the theory of elliptic curves and the negative solution of Hilbert's Tenth Problem concerning Diophantine equations. Along the way, we cover a diverse array of topics that should appeal to students and instructors, as well as casual readers simply wishing to learn about the mathematics of the natural numbers.

Our presentation of number theory features investigations, worked examples, and exercises for the reader to solve. We use the computer software systems *Mathematica*® and *Maple*™ to make large-number calculations. Readers with access to *Mathematica* or *Maple* will find it instructive to experiment with the given commands and see the mathematics in action. Readers without this software can follow the discussion, since all the calculations are included in the text. In the appendices, we give brief tutorials on *Mathematica* and *Maple* and supply references for pertinent Internet resources, including supplemental web pages containing *Mathematica* notebooks and *Maple* worksheets. The supplemental material may be found at

`www2.truman.edu/~erickson/introduction_to_number_theory/`.

Number theory is a vital and useful branch of mathematics. We make every attempt to show connections between number theory and other branches of mathematics, including algebra, analysis, and combinatorics. We also demonstrate applications of number theory to real life problems. For example, congruences are used to explain the ISBN system; modular arithmetic and Euler's theorem are employed to produce RSA encryption; and quadratic residues are utilized to construct round-robin tournaments. An entire chapter is devoted to cryptography.

We have found that the organization of the book allows for flexibility in teaching a number theory course. We recommend that instructors cover Chapters 1, 2, 3, and 5 (which serve as a foundation), plus a selection of other chapters of their choice. It should be noted that Chapter 12 (Logic and Number Theory) depends on Chapter 9 (Diophantine Equations), Section 9.7 (Continued fraction solution of Pell's equation) depends on Chapter 8 (Continued Fractions), and Section 11.6 (Encryption via elliptic curves) depends on Chapter 4 (Cryptography). In terms of rigor and prerequisites, the text is upper-level undergraduate, meaning that some previous experience with proof-based mathematics is assumed. The last three chapters demand a greater level of mathematical maturity, where some exposure to abstract algebra and analysis would be helpful (although we supply the relevant background information). The dependencies are shown in the following graph.

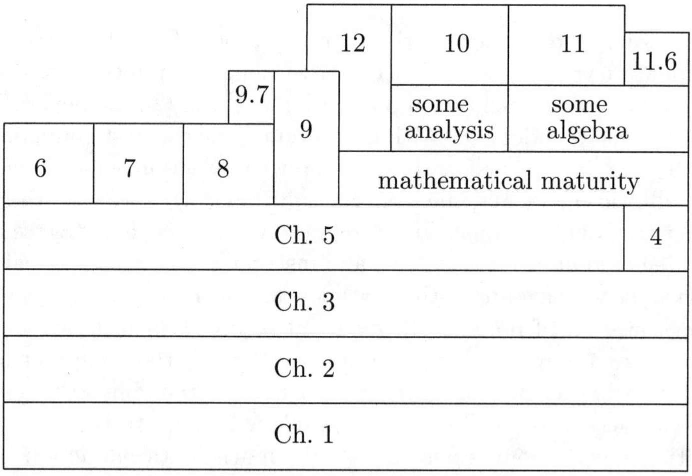

12 | 10 | 11 | 11.6
9.7 | 9 | some analysis | some algebra
6 | 7 | 8 | mathematical maturity
Ch. 5 | 4
Ch. 3
Ch. 2
Ch. 1

The exercise sets encompass a wide variety of problems, including many that relate number theory to other areas of mathematics or other fields (e.g., music). The problems range in difficulty from very easy and just-like-the-examples to quite challenging. Exercises designated with a star ($\star$) are particularly difficult or require advanced mathematical background; exercises designated with a diamond ($\diamond$) require the use of a calculator or computer; exercises designated with a dagger (†) are of special theoretical importance.

We thank those colleagues who provided suggestions about our work: Benjamin Braun, University of Kentucky; Robert Dobrow, Carleton College; Suren Fernando, Truman State University; Joe Flowers, St. Mary's University (San Antonio); David Garth, Truman State University; Joe Hemmeter, Ancor Corporation; Daniel R. Jordan, Columbia College of Chicago; Ken Price, University of Wisconsin–Oshkosh; H. Chad Lane, USC Institute for Creative Technologies; Chad Meiners, Michigan State University; Khang Tran, University of Illinois. We also thank the people at Wolfram Research and MapleSoft for providing the mathematical software that makes possible the computational exploration of number theory showcased here. Finally, we thank the people at CRC Press for their support during the realization of our project.

Part I

Core Topics

Chapter 1

Introduction

> Die ganzen Zahlen hat der liebe Gott gemacht, alles andere ist Menschenwerk.
>
> [The good Lord made the whole numbers; all else is the work of man.]
>
> LEOPOLD KRONECKER (1823–1891)

1.1 What is number theory?

The natural numbers (i.e., the positive integers) are the counting numbers

$$1, 2, 3, 4, 5, 6, 7, \ldots.$$

These numbers are one of the oldest, most universal concepts of mathematics. Number theory is the study of properties of the natural numbers.

One of the central issues of number theory is that of factorization and in particular prime numbers. A *prime number* is a natural number greater than 1 that is not a product of two smaller natural numbers. Thus, the prime numbers are

$$2, 3, 5, 7, 11, 13, 17, \ldots.$$

We will show that every positive integer greater than 1 can be (uniquely) written as the product of prime numbers. Therefore, understanding prime numbers is crucial.

A particularly appealing aspect of number theory is that one can start with a simple concept and quickly come upon deep, difficult-to-solve problems. Another attractive feature is that many interesting patterns are revealed through example calculations that are easy to carry out.

We illustrate these two points with a few questions about prime numbers. First, how many prime numbers are there? Over two thousand years ago, Euclid provided a simple, elegant proof that there are infinitely many. (We will give this proof in Section 2.6.)

Let's delve a little deeper. Apart from the number 2, all primes are odd. Consequently, any prime greater than 2 can be written in the form $4k+1$ or $4k+3$, for some integer k. For example, $13 = 4 \cdot 3 + 1$ and $19 = 4 \cdot 4 + 3$. One can easily work out representations for the first few primes, as shown below.

Prime	Representation
3	$4 \cdot 0 + 3$
5	$4 \cdot 1 + 1$
7	$4 \cdot 1 + 3$
11	$4 \cdot 2 + 3$
13	$4 \cdot 3 + 1$
17	$4 \cdot 4 + 1$
19	$4 \cdot 4 + 3$
23	$4 \cdot 5 + 3$
29	$4 \cdot 7 + 1$
31	$4 \cdot 7 + 3$

We see that four of the first ten odd primes are of the form $4k+1$ while the remaining six are of the form $4k+3$. With the aid of a computer one can easily make similar calculations for a much larger sample. (We will see how to make such computer calculations using the mathematical software systems *Mathematica* and *Maple*; try these calculations as you read the text!) The table below indicates how the first n odd primes are divided between the two sets.

n	Primes of the form $4k+1$	Primes of the form $4k+3$
10	4	6
100	47	53
1000	495	505
10000	4984	5016
100000	49950	50050

By modifying Euclid's proof one can show without substantial effort that there are an infinite number of primes of the form $4k+3$ (see Proposition 2.38). Strangely, it is not as easy to show that there are an infinite number of primes of the form $4k+1$. However, with the introduction of some mathematical machinery, we will be able to prove that there are an infinite number of such primes. Our data above suggest that there is more to the issue than the infinitude of both sets. For each value of n, approximately half of the primes are in each set. Moreover, the larger n is in our table, the closer the percentage of each type is to 50%. Developing even heavier machinery (which is beyond the scope of this book), one can show that this pattern continues. That is, the percentage of the first n primes of the form $4k+1$ approaches 50% as n grows larger.

One can ask similar questions about the number of primes of the form $ak+b$, for fixed integers a and b. Again, with a good deal of effort one can give a satisfactory description of what goes on. If we modify things a bit in a different direction, the problem becomes decidedly much more difficult. Consider the following question: Are there an infinite number of primes of the form $k^2 + 1$? For example, 5 is such a prime, as $5 = 2^2 + 1$. On the surface, this question doesn't seem much more difficult than the ones above, but at the present time, no one can provide an answer.

In addition to the issue of prime numbers, another central issue of number theory that we will visit repeatedly is that of solving Diophantine equations. A Diophantine equation is a polynomial equation in one or several variables with integer coefficients, for which we are interested only in integer solutions. The name is given in honor of the Greek mathematician Diophantus whose book *Arithmetica* contains a collection of problems of this type. Consider, as an example, the Diophantine equation

$$2x - 5y = 1. \tag{1.1}$$

This equation has solutions, for instance, $x = 3$, $y = 1$. On the other hand, the Diophantine equation

$$2x - 4y = 1 \tag{1.2}$$

has no integer solutions, because all integers of the form $2x - 4y$ are even and therefore cannot equal 1. Right away we see that Diophantine equations that look similar may behave dramatically differently. In fact, given a Diophantine equation, it is often a difficult problem to determine whether the equation has integer solutions. In general, given an equation that has solutions, we would like to know how many solutions there are and, if possible, describe the set of solutions completely. For *linear* Diophantine equations, such as (1.1) and (1.2) above, one can do just that. We will see that the equation (1.1) in fact has an infinite number of solutions, and we will show how to generate all solutions.

Another interesting Diophantine equation has a familiar look to it. The equation

$$x^2 + y^2 = z^2 \tag{1.3}$$

gives the relationship for the sides of a right triangle according to the famous Pythagorean theorem. One solution to this Diophantine equation is $x = 3$, $y = 4$, $z = 5$. With a little experimentation, it is easy to find several more solutions. With additional effort, we will completely describe the (infinite) set of all solutions to this Diophantine equation (see Section 9.2). Now let's modify the equation (1.3) slightly to

$$x^3 + y^3 = z^3. \tag{1.4}$$

Certainly, we can find solutions to this equation by taking $x = 0$ and setting $y = z$. Solutions of this type (where one of the variables is 0) are in some sense *trivial*, and so we ask are there any *nontrivial* solutions? Experimentation by hand (or computer) turns up nothing. It isn't immediately clear why the equation (1.3) has nontrivial solutions while the equation (1.4) would not, but this does prove to be the case. In fact, for any integer n greater than 2, the Diophantine equation

$$x^n + y^n = z^n$$

has no nontrivial solutions. This statement was first given by Pierre de Fermat in the 17th century. Much of Fermat's work comes to us in the form of marginal notes he made in his copy of Diophantus' book. These were published after his lifetime. The statement above, now known as Fermat's Last Theorem, was one such marginal note. Along with his observation that these equations have no nontrivial solutions, he added the following tantalizing statement. "I have discovered a truly marvelous proof of this which however the margin is not large enough to contain." No proof of the theorem was ever found among Fermat's papers except for the special case $n = 4$. The history of the effort to prove Fermat's Last Theorem is a colorful one. In addition, many important mathematical ideas were developed along the way. Finally, in 1995, the proof of the theorem was completed with work done by Andrew Wiles.

Number theory is a beautiful subject in its own right and needs no applications to justify its study. However, applications ranging from the simple to the sophisticated do exist and are used in the world all around us. Any sort of identification number one might encounter (e.g., ISBN codes on books, UPC symbols on merchandise, and ABA routing numbers on checks) is likely to have some number theory built into it. These applications help to detect and in some cases correct errors in transmission of such numbers. On a deeper level, using an easy-to-perform procedure, one can encrypt data without knowing how to decode it. As a result, a customer can encode a credit card number for safe passage to an online merchant over the Internet. At the same time, third parties are prevented from stealing credit card numbers from customers, because the method of decoding is known only to the merchant. This application demonstrates again the recurring theme we have developed above: elementary ideas live side by side among the complex in the world of number theory.

1.2 The natural numbers

The origins of number theory are found in simple observations about the natural numbers. We denote the set of natural numbers by $\mathbf{N}$; that is,

$$\mathbf{N} = \{1, 2, 3, 4, 5, 6, 7, \ldots\}.$$

Given a pair of natural numbers, we may compute their sum and product and obtain a natural number as a result. The operations of addition (+)

and multiplication ($\cdot$) conform to a list of familiar properties (associativity, commutativity, distributivity, etc.). These properties can all be proved from Peano's Axioms (see Chapter Notes).

***PROPOSITION* (Properties of the natural numbers)**
Assume that a, b, and c are arbitrary natural numbers.

- *Closure: $a + b$ and $a \cdot b$ are natural numbers.*
- *Commutative laws: $a + b = b + a$ and $a \cdot b = b \cdot a$.*
- *Associative laws: $(a + b) + c = a + (b + c)$ and $(a \cdot b) \cdot c = a \cdot (b \cdot c)$.*
- *Distributive law: $a \cdot (b + c) = a \cdot b + a \cdot c$.*
- *Identity element: $a \cdot 1 = a$.*
- *Cancelation laws: If $a + c = b + c$, then $a = b$. If $a \cdot c = b \cdot c$, then $a = b$.*

We typically write $a \cdot b$ as ab, and aa as a^2, aaa as a^3, etc.

Example 1.1 We will show that if a and b are natural numbers, then

$$(a + b)^2 = a^2 + 2ab + b^2.$$

Using properties of natural numbers, we have

$$\begin{aligned}
(a + b)^2 &= (a + b)(a + b) \\
&= (a + b)a + (a + b)b && \text{(distributive law)} \\
&= a(a + b) + b(a + b) && \text{(commutative law)} \\
&= a^2 + ab + ba + b^2 && \text{(distributive law)} \\
&= a^2 + ab + ab + b^2 && \text{(commutative law)} \\
&= a^2 + 2ab + b^2.
\end{aligned}$$

□

It is useful to endow the natural numbers with an order relation (which we call '$<$' or 'less than'). Properties of this relation can be proved from the axioms.

***DEFINITION* (Order relation)**
Let a and b be natural numbers. Then $a < b$ if there exists a natural number c such that $a + c = b$.

Example 1.2 We will show that if a, b, and c are natural numbers, with $a < b$ and $b < c$, then $a < c$. Assume that $a < b$ and $b < c$. Then there exist natural numbers x and y such that $a + x = b$ and $b + y = c$. Hence, $(a + x) + y = c$, which by the associative law we may write as $a + (x + y) = c$. Since x and y are natural numbers, $x + y$ is a natural number. Therefore, $a < c$. □

An especially important property concerning the order relation is known as the trichotomy law.

PROPOSITION (Trichotomy law)
If a and b are natural numbers, then exactly one of the following statements is true: $a = b$, $a < b$, or $b < a$.

In the course of our study of natural numbers, we will frequently work with negative integers and 0. We denote the set of all integers by **Z**; that is,

$$\mathbf{Z} = \{\ldots, -3, -2, -1, 0, 1, 2, 3, \ldots\}.$$

(The '**Z**' stands for 'Zahlen,' the German word for number.) Unless stated otherwise, the word "number" will mean either natural number or integer.

We will also have occasion to broaden our view to consider other sorts of numbers. A real number is *rational* if it can be expressed as a quotient of two integers. For instance, the number $2/3$ is a rational number. Every integer n is a rational number since we can write it in the form $n/1$. An *irrational* number is a real number that is not rational (i.e., a real number that cannot be expressed as a quotient of two integers). A real (or complex) number is *algebraic* if it is the root of some polynomial with integer coefficients. Since $\sqrt{2}$ is a root of $x^2 - 2$, it is an algebraic number. A rational number a/b is a root of the polynomial $bx - a$, and so all rational numbers are algebraic. A real (or complex) number that is not algebraic is said to be *transcendental*. The number π is transcendental, but it is difficult to prove this fact.

Exercises

1.1 Prove that if a and b are natural numbers, then

$$(a + b)^3 = a^3 + 3a^2b + 3ab^2 + b^3.$$

1.2 Let a, b, c be natural numbers.

(a) Prove that $a < b$ if and only if $a + c < b + c$.

(b) Prove that $a < b$ if and only if $ac < bc$.

1.3 Prove that if a and b are natural numbers, and $ab = 1$, then $a = 1$ and $b = 1$.

1.4 Prove that if a and b are natural numbers, then $a < b$ if and only if $a^2 < b^2$.

†**1.5** Of the properties of natural numbers given in this section, which are true for all integers (all elements of $\mathbf{Z}$)? What are some additional properties of the set $\mathbf{Z}$?

1.6 Show that if $ab = 0$, where $a, b \in \mathbf{Z}$, then $a = 0$ or $b = 0$.

1.7 How do we define the $<$ ('less than') relation on $\mathbf{Z}$?

1.3 Mathematical induction

We note a further important property of the set of natural numbers.

Principle of Mathematical Induction. Let S be a subset of $\mathbf{N}$ satisfying the following.

1. The number 1 is an element of S.

2. If $n \in S$ then $n + 1 \in S$.

Then $S = \mathbf{N}$.

The principle of mathematical induction is an axiom, rather than a fact to be proved. (See Chapter Notes.) The principle is equivalent to several other statements about the natural numbers, some of which we discuss in this section.

Example 1.3 Let S be the set of natural numbers n satisfying

$$1 + 2 + \cdots + n = \frac{n(n+1)}{2}.$$

We claim that $S = \mathbf{N}$. We see that S does indeed contain 1 since

$$1 = \frac{1(1+1)}{2}.$$

Now suppose that S contains a number n. We must check that S contains

$n + 1$. We have

$$\begin{aligned} 1 + 2 + \cdots + n + (n+1) &= \frac{n(n+1)}{2} + (n+1) \qquad \text{(since } n \in S\text{)} \\ &= \frac{n(n+1) + 2(n+1)}{2} \\ &= \frac{(n+1)(n+2)}{2} \\ &= \frac{(n+1)((n+1)+1)}{2}. \end{aligned}$$

Thus, we have shown that $n+1$ is also in S. By the principle of mathematical induction, it follows that $S = \mathbf{N}$. □

We refer to the $n = 1$ case of an induction proof as the *base case.* We call the assumption that n is a member of S the *inductive hypothesis* and the subsequent proof that $n + 1$ is also an element of S the *inductive step.* If it makes our notation less cumbersome, we may choose for our inductive hypothesis to assume that S contains $n-1$ and deduce that it also contains n.

In practice, we often use the principle of mathematical induction to prove that some formula or relation holds for all natural numbers (as in our example). In such situations, we typically dispense with the formality of introducing the set S (of natural numbers for which the result holds). For example, to prove that $1+2+\cdots+n = n(n+1)/2$ for all natural numbers n, we would first check that the formula is valid for $n = 1$ (the base case), assume that it holds for a particular integer n (the inductive hypothesis), and then deduce that the result holds when we replace n by $n + 1$ (the inductive step).

We mention a couple of variations of the principle of induction. First, it may be the case that we are attempting to prove a statement that holds only for natural numbers greater than or equal to some fixed number n_0. In this situation, we prove that the result holds for n_0 as our base case. Then we assume it to be true for n and deduce that it is also valid for $n + 1$. It follows that the statement is true for all $n \geq n_0$.

Example 1.4 Consider the inequality

$$2^n < n!.$$

(For $n \geq 1$, "n factorial" is defined as the product of the first n natural numbers: $n! = 1 \cdot 2 \cdot \cdots \cdot n$. In addition, $0! = 1$.)

The inequality does not hold for $n = 1$, 2, or 3, but in the $n = 4$ case we have $2^4 = 16 < 24 = 4!$. Assume that the inequality holds for some value of n

and, in particular, for a value of n greater than or equal to 4. Then

$$\begin{aligned} 2^{n+1} &= 2 \cdot 2^n \\ &< (n+1)2^n && \text{(since } n \geq 4) \\ &< (n+1)n! && \text{(by the inductive hypothesis)} \\ &= (n+1)!. \end{aligned}$$

Thus, the inequality $2^n < n!$ holds for all $n \geq 4$. □

We will also occasionally use a stronger version of the principle of mathematical induction. This version is equivalent to the (weak) principle of mathematical induction, in that each implies the other.

Strong Mathematical Induction. Let S be a subset of $\mathbf{N}$ satisfying the following.

1. The number 1 is an element of S.

2. If $1, 2, \ldots, n \in S$ then $n + 1 \in S$.

Then $S = \mathbf{N}$.

As with the (weak) principle of mathematical induction, we can start with any positive integer as our base case, and prove that the result holds for all integers greater than or equal to this starting number.

Example 1.5 The Fibonacci numbers f_n are defined recursively as follows:

$$\begin{aligned} f_0 &= 0, \quad f_1 = 1, \\ f_n &= f_{n-1} + f_{n-2}, \quad n \geq 2. \end{aligned}$$

Thus, the Fibonacci numbers are

$$0, 1, 1, 2, 3, 5, 8, 13, 21, 34, 59, 89, 144, \ldots.$$

The Fibonacci numbers were introduced by Leonardo of Pisa (also known as Fibonacci) (1180s–1250) in his book *Liber Abaci* (*Book of Calculation*).

We will prove that $f_n > (3/2)^n$, for $n \geq 11$. The inequality is true for $n = 11$, as $f_{11} = 89 > (3/2)^{11} \doteq 86.5$, and for $n = 12$, as $f_{12} = 144 > (3/2)^{12} \doteq 129.7$. Assume that the inequality holds for all k such that $11 \leq k \leq n$, where $n \geq 12$. We will show that the inequality holds for $n + 1$. Since the result holds for n and $n - 1$, we have $f_n > (3/2)^n$ and $f_{n-1} > (3/2)^{n-1}$. Using the recurrence relation for the Fibonacci numbers, we obtain

$$\begin{aligned} f_{n+1} &= f_n + f_{n-1} \\ &> (3/2)^n + (3/2)^{n-1} \\ &= (3/2)^{n-1}(3/2+1) \\ &= (3/2)^{n-1}(5/2) \\ &> (3/2)^{n-1}(3/2)^2 \\ &= (3/2)^{n+1}. \end{aligned}$$

We have shown that the inequality holds for $n+1$. Therefore, by the principle of strong mathematical induction, it holds for all $n \geq 11$. □

Another equivalent statement about the natural numbers, which we will use from time to time, is the well ordering principle.

Well Ordering Principle. Every nonempty set of natural numbers contains a least element.

Example 1.6 We will show that $n < 2^n$, for all $n \geq 1$. We could use induction but will instead use the well ordering principle. Let A be the set of natural numbers for which the inequality *does not* hold. We want to show that A is empty. If A is nonempty, then, by the well ordering principle, A contains a least element m. Clearly, $m \neq 1$ because $1 < 2^1$. So $m \geq 2$ and $m \geq 2^m$. It follows that $2^{m-1} = 2^m/2 \leq m/2 \leq m - 1$. But this means that $m - 1 \in A$, contradicting the assumption that m is the least element of A. Hence, A is empty, and the inequality $n < 2^n$ holds for all $n \geq 1$. □

Exercises

1.8 Use the principle of mathematical induction to prove that

$$1 + 3 + 5 + \cdots + (2n - 1) = n^2, \text{ for } n \geq 1.$$

1.9 Use the principle of mathematical induction to prove that

$$1^2 + 2^2 + 3^2 + \cdots + n^2 = \frac{n(n+1)(2n+1)}{6}, \text{ for } n \geq 1.$$

1.10 Prove that

$$\sum_{k=0}^{n} 2^k = 2^{n+1} - 1, \text{ for } n \geq 0.$$

1.11 Use mathematical induction to prove that $n < 2^n$, for $n \geq 0$.

1.12 Prove Bernoulli's inequality: for all real numbers $x > -1$ and all positive integers n,

$$(1+x)^n \geq 1+nx.$$

1.13 Let $a_1 = 1$, $a_2 = 5$, and $a_n = a_{n-1} + 2a_{n-2}$, for $n \geq 3$. Prove that $a_n = 2^n + (-1)^n$, for $n \geq 1$.

1.14 Prove that the Fibonacci numbers f_n satisfy the inequality

$$f_n < (5/3)^n, \quad n \geq 0.$$

†1.15 Prove that the Fibonacci numbers f_n satisfy Cassini's identity:

$$f_n^2 - f_{n-1}f_{n+1} = (-1)^{n+1}, \quad n \geq 1.$$

†1.16 For $0 \leq k \leq n$, the "binomial coefficient" $\binom{n}{k}$ is defined as $n!/(k!(n-k)!)$. Prove the following:

(a) $\binom{n}{k} = \binom{n-1}{k} + \binom{n-1}{k-1}$, for $1 \leq k \leq n$;
(b) (The Binomial Theorem)

$$(a+b)^n = \sum_{k=0}^{n} \binom{n}{k} a^k b^{n-k}, \text{ for } n \geq 0.$$

†1.17 Prove the following formula for the sum of a geometric series:

$$a + ar + ar^2 + \cdots + ar^n = a(r^{n+1} - 1)/(r-1).$$

1.18 (a) The "floor" of x, denoted $\lfloor x \rfloor$, is the greatest integer less than or equal to x. Define $a_0 = 1$ and $a_n = \lfloor 5a_{n-1}/2 \rfloor$, for $n \geq 1$. Given any $r < 5/2$, prove that $a_n > r^n$, for all sufficiently large n.
(b) The "ceiling" of x, denoted $\lceil x \rceil$, is the least integer greater than or equal to x. Define $b_0 = 1$ and $b_n = \lceil 5b_{n-1}/2 \rceil$, for $n \geq 1$. Given any $r > 5/2$, prove that $b_n < r^n$, for all sufficiently large n.

1.19 Show that the set $A = \{(5n^5 - 100n^4 + 7n^3 + 2n^2 - 10n + 4)^2 : n \in \mathbf{N}\}$ contains a least element, which is a nonnegative integer.

⋆1.20 For each integer $k \geq 0$, define

$$S_k(n) = \sum_{i=1}^{n} i^k.$$

Prove that $S_k(n)$ is a polynomial in n of degree $k+1$ and leading coefficient $1/(k+1)$.

⋆1.21 Prove that every positive rational number is the sum of a finite number of distinct terms from the sequence $1, 1/2, 1/3, 1/4, \ldots$.

Numbers of the form $1/q$ are called *Egyptian fractions*. The ancient Egyptians used them to represent all fractions (except $2/3$, for which they had a special hieroglyph).

1.4 Notes

The Peano Axioms

In 1889 Giuseppe Peano (1858–1932) gave a list of axioms for the natural numbers in his book *Arithmetices principia, nova methodo exposita* (*The principles of arithmetic, presented by a new method*). These axioms may be stated informally as follows.

Peano Axioms. The set of natural numbers $\mathbf{N}$ has the following properties:

- 1 is a natural number.
- Every natural number n has a successor, denoted by $s(n)$.
- No natural number has 1 as its successor.
- Distinct natural numbers have distinct successors: $s(m) = s(n)$ only if $m = n$.
- If a subset S of $\mathbf{N}$ contains 1 and contains the successor of every element that it contains, then $S = \mathbf{N}$. (This axiom is the principle of mathematical induction.)

The properties of the natural numbers given in Section 1.2 can be proved from Peano's Axioms.

In practice, we define $2 = s(1)$, $3 = s(2)$, $4 = s(3)$, etc.

Chapter 2

Divisibility and Primes

> A number is a part of a number, the less of the greater, when it measures the greater.
>
> EUCLID, *The Elements* (c. 300 B.C.E.)

In this chapter we introduce two fundamental concepts of number theory: divisibility and primality. We also investigate various other mathematical topics, including greatest common divisor, Euclidean algorithm, and Fibonacci numbers. In the course of our explorations, we use *Mathematica* and *Maple* to perform numerical calculations that illustrate the theory. Some of the calculations would be difficult or impossible to perform by hand.

2.1 Basic definitions and properties

Notice that 6 and 7 are very different kinds of numbers. The number 6 "factors" as $2 \cdot 3$, but 7 does not "factor" into smaller positive integers. This observation leads to the notions of divisibility and primality.

DEFINITION 2.1 *A nonzero integer a* divides *an integer b if $b = ak$, for some integer k. If a divides b, then a is a* divisor *of b and b is a* multiple *of a.*

If a divides b, we write $a \mid b$. If a does not divide b, we write $a \nmid b$.

Example 2.2 Since $6 = 2 \cdot 3$, we have

$$2 \mid 6.$$

On the other hand,

$$2 \nmid 7.$$

□

Note: Let n be a positive integer. A "proper divisor" of n is a positive divisor of n that is less than n. A "proper multiple" of n is a multiple of n that is greater than n.

DEFINITION 2.3 *An integer that is divisible by* 2 *is* even. *An integer that is not divisible by* 2 *is* odd.

Example 2.4 The integers 2, 10, 0, and -400 are even. The integers 1, 13, -1, and -137 are odd. □

DEFINITION 2.5 *A* prime number *is an integer* $n > 1$ *that has no positive divisors other than* 1 *and* n. *A* composite number *is an integer greater than* 1 *that is not a prime number.*

Example 2.6 The number 17 is prime, as it has no positive divisors other than 1 and 17. The number 16 is composite, because 16 has positive divisors other than 1 and 16, namely, 2, 4, and 8. □

We state and prove a few simple results on divisibility.

PROPOSITION 2.7 (Properties of divisibility)
Let a, b, and c be integers.

(i) If $a \mid b$, then $a \mid bc$.

(ii) If $a \mid b$ and $b \mid a$, then $a = \pm b$.

(iii) If $c \neq 0$, then $a \mid b$ if and only if $ac \mid bc$.

(iv) If $a \mid b$ and $b \mid c$, then $a \mid c$.

(v) If $a \mid b$ and $a \mid c$, then $a \mid (bx + cy)$, for all integers x and y.

PROOF (i) Suppose that $a \mid b$. By Definition 2.1, there exists an integer k such that $b = ak$. It follows that $bc = a(kc)$, and so a divides bc.

(ii) Suppose that $a \mid b$ and $b \mid a$. As a consequence of Definition 2.1, there exist integers k and l with $b = ak$ and $a = bl$. Hence $a = akl$. Since, by definition, $a \neq 0$, we have $kl = 1$. Therefore $k = 1$ and $l = 1$, in which case $a = b$, or $k = -1$ and $l = -1$, in which case $a = -b$.

We leave the proofs of statements (iii)–(v) as exercises. □

Example 2.8 Prove that $n^5 - n$ is divisible by 5, for all positive integers n. We will give a proof by mathematical induction on n. The result is true for the case $n = 1$, since $5 \mid 1^5 - 1$. Assume that the result holds for n; that is, $5 \mid n^5 - n$. We must show that the result also holds for $n + 1$. We find that

$$\begin{aligned}(n+1)^5 - (n+1) &= n^5 + 5n^4 + 10n^3 + 10n^2 + 5n + 1 - n - 1 \\ &= (n^5 - n) + 5(n^4 + 2n^3 + 2n^2 + n).\end{aligned}$$

By assumption, $n^5 - n$ is divisible by 5. Clearly, $5(n^4+2n^3+2n^2+n)$ is divisible by 5. Hence, by Proposition 2.7 (v), the quantity $(n+1)^5 - (n+1)$ is divisible by 5. Therefore, by the principle of mathematical induction, $5 \mid n^5 - n$, for all positive integers n. ∎

Exercises

2.1 Find all proper divisors of the following numbers:

$$99, 100, 101, 102.$$

2.2 List ten prime numbers and ten composite numbers.

2.3 Determine (by hand) which of the following numbers are primes:

$$79, 121, 739, 1999, 2001, 10201.$$

2.4 Show that if $a \mid b$ and $c \mid d$, then $ac \mid bd$.

2.5 Show that $n^3 - n$ is divisible by 3 for every positive integer n.

2.6 Show that $n^7 - n$ is divisible by 7 for every positive integer n.

†**2.7** Prove Proposition 2.7 (iii)–(v).

2.2 The division algorithm

Whether or not a positive integer b divides an integer a, we can divide b into a and obtain a quotient and a remainder.

THEOREM 2.9 (Division algorithm)
For any integers a and b, with $b > 0$, there exist unique integers q and r such that

$$a = bq + r \quad \text{and } 0 \le r < b. \tag{2.1}$$

We call q the "quotient" and r the "remainder."

Note: The case $r = 0$ is when $b \mid a$. A variety of generalizations of the division algorithm have applications in many areas of mathematics, for example, in the theory of partial fractions used in integral calculus.

PROOF First we demonstrate existence. Let

$$S = \{a - bk \colon k \in \mathbf{Z} \text{ and } a - kb \geq 0\}.$$

We first note that S is nonempty as a is an element of S if $a \geq 0$, and $a - b(a) = (-a)(-1 + b)$ is an element of S if $a < 0$. If S contains 0, then we have an integer q such that $a = bq$. In this case we take $r = 0$ and we have the desired relation. If S does not contain 0, then we can apply the well ordering principle. Let r be the smallest positive integer in S and choose q so that $r = a - bq$. If $r \geq b$, then

$$r - b = a - bq - b = a - (q + 1)b,$$

which is a number in S and a smaller value than r. This contradicts our choice of r. Hence, in either case our choice of r satisfies $0 \leq r < b$.

Now we verify uniqueness. Suppose that

$$a = bq_1 + r_1 = bq_2 + r_2,$$

where $0 \leq r_1 < b$ and $0 \leq r_2 < b$. Without loss of generality, assume that $r_2 \geq r_1$. Then

$$b(q_1 - q_2) = r_2 - r_1 \geq 0. \tag{2.2}$$

Since $r_2 < b$, we know that $r_2 - r_1 < b$, and so we have

$$0 \leq b(q_1 - q_2) < b.$$

It follows that $q_1 - q_2 = 0$, which leads us to the conclusion that $q_1 = q_2$ and $r_1 = r_2$. ∎

From the definition of even numbers, we may characterize even numbers as those that can be expressed in the form $2n$. As a consequence of the division algorithm, we may characterize odd numbers as those that can be expressed in the form $2n+1$. It follows that the product of two even numbers is even, the product of two odd numbers is odd, and the product of an even number and an odd number is even. Let's represent arbitrary even numbers as $2a$ and $2b$, and arbitrary odd numbers as $2c + 1$ and $2d + 1$. We find that $(2a)(2b) = 2(2ab)$, $(2c + 1)(2d + 1) = 2(2cd + c + d) + 1$, and $(2a)(2c + 1) = 2(2ac + a)$.

Example 2.10 The number $\sqrt{2}$ is irrational. Proof by contradiction: Suppose that $\sqrt{2}$ is rational, say, equal to a/b, with a and b not both even. (If a and b were both even, we could replace them with $a/2$ and $b/2$; and if *these* numbers were both even, we could use $a/4$ and $b/4$, and so on.) Then $2b^2 = a^2$, and hence a^2 is even. By the characterization of odd numbers given above, we see that a must be even too. (Odd times odd is odd and even times even is even.) Write $a = 2c$. It follows that $b^2 = 2c^2$, and hence b^2 is even, and also b is even. Thus, we have a contradiction to the assumption that a and b are not both even. Therefore, $\sqrt{2}$ is irrational. □

The *Mathematica* command `Mod[a,b]` calculates the remainder when a is divided by b. For example, the following computation shows that when 20 is divided by 7, the remainder is 6.

Mathematica

```
Mod[20,7]

6
```

Note: The terminology "mod" is used to indicate the remainder when one number is divided by another. For example, the above calculation is written $20 \bmod 7 = 6$.

We can also use the `Mod` command to determine the quotient that results when one number is divided by another.

Mathematica

```
Mod[23312, 232]

112

(23312 - 112)/232

100
```

This tells us that 23312 divided by 232 gives a quotient of 100 (and a remainder of 112). We check: $232 \cdot 100 + 112 = 23312$.

The corresponding *Maple* command for the remainder is `modp(a,b)`.

Maple

```
modp(20,7);

6

modp(23312, 232);

112

(22312-112)/232;

100
```

Exercises

◇**2.8** Use a computer to find the quotient and remainder when 171316247 is divided by 631.

◇**2.9** Use a computer to find the smallest natural number n such that 11^n divided by 10^n gives a quotient of 2. What is the remainder in this case?

2.10 Prove that the sum of two even integers is even, the sum of two odd integers is even, and the sum of an even integer and an odd integer is odd.

2.11 Let a and b be integers.

(a) Show that if a is not divisible by 3, then $a = 3k + 1$ or $a = 3k + 2$, for some integer k.

(b) Show that if ab is divisible by 3, then either a or b is divisible by 3.

2.12 Prove that $\sqrt{3}$ is an irrational number.

2.13 Show that there are an infinite number of positive integers n such that $n^4 - n$ is *not* divisible by 4.

†**2.14** (Base-b representation) Let b be an integer greater than 1. Prove that every positive integer n can be written uniquely in the form

$$n = d_k b^k + d_{k-1} b^{k-1} + \cdots + d_1 b + d_0,$$

where $k \geq 0$, each d_j is an integer (called a base-b "digit") such that $0 \leq d_j \leq b - 1$, and $d_k \neq 0$.

2.15 Show that in the base-b representation of n in the previous exercise, the number of base-b digits is $\lfloor \log_b n \rfloor + 1$.

$\star$**2.16** Let c be an integer greater than 1. Prove that there are infinitely many positive integers b such that, for infinitely many positive integers n, division of c^n by b leaves a remainder of 1.

$\star$**2.17** Let k and m be positive integers such that $0 \leq m \leq 2^k - 1$. Prove that the binomial coefficient $\binom{2^k-1}{m}$ is odd.

2.3 Greatest common divisor

DEFINITION 2.11 *The* greatest common divisor *(*gcd*) of two integers (not both* 0*) is the largest integer that divides them both.*

We write $\gcd(a, b)$ for the gcd of the integers a and b. By definition, $\gcd(a, 0) = |a|$, for any integer $a \neq 0$. Since $\gcd(a, b) = \gcd(|a|, |b|)$, we may as well assume that a and b are nonnegative numbers when calculating the gcd.

Example 2.12 We have $\gcd(12, 18) = 6$, since 6 is the largest number that divides both 12 and 18. $\square$

Note: We define $\gcd(a_1, a_2, \ldots, a_n)$, where not all the a_i are 0, to be the largest integer that divides all the a_i. For instance, $\gcd(42, 105, 70) = 7$.

DEFINITION 2.13 *Two or more integers are* relatively prime *(also called* coprime*) if their greatest common divisor is* 1*. The numbers in a list are* pairwise relatively prime *if each pair of integers in the list is relatively prime.*

Example 2.14 The numbers 9 and 10 are relatively prime. The numbers 6, 10, and 15 are relatively prime but not pairwise relatively prime. $\square$

PROPOSITION 2.15
If a and b are integers (not both 0*) and $g = \gcd(a, b)$, then $\gcd(a/g, b/g) = 1$.*

PROOF Let $h = \gcd(a/g, b/g)$. There exist integers k and l such that $a/g = hk$ and $b/g = hl$. Multiplying through by g in both equations, we obtain

$$\begin{aligned} a &= (gh)k \\ b &= (gh)l. \end{aligned}$$

Thus, we see that gh is a common divisor of a and b. If h is any number other than 1, we have a contradiction of the assumption that $g = \gcd(a, b)$. □

A "linear combination" of a and b is an expression of the form $ax + by$, where x and y are integers. As a result of Proposition 2.7 (v), any common divisor of a and b is also a divisor of any linear combination of a and b. We now show that the greatest common divisor of a and b is the smallest positive linear combination of the two numbers.

THEOREM 2.16 (GCD as a linear combination)
Let a and b be integers (not both 0). Then $\gcd(a, b)$ is the smallest positive value of $ax + by$, where x and y are integers. That is, $\gcd(a, b)$ is the smallest positive linear combination of a and b.

Note: In particular, if a and b are relatively prime, then there exist integers x and y such that $ax + by = 1$.

PROOF By taking $x = a$ and $y = b$, we see that positive linear combinations of a and b do exist. By the well ordering principle, there exists a smallest positive linear combination of a and b. Let g be this value, and write $g = ax_0 + by_0$. (The pair (x_0, y_0) need not be unique.) By the division algorithm (Theorem 2.9),

$$a = gq + r = (ax_0 + by_0)q + r,$$

where $0 \leq r < g$. By writing $r = a(1 - qx_0) + b(-qy_0)$, we see that r is a linear combination of a and b. Therefore, r must not be positive (since $r < g$). Hence, $r = 0$ and $g \mid a$. By a similar argument, $g \mid b$. We have shown that g is a common divisor of a and b.

We now show that g is the *greatest* common divisor of a and b. Suppose that d is any common divisor of a and b. Then by Proposition 2.7 (v), $d \mid (ax_0 + by_0) = g$. Hence, $d \leq g$ and so g is the greatest common divisor of a and b. □

Example 2.17 With $a = 5$ and $b = 7$, we have $\gcd(5, 7) = 1$, and

$$5 \cdot 3 + 7 \cdot (-2) = 1.$$

□

We are typically introduced to prime numbers and factorization early on in school, and in this early exposure there is an implicit assumption that every integer can be written as a product of prime numbers in an essentially unique way. This fact has a nontrivial proof (which we will give in Theorem 2.34). However, given that most of us are in possession of this fact at an early age, it plays a large role in guiding our intuition about primes and factorization. The following corollary is an example of a statement that seems obvious in light of our knowledge of unique factorization. However, its proof is, somewhat surprisingly, an application of Theorem 2.16, and in fact the corollary is a key step in the proof of unique factorization.

COROLLARY 2.18 (Euclid's lemma)
If p is a prime and $p \mid ab$, then $p \mid a$ or $p \mid b$. More generally, if p divides $a_1 a_2 \ldots a_n$, then $p \mid a_i$ for some i.

Note: Observe that this result is, in general, false if p is not a prime. For example, $4 \mid (6 \cdot 10)$ but 4 divides neither 6 nor 10.

PROOF To begin, suppose that $p \mid ab$, but $p \nmid a$. We must show that $p \mid b$. Since p is a prime, $\gcd(a, p) = 1$. So by Theorem 2.16, there exist integers x and y with $ax + py = 1$. Since we must show that p divides b, we need to somehow introduce b into the situation. We achieve this by multiplying the equation through by b to obtain

$$abx + pby = b.$$

Since p divides the left side of this relation (by Proposition 2.7 (v)), p divides b.

We will use induction to extend the result to products of arbitrary length. The argument given above serves as the base case (with $n = 2$). Now we suppose that the result holds for a product of length n. Suppose that p divides $a_1 a_2 \ldots a_n a_{n+1}$. By grouping the terms of the product together as

$$(a_1 a_2 \ldots a_n)(a_{n+1}),$$

we conclude from our previous work that either p divides $a_1 a_2 \ldots a_n$ or p divides a_{n+1}. If the latter holds, we are done. In the former case, we may apply the inductive hypothesis to conclude that p divides one of the a_i. ∎

COROLLARY 2.19
If $d \mid a$ and $d \mid b$, then $d \mid \gcd(a, b)$.

The proof of this corollary is called for in the exercises.

DEFINITION 2.20 *The* least common multiple *(*lcm*) of a and b (not both 0) is the smallest positive integer that is a multiple of both a and b.*

We write $\text{lcm}(a, b)$ for the lcm of a and b.

Note: We define $\text{lcm}(a_1, a_2, \ldots, a_n)$ to be the smallest positive integer that is a multiple of all the a_i.

Example 2.21 We have $\text{lcm}(15, 50) = 150$, because 150 is the smallest positive integer that is a multiple of both 15 and 50. □

PROPOSITION 2.22
Let a, b, and c be positive integers. Then

(i) $\gcd(ac, bc) = c \cdot \gcd(a, b)$ *and* $\text{lcm}(ac, bc) = c \cdot \text{lcm}(a, b)$;

(ii) $\gcd(a, b) = 1$ *if and only if* $\text{lcm}(a, b) = ab$;

(iii) $\gcd(a, b) \cdot \text{lcm}(a, b) = ab$

PROOF Proofs of statements (i) and (ii) are called for in the exercises. Using these, we will prove statement (iii). Let $g = \gcd(a, b)$. By Proposition 2.15, $\gcd(a/g, b/g) = 1$. Applying (ii), we have

$$\text{lcm}(a/g, b/g) = ab/g^2.$$

Multiplying by g^2 and applying (i), we obtain

$$g \cdot \text{lcm}(g \cdot a/g, g \cdot b/g) = ab.$$

The result now follows. □

The *Mathematica* command for gcd is `GCD`.

Mathematica

```
GCD[81959,73963]

1999
```

The corresponding *Maple* command is `gcd`.

Maple

```
gcd(81959,73963);

1999
```

Although commands for least common multiple exist in *Mathematica* and *Maple*, we can program our own command, called `ourlcm`, using Proposition 2.22 (iii).

Mathematica

```
ourlcm[a_, b_] := a b /GCD[a, b]

ourlcm[18, 24]

72
```

Maple

```
ourlcm := (a,b) -> a*b / gcd(a,b):

ourlcm(18,24);

72
```

Exercises

2.18 Prove the following.

(a) The greatest common divisor of two even numbers is even.

(b) The greatest common divisor of two odd numbers is odd.

(c) The greatest common divisor of an even number and odd number is odd.

2.19 Show that $\gcd(12m+6, 15n+9) > 1$, for all m and n.

2.20 Show that $\gcd(2n+1, 3n+1) = 1$, for all n.

2.21 Prove that if $a \mid bc$ and $\gcd(a, b) = 1$, then $a \mid c$.

2.22 Prove that if $a \mid c$, $b \mid c$, and $\gcd(a, b) = 1$, then $ab \mid c$.

2.23 Prove that if $\gcd(a, b) = 1$, then $\gcd(a, bc) = \gcd(a, c)$.

†**2.24** Prove Corollary 2.19.

2.25 (a) Show that if a and b are relatively prime, then so are a^2 and b^2.

(b) Given that n is a positive integer, prove that $\sqrt{n}$ is a rational number if and only if n is a perfect square.

2.26 Let a, b, and c be nonzero integers. Show that

$$\gcd(a, b, c) = \gcd(\gcd(a, b), c).$$

2.27 (a) Prove that for any integers a_1, a_2, $\ldots$, a_n, not all 0, there exist integers x_1, x_2, $\ldots$, x_n with

$$\gcd(a_1, a_2, \ldots, a_n) = a_1x_1 + a_2x_2 + \cdots + a_nx_n.$$

(b) Show that $\gcd(a_1, a_2, \ldots, a_n)$ is even if and only if a_1, a_2, $\ldots$, a_n are all even.

†**2.28** Prove Proposition 2.22 (i) and (ii).

2.29 Prove that if λ is a multiple of a and b, then $\operatorname{lcm}(a, b) \mid \lambda$.

◇**2.30** Use a computer to find $\gcd(1212121212121212, 2323232323232323)$ and $\operatorname{lcm}(1212121212121212, 2323232323232323)$.

⋆**2.31** Show that there is no time other than at 12:00 when the hour hand, minute hand, and second hand of a clock all coincide. Assume that the hands of the clock move at constant rates.

⋆**2.32** Let $H_n = 1 + 1/2 + \cdots + 1/n$, for $n \geq 1$. Show that H_n is not an integer for any $n > 1$. The numbers H_n are called "harmonic sums."

⋆**2.33** (Paul Erdös and George Szekeres) Prove that every two entries in a row of Pascal's triangle (except the two 1's) have a factor in common.

Hint: Prove and use the "subcommittee identity" for binomial coefficients:

$$\binom{n}{a}\binom{a}{b} = \binom{n}{b}\binom{n-b}{a-b}.$$

This is called the subcommittee identity because it counts the ways a committee and a subcommittee may be chosen from a group of people.

⋆**2.34** (For those who are familiar with groups) Let G be a finite group of order n and k a positive integer relatively prime to n. Show that the map $f\colon G \to G$ defined by $f(x) = x^k$ is a bijection.

2.4 The Euclidean algorithm

Most of us have been taught to find the gcd of two numbers by factoring the numbers and multiplying the common prime power factors together. For example, to find $\gcd(306, 252)$, we factor 306 and 252:

$$306 = 2 \cdot 3^2 \cdot 17, \quad 252 = 2^2 \cdot 3^2 \cdot 7.$$

The common prime power factors are 2 and 3^2. Hence,

$$\gcd(306, 252) = 2 \cdot 3^2 = 18.$$

This procedure always works in theory (as a result of the forthcoming Fundamental Theorem of Arithmetic), but it is only practical when the two given numbers are small or otherwise easily factored. It would be difficult to use the procedure to find the gcd of, say, 73963 and 81959, because these numbers are not easily factored by hand. Factoring numbers with hundreds of digits is often a practical impossibility, even with the aid of a computer.

However, the *Mathematica* command `GCD` and the *Maple* command `gcd` calculate gcd very quickly (even for numbers with hundreds of digits). How do these commands work? They use a highly efficient procedure called the Euclidean algorithm.

The Euclidean algorithm is a procedure for finding $\gcd(a, b)$, where a and b are any given positive integers. Recall that the division algorithm (Theorem 2.9) allows us to write

$$a = bq + r, \tag{2.3}$$

where $0 \leq r < b$. The basis of the Euclidean algorithm is the observation that $\gcd(a, b) = \gcd(b, r)$ in (2.3). This observation reduces the problem of finding $\gcd(a, b)$, where $a \geq b$, to the simpler problem of finding $\gcd(b, r)$. (The problem is simpler because r is less than a.)

LEMMA 2.23
If $a = bq + r$, *then* $\gcd(a, b) = \gcd(b, r)$.

PROOF Suppose that d divides both a and b. Since $r = a - bq$, it follows that d also divides r; hence d divides b and r. Conversely, if d divides both b and r, then d divides a and b. We have shown that the common divisors of a and b are the same as the common divisors of b and r. As the gcd of two numbers is the greatest of their common divisors, $\gcd(a, b) = \gcd(b, r)$. ☐

The Euclidean algorithm is the repeated application of the division algorithm, starting with the numbers a and b and terminating when a remainder of 0 occurs.

Example 2.24 With $a = 306$ and $b = 252$, the Euclidean algorithm produces the following computations:

$$\begin{aligned} 306 &= 252 \cdot 1 + 54 \\ 252 &= 54 \cdot 4 + 36 \\ 54 &= 36 \cdot 1 + 18 \\ 36 &= 18 \cdot 2 + 0. \end{aligned}$$

Applying Lemma 2.23 repeatedly, we find that $\gcd(306, 252) = \gcd(252, 54) = \gcd(54, 36) = \gcd(36, 18) = \gcd(18, 0) = 18$. ∎

In general, the Euclidean algorithm runs as follows:

$$\begin{aligned}
a &= bq_1 + r_1 & 0 &\le r_1 < b \\
b &= r_1 q_2 + r_2 & 0 &\le r_2 < r_1 \\
r_1 &= r_2 q_3 + r_3 & 0 &\le r_3 < r_2 \\
&\vdots \\
r_i &= r_{i+1} q_{i+2} + r_{i+2} & 0 &\le r_{i+2} < r_{i+1} \\
&\vdots \\
r_{n-2} &= r_{n-1} q_n + r_n & 0 &\le r_n < r_{n-1} \\
r_{n-1} &= r_n q_{n+1} + 0.
\end{aligned}$$

The algorithm must terminate because of the inequalities on the r_i.

THEOREM 2.25
With notation as above, $\gcd(a, b) = r_n$.

PROOF Applying Lemma 2.23,

$$\gcd(a, b) = \gcd(b, r_1) = \gcd(r_1, r_2) = \cdots = \gcd(r_{n-1}, r_n) = \gcd(r_n, 0) = r_n.$$

∎

Although `GCD` already exists in *Mathematica*, we can program our own gcd procedure with the following commands.

Mathematica

```
ourgcd[a_, b_] := (
  {atemp, btemp} = {a, b};
  While[
    btemp > 0,
    {atemp, btemp} = {btemp, Mod[atemp, btemp]}
  ];
  atemp
)
```

In our variable names, the suffix 'temp' stands for 'temporary.' Applying Lemma 2.23, we repeatedly replace the pair `{atemp,btemp}` with the pair `{btemp,Mod[atemp,btemp]}`, where `Mod[atemp,btemp]` is the remainder from the division algorithm on `atemp` and `btemp`. This reduction happens until `btemp` equals 0, and then `atemp` is reported as the gcd.

We test our program by comparing results with the one obtained by `GCD` in the previous section.

Mathematica

```
ourgcd[81959,73963]

1999
```

We can modify our Euclidean algorithm to report the number of steps it takes to produce the gcd. To keep track of the number of steps in the computation, we insert a counter, `c`, as follows.

Mathematica

```
ourgcdcount[a_, b_] := (
  {atemp, btemp} = {a, b};
  c = 0;
  While[
    btemp > 0,
    {atemp, btemp} = {btemp, Mod[atemp, btemp]};
    c++;
  ];
  {atemp, c}
)
```

At each step in the computation, the counter `c` is incremented by 1 (by `c++`). The value of `c` in `{atemp,c}` is the total number of steps in the computation.

We test our program.

Mathematica

```
ourgcdcount[81959,73963]

{1999,3}
```

The computation of $\gcd(81959, 73963)$ takes three steps. To check this, we perform the Euclidean algorithm "by hand":

$$\begin{aligned} 81959 &= 73963 \cdot 1 + 7996 \\ 73963 &= 7996 \cdot 9 + 1999 \\ 7996 &= 1999 \cdot 4 + 0. \end{aligned}$$

Indeed, the computation requires three steps.

Here are the corresponding *Maple* commands for the gcd algorithms.

Maple

```
ourgcd := proc(a,b)
  local atemp, btemp;
  (atemp, btemp) := (a, b);
  while btemp > 0 do
    (atemp, btemp) := (btemp,modp(atemp,btemp));
  end do;
  atemp;
end proc:

ourgcd(81959,73963);

1999

ourgcdcount := proc(a,b)
  local atemp, btemp, c;
  (atemp, btemp) := (a, b); c := 0;
  while btemp > 0 do
    (atemp, btemp) := (btemp, modp(atemp,btemp));
    c := c+1;
  end do;
  [atemp, c];
end proc:

ourgcdcount(81959,73963);

[1999, 3]
```

Let's find the gcd of two very large numbers.

Mathematica

```
ourgcdcount[124131414142342424234l3412,123123123909023233211]

{1,37}
```

Maple

```
ourgcdcount(124131414142342424234134l2,1231231239090232332l1)

[1,37]
```

This computation, determining that the gcd is 1, takes 37 steps—not bad given the size of the numbers! What is the relationship between the size of the inputs and the number of steps required by the Euclidean algorithm? What inputs make the Euclidean algorithm take the most steps? Consider a specific question: Can you find two positive numbers both less than 1000 for which the Euclidean algorithm requires 14 steps? With some trial and error, it can be done.

It seems reasonable that the Euclidean algorithm takes the most steps when the quotients q_i are as small as possible, that is, equal to 1. In this case, "working the algorithm backwards" yields the following computations.

$$\begin{aligned} 2 &= 1 \cdot 2 + 0 \\ 3 &= 2 \cdot 1 + 1 \\ 5 &= 3 \cdot 1 + 2 \\ 8 &= 5 \cdot 1 + 3 \\ 13 &= 8 \cdot 1 + 5 \\ 21 &= 13 \cdot 1 + 8 \\ 34 &= 21 \cdot 1 + 13 \\ 55 &= 34 \cdot 1 + 21 \\ &\vdots \end{aligned}$$

(The first quotient is 2, rather than 1, because of the condition that the remainder is less than the number we are dividing by.)

Are the entries in the left column familiar? They are the Fibonacci numbers (which we saw in Chapter 1). Recall that the sequence of Fibonacci numbers, f_n, is defined recursively as follows:

$$\begin{array}{l} f_0 = 0, \quad f_1 = 1, \\ f_n = f_{n-1} + f_{n-2}, \quad n \geq 2. \end{array} \tag{2.4}$$

The Fibonacci numbers are

$$0, 1, 1, 2, 3, 5, 8, 13, 21, 34, 55, 89, 144, 233, 377, 610, 987, \ldots.$$

The numbers f_0 and f_1 are the *initial values* of the sequence $\{f_n\}$; the equation $f_n = f_{n-1} + f_{n-2}$ is the *recurrence relation* of the sequence.

If our reasoning is correct, then a pair of consecutive Fibonacci numbers are the "worst-case" inputs in the algorithm. Thus, to solve the puzzle posed earlier in this section, let $a = f_{16} = 987$ and $b = f_{15} = 610$. The Euclidean algorithm takes 14 steps for these inputs.

Mathematica

```
ourgcdcount[987,610]

{1,14}
```

Maple

```
ourgcdcount(987,610)

[1,14]
```

By comparing any pair of inputs to suitable Fibonacci numbers, we can estimate the number of steps required for a Euclidean algorithm computation. Given inputs a and b, with $b \leq a$, the Euclidean algorithm takes fewer than $\log_{\alpha} a$ steps, where $\alpha = (1 + \sqrt{5})/2$. We provide a proof of this fact in the Chapter Notes.

Exercises

2.35 Use the Euclidean algorithm to find the greatest common divisors of the following pairs of numbers.

(a) $(12, 34)$
(b) $(13, 31)$
(c) $(99, 51)$
(d) $(123, 456)$
(e) $(1234, 567)$
(f) $(121212, 343434)$

2.36 Compute the least common multiple of the following pairs of numbers.

(a) $(12, 34)$
(b) $(99, 51)$
(c) $(123, 456)$
(d) $(121212, 343434)$

2.37 Use the Euclidean algorithm (and Exercise 2.26) to compute the following.

(a) $\gcd(70, 182, 455)$

(b) $\gcd(120, 144, 156, 171)$

2.38 Find a pair of numbers a and b such that the Euclidean algorithm requires exactly five steps when run on a and b.

2.39 Use Lemma 2.23 to show that for any integer n,

$$\gcd(n+1, 3n^2+4n+2) = 1.$$

2.40 Show that $\gcd(8n^2+14n+6, 2n+3) = 1$ or 3 for any integer n. Show that both 1 and 3 occur as the gcd for appropriate choices of n.

2.41 Show that for any integers a and b,

$$\gcd(13a+4b, 3a+b) = \gcd(a, b).$$

2.42 Let $\{f_n\}$ be the Fibonacci sequence.

(a) Show that the number of steps required in the Euclidean algorithm to compute $\gcd(f_n, f_{n-1})$ is $n-2$, for $n \geq 3$.

(b) Show that if $b < a \leq f_n$, then the number of steps required to compute $\gcd(a, b)$ is at most $n-2$, for $n \geq 3$.

$\diamond$**2.43** With a computer, use the Euclidean algorithm to the find the gcd of a pair of 100-digit numbers and count the number of steps. Generate a pair of consecutive Fibonacci numbers with 100 digits each and find their gcd, counting the number of steps.

2.44 Describe a procedure for computing the gcd of two numbers that uses subtraction instead of division.

$\star$**2.45** Let a be a positive integer greater than 1, and x and y arbitrary positive integers. Prove that

$$\gcd(a^x - 1, a^y - 1) = a^{\gcd(x,y)} - 1.$$

2.5 Linear Diophantine equations

As a result of Theorem 2.16, we know that for positive integers a and b, there exist integers x and y such that $ax + by = \gcd(a, b)$. This leads us to the question: Given integers a, b, c, can we find a solution to the equation

$$ax + by = c, \tag{2.5}$$

where x and y are integers? An equation of this form, a polynomial equation in which we are interested only in integer solutions, is called a *Diophantine equation*. We will study a variety of Diophantine equations in Chapter 9.

Using Theorem 2.16, we are in a position to deal with linear Diophantine equations in the form of (2.5) right now. If $c = \gcd(a, b)$, then a solution exists. Now consider the equation

$$6x + 4y = 10.$$

We know that we can find numbers x_0 and y_0 such that

$$6x_0 + 4y_0 = 2,$$

and so if we multiply through by 5,

$$6(5x_0) + 4(5y_0) = 10$$

and we obtain a solution to the original equation. In general, if c is a multiple of $\gcd(a, b)$, say $c = k \gcd(a, b)$, then we can find numbers x_0 and y_0 such that $ax_0 + by_0 = \gcd(a, b)$ and then multiply through by k to obtain a solution to (2.5).

Now consider the equation

$$6x + 4y = 15.$$

No matter what values we choose for x and y, the left side of this equation will always produce an even number. Thus, this equation has no solutions. In general, if c is not a multiple of $\gcd(a, b)$, then Proposition 2.7 (v) implies that the left side of (2.5) is divisible by $\gcd(a, b)$, and so the equation cannot have a solution in this case. We have proved the following theorem.

THEOREM 2.26

The Diophantine equation $ax + by = c$ has a solution if and only if $\gcd(a, b)$ *divides c.*

Given values of a, b, and c such that $\gcd(a, b)$ divides c, how do we actually find the solutions to (2.5)? In light of our discussion above, it comes down to finding solutions to the equation

$$ax + by = \gcd(a, b).$$

Let's consider a simple example,

$$15x + 11y = 1.$$

Since $\gcd(15, 11) = 1$, we know there is a solution to the equation. The Euclidean algorithm yields the following computations:

$$\begin{aligned} 15 &= 11 \cdot 1 + 4 \\ 11 &= 4 \cdot 2 + 3 \\ 4 &= 3 \cdot 1 + 1 \\ 3 &= 1 \cdot 3 + 0. \end{aligned}$$

The first remainder, 4, is a linear combination of 11 and 4; thus,

$$4 = 15 \cdot 1 - 11 \cdot 1.$$

The second remainder, 3, is a linear combination of 4 and 3; thus,

$$3 = 11 \cdot 1 - 4 \cdot 2 = 11 \cdot 1 - (15 \cdot 1 - 11 \cdot 1) \cdot 2 = 15 \cdot (-2) + 11 \cdot 3.$$

The third remainder (which is the gcd), 1, is a linear combination of 15 and 11; thus,

$$1 = 4 \cdot 1 - 3 \cdot 1 = 15 \cdot 1 - 11 \cdot 1 - (15 \cdot (-2) + 11 \cdot 3) = 15 \cdot 3 + 11 \cdot (-4).$$

Hence, we have a solution to our equation: $(x, y) = (3, -4)$.

In general, suppose that the Euclidean algorithm operates on a and b:

$$\begin{aligned} a &= bq_1 + r_1 \\ b &= r_1 q_2 + r_2 \\ r_1 &= r_2 q_3 + r_3 \\ &\vdots \\ r_i &= r_{i+1} q_{i+2} + r_{i+2} \\ &\vdots \\ r_{n-2} &= r_{n-1} q_n + r_n \\ r_{n-1} &= r_n q_{n+1} + 0. \end{aligned}$$

From the first step of the algorithm, it is apparent that r_1 is equal to a linear combination of a and b. Specifically,

$$r_1 = a \cdot 1 + b \cdot (-q_1).$$

The second equation gives

$$r_2 = b - q_2 \cdot r_1 = a \cdot (-q_2) + b \cdot (1 + q_1 q_2),$$

so that r_2 is a linear combination of a and b. We will prove that every r_i is a linear combination of a and b. In particular, $r_n = \gcd(a, b)$ will be such a combination (which is consistent with our earlier result about gcd). We want to define sequences $\{x_i\}_{i=1}^n$ and $\{y_i\}_{i=1}^n$ recursively so that

$$r_i = ax_i + by_i,$$

for $1 \le i \le n$. We take $x_1 = 1$, $x_2 = -q_2$, $y_1 = -q_1$, $y_2 = 1 + q_1q_2$. Since $r_i = r_{i+1}q_{i+2} + r_{i+2}$, for $1 \le i \le n-1$, we require that

$$\begin{aligned} r_{i+2} &= ax_i + by_i - (ax_{i+1} + by_{i+1})q_{i+2} \\ &= a(x_i - x_{i+1}q_{i+2}) + b(y_i - y_{i+1}q_{i+2}). \end{aligned}$$

Therefore, we obtain the desired linear combination if we set

$$\begin{aligned} x_{i+2} &= x_i - x_{i+1}q_{i+2} \\ y_{i+2} &= y_i - y_{i+1}q_{i+2}, \end{aligned} \tag{2.6}$$

for $1 \le i \le n-2$. Now we have $\gcd(a, b) = r_n = ax_n + by_n$.

As an aid to computation, we use the table below.

		q_1	q_2	$\ldots$	q_i	q_{i+1}	q_{i+2}	$\ldots$	q_n
1	0	x_1	x_2	$\ldots$	x_i	x_{i+1}	x_{i+2}	$\ldots$	x_n
0	1	y_1	y_2	$\ldots$	y_i	y_{i+1}	y_{i+2}	$\ldots$	y_n

The two leftmost columns of the table are always set to $\{1, 0\}$ and $\{0, 1\}$. For each $i = 1, \ldots, n$, we compute x_i by multiplying the entry to the left of x_i by q_i, and subtracting the product from the entry two columns to the left of x_i. We compute y_i similarly.

Example 2.27 Let's compute $\gcd(817, 615)$ and find a linear combination of 817 and 615 equal to this gcd.

First, we use the Euclidean algorithm to find the gcd and the q_i:

$$\begin{aligned} 817 &= 615 \cdot 1 + 202 \\ 615 &= 202 \cdot 3 + 9 \\ 202 &= 9 \cdot 22 + 4 \\ 9 &= 4 \cdot 2 + 1 \\ 4 &= 1 \cdot 4 + 0. \end{aligned}$$

Hence, $\gcd(817, 615) = 1$, $q_1 = 1$, $q_2 = 3$, $q_3 = 22$, and $q_4 = 2$.

Second, we construct the table.

		1	3	22	2
1	0	1	−3	67	−137
0	1	−1	4	−89	182

We have found a linear combination equal to the gcd:

$$1 = 817 \cdot (-137) + 615 \cdot 182.$$

□

By modifying our procedure implementing the Euclidean algorithm, we can use *Mathematica* to produce the linear combination. (In fact, the *Mathematica* command `ExtendedGCD` performs a similar task.)

Mathematica

```
ourextendedgcd[a_, b_] := (
  {atemp, btemp} = {a, b};
  {xx, x} = {1, 0}; {yy, y} = {0, 1};
  While[
    btemp > 0,
    r = Mod[atemp, btemp]; q = (atemp - r)/btemp;
    {atemp, btemp} = {btemp, r};
    {xx, x} = {x, xx - x q}; {yy, y} = {y, yy - y q}
  ];
  {atemp, {xx, yy}}
)
```

The variables `xx` and `x` represent x_{k-1} and x_k, respectively. Thus, when executing the function, we set `xx`=1 and `x`=0. Similarly, `yy` and `y` represent y_{k-1} and y_k, respectively, and we set `yy`=0 and `y`=1. The output of the function is the list of the quantities $\gcd(a, b)$, x_n, and y_n. We demonstrate with the numbers from Example 2.27.

Mathematica

```
ourextendedgcd[817,615]

{1,{-137,182}}
```

Here is the corresponding *Maple* code.

Maple

```
ourextendedgcd := proc(a,b)
  local atemp, btemp, x, xx, y, yy, r, q;
  (atemp, btemp) := (a, b);
  (xx, x) := (1, 0); (yy, y) := (0, 1);
  while btemp > 0 do
    r := modp(atemp, btemp); q := (atemp-r)/btemp;
    (atemp, btemp) := (btemp, r);
    (xx, x) := (x, xx-x*q); (yy, y) := (y, yy-y*q);
  end do;
  [atemp, (xx,yy)];
end proc:

ourextendedgcd(817,615);

[1, [-137, 182]]
```

Example 2.28 Let's find a solution to the Diophantine equation

$$12345x + 54321y = 99. \tag{2.7}$$

With our extended gcd algorithm, we find that $\gcd(54321, 12345) = 3$ (so that our equation does in fact have a solution, since $3 \mid 99$), and

$$(12345)(3617) + (54321)(-822) = 3.$$

Thus, if we take

$$x = (33)(3617) = 119361$$
$$y = (33)(-822) = -27126$$

we obtain a solution to (2.7). □

Example 2.29 Consider the equation

$$6x + 4y = 10.$$

Proceeding as in the previous example (or by inspection), we obtain the solution $x = 5$, $y = -5$. Clearly, there are other solutions. In fact, if we can find a solution to the equation $6\hat{x} + 4\hat{y} = 0$, we can simply add the two equations to obtain a new solution. We could take $\hat{x} = 2$ and $\hat{y} = -3$, and obtain the solution $x = 5 + 2 = 7$ and $y = -5 - 3 = -8$. We see that for every integer m, the pair $(5 + 2m, -5 - 3m)$ provides a solution, as

$$6(5 + 2m) + 4(-5 - 3m) = 6(5) + 12m + 4(-5) - 12m = 10.$$

The following theorem tells us that all solutions are obtained in this way. □

THEOREM 2.30
Let $g = \gcd(a, b)$, *and suppose that* $ax_0 + by_0 = c$. *Then* $x = x_0 + mb/g$ *and* $y = y_0 - ma/g$ *is a solution to the equation*

$$ax + by = c,$$

for all integers m. *Moreover, every solution to the equation has this form.*

PROOF It is easy to verify that $(x_0 + mb/g, y_0 - ma/g)$ is a solution to the equation:

$$a(x_0 + mb/g) + b(y_0 - ma/g) = ax_0 + by_0 = c.$$

Now suppose that (x_1, y_1) is any solution. We have

$$ax_1 + by_1 = ax_0 + by_0,$$

which we can rewrite as

$$a(x_1 - x_0) = b(y_0 - y_1). \tag{2.8}$$

We see that b/g divides $(a/g)(x_1 - x_0)$. Since b/g and a/g are relatively prime, Exercise 2.21 tells us that b/g divides $x_1 - x_0$. Hence, we can write

$$x_1 - x_0 = mb/g,$$

for some integer m. Therefore,

$$x_1 = x_0 + mb/g,$$

as desired. Substituting back into equation (2.8) and solving, we also find that $y_1 = y_0 - ma/g$. □

Exercises

2.46 Decide which of the equations below can be solved in integers. Find a solution for each equation that is solvable.

(a) $12x + 22y = 1$

(b) $12x + 22y = 2$

(c) $12x + 22y = 3$

(d) $12x + 22y = 4$

(e) $12x + 21y = 3$

(f) $12x + 23y = 1$

2.47 (a) Find an integer solution to the equation

$$561x + 379y = 1.$$

(b) Find a second solution to the equation with $x > 0$.

2.48 (a) Find an integer solution to the equation

$$121x + 323y = 1.$$

(b) Find an integer solution to the equation

$$121x + 323y = 5.$$

2.49 Find the integer solution to the equation

$$237x + 81y = 21$$

with the smallest positive value of y.

2.50 Find an integer solution to the equation

$$111x + 159y = 3.$$

Also, describe the complete set of integer solutions to this equation.

◇**2.51** (a) Use a computer to find an integer solution to

$$101234567891x + 98765432173123y = 1.$$

(b) Modify the `ourextendedgcd` program to display a calculation table as shown on p. 36. For a calculation as in part (a), how many quotients are equal to 1? Can you explain why there are so many?

2.52 Find all integer solutions to the equation

$$6x + 10y + 15z = 1.$$

◇**2.53** Use a computer to find an integer solution to the equation

$$342342342x + 345345345y + 1999z = 1.$$

Find all integer solutions.

2.6 Primes and the Fundamental Theorem of Arithmetic

Recall that a prime number is an integer $n > 1$ that has no proper divisors other than 1, and a composite number is an integer greater than 1 that is not a prime number. For example, 2, 3, 5, 7, and 11 are prime numbers, while 4, 6, 8, 9, 10, and 12 are composite numbers. The number 1 is neither prime nor composite; it is sometimes called a "unit."

Note: We do not call 1 a prime number or a composite number because otherwise the uniqueness part of the (forthcoming) Fundamental Theorem of Arithmetic would fail. If 1 were considered prime, then 6 would have several distinct representations as a product of prime numbers, such as $2 \cdot 3$ and $1 \cdot 2 \cdot 3$.

How can we produce a list of prime numbers? A simple, efficient method for listing primes up to a given number was discovered over two thousand years ago by Eratosthenes (276–194 B.C.E.). Eratosthenes' procedure starts with a list of all numbers from 2 to n. Then multiples of primes are "sieved out." The remaining numbers are primes. A particularly nice feature of the procedure is that it requires no multiplications or divisions. Before describing the algorithm we make a small but helpful observation.

PROPOSITION 2.31
If a number n has a proper divisor, then it has one less than or equal to $\sqrt{n}$.

PROOF If d is a proper divisor of n, then n/d is also a proper divisor of n. If $d > \sqrt{n}$ and $n/d > \sqrt{n}$, then $n = d \cdot (n/d) > \sqrt{n} \cdot \sqrt{n} = n$, a contradiction. Hence, $d \leq \sqrt{n}$ or $d/n \leq \sqrt{n}$. ∎

We are now ready for the algorithm.

ALGORITHM 2.32 (Sieve of Eratosthenes)
Given n, the prime numbers up to n are identified.

(1) *Let n be a positive integer; set $S = \{2, 3, \ldots, n\}$.*

(2) *For $2 \leq i \leq \lfloor \sqrt{n} \rfloor$, do:*

If i has not been crossed out in S, then cross out all proper multiples of i in S.

(3) *The elements of S that are not crossed out are the prime numbers up to n.*

We demonstrate Algorithm 2.32 with $n = 16$:

(1) $n = 16$; $S = \{2, 3, 4, 5, 6, 7, 8, 9, 10, 11, 12, 13, 14, 15, 16\}$

(2) $i = 2$: $S = \{2, 3, \not{4}, 5, \not{6}, 7, \not{8}, 9, \not{1}0, 11, \not{1}2, 13, \not{1}4, 15, \not{1}6\}$

$i = 3$: $S = \{2, 3, \not{4}, 5, \not{6}, 7, \not{8}, \not{9}, \not{1}0, 11, \not{1}2, 13, \not{1}4, \not{1}5, \not{1}6\}$

$i = 4$: 4 has been crossed out

(3) The primes up to 16 are 2, 3, 5, 7, 11, and 13.

We will give *Mathematica* and *Maple* implementations of Algorithm 2.32 in Section 7.1.

We now take up the issue of the uniqueness of factorization of integers. We begin with the simple, but important observation that all integers can, in fact, be written as a product of prime numbers.

LEMMA 2.33

Every integer $n > 1$ is either prime or can be written as a product of prime factors.

PROOF We will use strong mathematical induction. The statement holds in the base case, $n = 2$, since 2 is a prime number. Now suppose that the result holds for numbers 2, 3, ..., $n - 1$. Consider the number n. If n is prime, then there is nothing to prove. If n is composite, say, $n = ab$, with $1 < a, b < n$, then we may apply our strong inductive hypothesis to write a and b as products of primes. It follows that n is a product of primes. ▯

In our discussion of Euclid's lemma (Corollary 2.18), we noted that the uniqueness of factorization of integers is a fact that we often take for granted given the way it is introduced in school. For example, as young students we may be asked to write the number 12 as a product of prime numbers. It is implicit in the assignment of this task that all students will get the same answer (if we disregard the order of the prime factors). Suppose that "Andrew" begins to solve the problem by factoring 12 as $2 \cdot 6$. On the other hand, "Beth" begins by breaking down 12 as $3 \cdot 4$. We can easily see that after Andrew factors 6 and Beth factors 4, they will both arrive at the factorization $2 \cdot 2 \cdot 3$. However, nothing we have proved so far guarantees that if two students start factoring a given number in different ways, then they will both eventually arrive at the same prime factorization. As we have mentioned, the key to the proof of unique factorization of integers is Euclid's lemma. The first explicit statement and subsequent proof of unique factorization was not until 1801 when it appeared in Gauss's *Disquisitiones Arithmeticae.* For a discussion of several examples of noninteger arithmetic in which unique factorization fails, see the Chapter Notes.

THEOREM 2.34 (Fundamental Theorem of Arithmetic)
Every integer $n > 1$ is either prime or can be written as a product of prime factors. Furthermore, such a factorization is unique except for rearrangement of the factors.

PROOF Having already proved Lemma 2.33, it remains for us to verify the uniqueness. We will again use strong induction. Uniqueness is clear in the base case $n = 2$, since 2 is prime. Now suppose that uniqueness of factorization holds for numbers 2, 3, ..., $n - 1$. Suppose that n can be represented as a product of primes in two ways:

$$n = p_1 p_2 p_3 \dots p_s = q_1 q_2 q_3 \dots q_t,$$

where the p_i and q_j are primes. We must show that $s = t$ and that the q_j are just a rearrangement of the p_i. If $s = 1$, then n is prime and so $t = 1$ and $p_1 = q_1$.

Suppose that $s > 1$. Since p_1 divides $p_1 p_2 p_3 \dots p_s$, it also divides $q_1 q_2 q_3 \dots q_t$. It follows from Euclid's lemma (Corollary 2.18) that $p_1 \mid q_k$, and so $p_1 = q_k$ for some k. Because we are not concerned with the order of the prime factors, we can reorder the q_j, if necessary, and assume that $k = 1$. Now write

$$n/p_1 = p_2 p_3 \dots p_s = q_2 q_3 \dots q_t.$$

Since $p_1 > 1$, we have a factorization of a number, n/p_1, that is smaller than n but greater than 1. Our strong inductive hypothesis applies to this number, and so we may conclude that $s = t$ and the q_j are just a rearrangement of the p_i. ☐

Example 2.35 We easily find the prime factorization of 100:

$$100 = 2^2 \cdot 5^2.$$

☐

As a consequence of the Fundamental Theorem of Arithmetic, every integer $n > 1$ can be written uniquely in the form

$$n = \prod_{i=1}^{k} p_i^{\alpha_i}, \qquad (2.9)$$

where $p_1, \dots, p_k$ are distinct primes, $p_1 < \dots < p_k$, and $\alpha_1, \dots, \alpha_k$ are positive integers. This is called the "canonical factorization" of n.

The commands `FactorInteger` (*Mathematica*) and `ifactor` (*Maple*) give the canonical factorization of an integer. These commands work quickly when the number to be factored is not extremely large, but they don't return an answer in a reasonable amount of time for most numbers greater than 10^{40}. Try it! In fact, there is no known efficient procedure for factoring large numbers. (See Sections 4.4 and 11.7.)

Given that n has the prime factorization in (2.9), suppose that d divides n, so that $n = dk$ for some number k. Then the prime factorization of dk is the product of the prime factorizations of d and k. Since this factorization is unique, it must be the case that

$$d = \prod_{i=1}^{k} p_i^{\alpha'_i},$$

where $0 \leq \alpha'_i \leq \alpha_i$, for all i. Conversely, any number d of this form is a divisor of n.

Now suppose that a and b are two numbers greater than 1, and $p_1, \ldots, p_k$ are the primes that divide either a or b. Then we may write

$$a = \prod_{i=1}^{k} p_i^{\alpha_i} \quad \text{and} \quad b = \prod_{i=1}^{k} p_i^{\beta_i}.$$

(Here the α_i and β_i are nonnegative integers.) It follows that

$$\gcd(a, b) = \prod_{i=1}^{k} p_i^{\min\{\alpha_i, \beta_i\}} \tag{2.10}$$

and

$$\operatorname{lcm}(a, b) = \prod_{i=1}^{k} p_i^{\max\{\alpha_i, \beta_i\}}. \tag{2.11}$$

Formulas (2.10) and (2.11) confirm Proposition 2.22 (iii), as $\max\{\alpha_i, \beta_i\} + \min\{\alpha_i, \beta_i\} = \alpha_i + \beta_i$.

We now proceed to give some partial answers to the question, "What does the set of prime numbers look like?" The most fundamental fact about the set of primes is proved simply and elegantly in Euclid's *The Elements*.

THEOREM 2.36 (Euclid)
There are infinitely many prime numbers.

PROOF The proof is by contradiction. Assume that there are only finitely many prime numbers, say, $p_1, p_2, \ldots, p_n$. Let

$$N = p_1 p_2 \ldots p_n + 1.$$

Let q be any prime factor of N. It is not possible that $q = p_i$ for some p_i, because $q \mid N$ and $q \nmid 1$ (recall Proposition 2.7). Hence, q is different from all the p_i. This is a contradiction. Therefore, there are infinitely many prime numbers. ∎

Next we note that the set of primes contains arbitrarily long "gaps."

PROPOSITION 2.37
Given any $n > 1$, there exist n consecutive composite numbers.

PROOF Let $a = (n+1)! + 2$. Then each of the numbers a, $a+1$, $\ldots$, $a+n-1$ is composite, because $(k+2) \mid (a+k)$, for each $k = 0, \ldots, n-1$. ∎

The numbers produced by the argument in the proof are rather large in comparison to n, and certainly we would not expect this gap to be the first of length n. For instance, when $n = 10$, the proof provides the list of composite numbers 39916802, 39916803, $\ldots$, 39916811. However, one finds 10 (in fact 13) consecutive composite numbers between the primes 113 and 127.

What is the smallest possible gap between primes? Apart from the numbers 2 and 3, the smallest difference between two consecutive primes is 2 (since the primes must both be odd). A glance at the list of primes less than 100 reveals several pairs of primes that differ by 2. For example, two such pairs are $\{5, 7\}$ and $\{11, 13\}$.

CONJECTURE (Twin primes)
There are infinitely many primes p for which $p + 2$ is also prime. (Primes p, $p + 2$ are called "twin primes.")

While no one has been able to prove this statement, it is known that there are an infinite number of primes p for which $p + 2$ has at most two factors. See [7].

Another famous conjecture on primes that is also easy to state is due to Christian Goldbach (1690–1764).

CONJECTURE (Goldbach, 1742)
Every even number greater than 2 is the sum of two primes.

For example, $18 = 7 + 11$, and $100 = 3 + 97$. It is known that every *sufficiently large* even number is the sum of a prime and a number with at most two factors. (The proof of this fact produces a number n_0 such that every even number greater than n_0 is the sum of a prime and a number with at most two factors. Unfortunately, n_0 is so large that there is no hope of checking the finitely many remaining cases, even with the aid of a computer.)

x	$\pi(x)$	$x/\log(x)$	$\frac{\pi(x)}{x/\log x}$
10	4	4.3	0.921
100	25	21.7	1.151
1000	168	144.8	1.161
10000	1229	1085.8	1.132
100000	9592	8685.9	1.104
1000000	78498	72382.4	1.084

TABLE 2.1: Comparison of $\pi(x)$ with approximations.

Notice that the conjecture does not hold in general for odd numbers. For example, 27 is not the sum of two prime numbers. However, if n is any odd number, then $n-3$ is even, and if Goldbach's conjecture is true (and $n > 6$), then $n-3$ can be written as a sum of two primes. Thus Goldbach's conjecture implies the "weak Goldbach conjecture," which predicts that every odd number greater than 6 can be written as the sum of three primes. This conjecture is known to hold for sufficiently large odd numbers, and so there are finitely many odd numbers to check. Unfortunately, the list of numbers to check is still too large to be done by a computer in a reasonable amount of time. See [11].

Let $\pi(x)$ be the number of primes less than or equal to x. For example, $\pi(6) = 3$ and $\pi(7) = 4$. How does the function $\pi(x)$ behave in general? The Prime Number Theorem, proved independently by Jacques Hadamard (1865–1963) and Charles de la Vallée Poussin (1866–1962) in 1896, provides an answer. The theorem states that the function $x/\log x$ provides a good approximation for $\pi(x)$. (Logarithms are base e in this discussion.) Table 2.1 gives some comparisons of the values of $\pi(x)$ with the approximations.

Notice that while the difference between the actual values and the approximations is getting larger, the quotient of these two quantities seems to be approaching 1. This is precisely what the theorem predicts.

THEOREM (Prime Number Theorem)

$$\lim_{x\to\infty} \frac{\pi(x)}{x/\log x} = 1.$$

The proof given by Hadamard and De la Vallée Poussin makes use of complex analysis. The result is such a fundamental one in number theory that many continued to search for an *elementary* proof after they published theirs. (An elementary proof is one that uses only elementary mathematical ideas, i.e., at the level of calculus.) In 1949 Atle Selberg made a breakthrough that enabled him and Paul Erdős (1913–1996) to provide such a proof. For his work, Selberg was awarded the Fields Medal in 1950. (The Fields Medal, presented once every four years at the International Congress of Mathematics, is perhaps the highest honor a mathematician can receive for his or her work.) We will have more to say about the Prime Number Theorem in Chapter 10.

Here is a proposition similar in spirit to Theorem 2.36.

PROPOSITION 2.38

There are infinitely many primes of the form $4n + 3$.

PROOF Suppose that there are only finitely many primes of the form $4n+3$. One of them is obviously 3. Let $p_1, p_2, \ldots, p_k$ be the primes of the form $4n+3$, other than 3. Define $N = 4p_1p_2 \ldots p_k + 3$. Note that 2 is not a factor of N, since N is odd. Also, 3 is not a factor of N, since 3 would have to divide one of the p_i, which is impossible. Not all of the prime factors of N may be of the form $4n + 1$, since, if so, their product would be of the form $4n + 1$ instead of $4n + 3$. Therefore, N must contain at least one prime factor of the form $4n + 3$. This prime must be different from all the p_i, since clearly no p_i divides N. Hence, there exists a prime of the form $4n + 3$ different from all the p_i. But this contradicts our assumption that the number of primes of the form $4n + 3$ is finite. Therefore, there are infinitely many primes of the form $4n + 3$. ☐

Note: The techniques in the proof of Proposition 2.38 cannot be used to demonstrate that there are an infinite number of primes of the form $4n + 1$. More subtle proof techniques are needed (see Proposition 3.38).

The result of this theorem is a special case of a theorem of Johann Peter Gustav Lejeune Dirichlet (1805–1859). We will discuss the general theorem in Chapter 10.

THEOREM (Dirichlet)

If $\gcd(a, b) = 1$, *then there are infinitely many primes of the form* $a + bn$.

An "arithmetic progression of length l" (or l-AP) is a sequence

$$a,\ a+d,\ a+2d,\ \ldots,\ a+(l-1)d$$

of l numbers, each consecutive pair of which differ by a constant number d. For example, the sequence 20, 30, 40, 50, 60 is a 5-AP. Is there a 5-AP of prime numbers? Yes: 5, 11, 17, 23, 29. In 2007 Jaroslaw Wroblewski discovered 24 primes in arithmetic progression: $468395662504823 + 45872132836530n$, for $0 \leq n \leq 23$.

In 2004 Ben Green and Terence Tao proved that there exist arbitrarily long arithmetic progressions of prime numbers. Their proof, which uses ideas from ergodic theory (a branch of analysis) and Ramsey theory (a branch of combinatorics), is non-constructive: it does not show how to find long arithmetic progressions of primes.

THEOREM (Primes in arithmetic progression)
For every positive integer n, there exist n primes in arithmetic progression.

Exercises

2.54 Find the prime factorizations of the following numbers.

(a) 120
(b) 121
(c) 124
(d) 131
(e) 1234
(f) 4321

2.55 Use the Sieve of Eratosthenes to find all prime numbers less than 100.

2.56 Compute $\gcd(5389, 7633)$

(a) using the Euclidean algorithm
(b) using equation (2.10).

2.57 Show that the binomial coefficient $\binom{p}{k}$ is divisible by p if p is a prime and $1 \leq k \leq p-1$.

2.58 A piano cannot be tuned to play both octaves and fifths correctly. Two notes an octave apart should have frequencies in the ratio $2\colon 1$ and two notes a fifth apart should have frequencies in the ratio $3\colon 2$. But these two ratios are incommensurable: there are no positive integers a and b such that

$$\left(\frac{3}{2}\right)^a = \left(\frac{2}{1}\right)^b.$$

(a) Prove this.
(b) Calculate how far "off" the frequencies on a piano are when two notes 84 steps apart are compared by octaves and by fifths. (An octave is 12 steps and a fifth is 7 steps.)

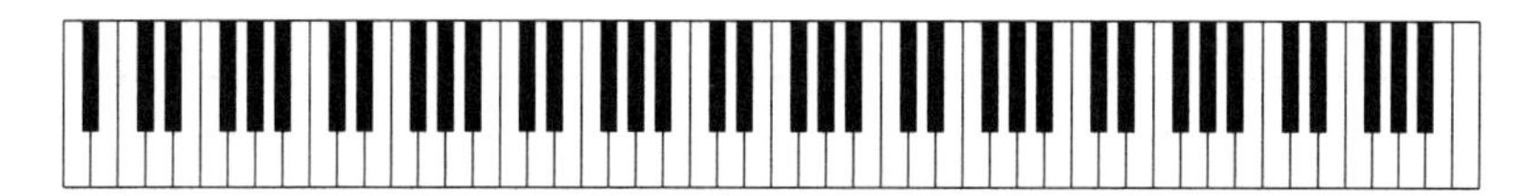

See [5] for a discussion of the relationship of number theory to music.

2.59 Let p/q be a nonzero rational number. Suppose that the canonical factorization of q is

$$q = \prod_{i=1}^{k} q_i^{\alpha_i}.$$

Show that there exist integers $a_1, \ldots, a_k$ for which

$$\frac{p}{q} = \frac{a_1}{q_1^{\alpha_1}} + \cdots + \frac{a_k}{q_k^{\alpha_k}}.$$

†2.60 (a) Show that there are infinitely many prime numbers of the form $3n + 2$.

(b) Show that there are infinitely many prime numbers of the form $6n + 5$.

2.61 Prove that for every integer $n \geq 1$, the number $2^{2^n} - 1$ is divisible by at least n distinct primes.

2.62 Let p_n be the nth prime number.

(a) Show that $p_n \leq p_1 p_2 \ldots p_{n-1} + 1$.

(b) Use part (a) to show that $p_n < 2^{2^n}$.

⋄2.63 Use a computer to find a pair of twin primes greater than 10^6.

⋆2.64 (For those who have studied field theory) Use the idea in Euclid's proof of the infinitude of prime numbers to prove that no finite field is algebraically closed (contains a root of every polynomial with coefficients in the field).

2.65 Show that 3 is the only prime such that p, $p+2$, and $p+4$ are all primes.

⋄2.66 Use a computer to find a 10-AP of prime numbers.

2.67 Prove or disprove: There exists an infinite arithmetic progression of prime numbers.

2.7 Notes

Euclid

Perhaps the most famous text in mathematics ever written is *The Elements* ([13], [14], [15]) by Euclid (325?–265? B.C.E.). Some have suggested that it ranks second only to the Bible as an object of study and in overall worldwide circulation [16]. In *The Elements*, Euclid organized in a deductive system much of the mathematical knowledge of his time. While geometry is the central focus of this work, some elementary number theory is developed in three of the thirteen books. The number theory is presented within a geometric context. For instance, the definition for part (factor) of a number (see the quote at the beginning of the chapter) suggests that we are to picture a line segment broken up into a number of smaller line segments all of the same length. In this chapter of our book, we have seen two of the most famous results on number theory from *The Elements* (the Euclidean algorithm and the infinitude of the prime numbers). In addition to these, Euclid also derives formulas for summing a finite geometric series and for producing Pythagorean triples. The book stops short of proving the Fundamental Theorem of Arithmetic, though this result seems likely to have been known to Euclid [32].

Little is known about the life of Euclid. Most scholars believe he lived in Alexandria. One of the few stories we are told of Euclid seems strangely familiar to mathematics teachers today. According to the historian Joannes Stobaeus (c. 5th century), a new student of Euclid asks, upon learning the first theorem, "But what is the good of this, and what shall I get by learning these things?" In reply, Euclid calls for a slave saying, "Give this fellow a penny, since he must make gain from what he learns."

The number of steps in the Euclidean algorithm

How fast is the Euclidean algorithm? We now prove the assertion that the inputs that make the Euclidean algorithm take the most steps are consecutive Fibonacci numbers. We will also be able to estimate the number of steps that the Euclidean algorithm takes, as a function of the size of its inputs.

The first step is an explicit formula for the Fibonacci numbers known as Binet's formula, named after Jacques Binet (1786–1856).

THEOREM (Binet's formula)
The Fibonacci numbers are given by the formula

$$f_n = \frac{1}{\sqrt{5}}\left(\frac{1+\sqrt{5}}{2}\right)^n - \frac{1}{\sqrt{5}}\left(\frac{1-\sqrt{5}}{2}\right)^n, \quad n \geq 0. \tag{2.12}$$

Isn't it remarkable that the right side of (2.12) is an integer for all n?

PROOF Recall that

$$f_n = f_{n-1} + f_{n-2}, \quad n \geq 2; \quad f_0 = 0, \ f_1 = 1.$$

Assume that there exists a value of x such that the sequence x^n satisfies the recurrence relation. Then

$$x^n = x^{n-1} + x^{n-2},$$

and hence,

$$x^2 = x + 1.$$

This relation is called the "characteristic equation" and its roots are called "characteristic roots."

From the quadratic formula, we find that the roots are

$$\alpha = \frac{1+\sqrt{5}}{2} \doteq 1.618, \quad \beta = \frac{1-\sqrt{5}}{2} \doteq -0.618.$$

Hence, the sequences

$$\alpha^n, \quad \beta^n$$

satisfy the recurrence relation. It is straightforward to demonstrate that if two sequences u and v satisfy the recurrence relation, then any linear combination of them, $c_1 u + c_2 v$, with c_1 and c_2 constants, also satisfies the relation. Hence, the sequence

$$c_1 \alpha^n + c_2 \beta^n$$

satisfies the relation. From the initial conditions, we find that

$$c_1 = \frac{1}{\sqrt{5}}, \quad c_2 = -\frac{1}{\sqrt{5}},$$

and so the sequence

$$\frac{1}{\sqrt{5}}\left(\frac{1+\sqrt{5}}{2}\right)^n - \frac{1}{\sqrt{5}}\left(\frac{1-\sqrt{5}}{2}\right)^n$$

satisfies the recurrence relation and the initial conditions. But there is only one sequence that does both. Therefore,

$$f_n = \frac{1}{\sqrt{5}}\left(\frac{1+\sqrt{5}}{2}\right)^n - \frac{1}{\sqrt{5}}\left(\frac{1-\sqrt{5}}{2}\right)^n, \quad n \geq 0.$$

$\square$

Recall the notation of the Euclidean algorithm:

$$\begin{aligned}
a &= bq_1 + r_1 & 0 &\le r_1 < b\\
b &= r_1q_2 + r_2 & 0 &\le r_2 < r_1\\
r_1 &= r_2q_3 + r_3 & 0 &\le r_3 < r_2\\
&\vdots\\
r_i &= r_{i+1}q_{i+2} + r_{i+2} & 0 &\le r_{i+2} < r_{i+1}\\
&\vdots\\
r_{n-2} &= r_{n-1}q_n + r_n & 0 &\le r_n < r_{n-1}\\
r_{n-1} &= r_nq_{n+1} + 0.
\end{aligned}$$

We fix n and determine lower bounds for a and b. Working from the bottom up, since $q_i \ge 1$, we see that

$$\begin{aligned}
a &\ge f_{n+2} \cdot 1 + f_{n+1} = f_{n+3}\\
b &\ge f_{n+1} \cdot 1 + f_n = f_{n+2}\\
r_1 &\ge f_n \cdot 1 + f_{n-1} = f_{n+1}\\
&\vdots\\
r_i &\ge f_{n-i+1} \cdot 1 + f_{n-i} = f_{n-i+2}\\
&\vdots\\
r_{n-2} &\ge 2 \cdot 1 + 1 = 3 = f_4\\
r_{n-1} &\ge 2 = f_3.
\end{aligned}$$

The last step holds since $r_{n-1} > r_n \ge 1$.

Thus, we have found that if the Euclidean algorithm takes $n+1$ steps for inputs a, b, with $a \ge b$, then

$$a \ge f_{n+3}, \quad b \ge f_{n+2}.$$

We retain the notation $\alpha = (1+\sqrt{5})/2$ and $\beta = (1-\sqrt{5})/2$.

PROPOSITION

Given inputs a and b, with $a \le b$, the Euclidean algorithm takes fewer than $\log_\alpha a$ steps.

PROOF As the Euclidean algorithm takes $n+1$ steps (following our previous notation), we need to show that $n+1 < \log_\alpha a$.

Since $|\alpha| > 1$ and $|\beta| < 1$, we obtain

$$\left|\frac{\beta}{\alpha}\right|^{n+3} \le \left|\frac{\beta}{\alpha}\right|^3 < 0.06,$$

and hence,

$$|\beta^{n+3}| < 0.06\alpha^{n+3}.$$

Therefore,

$$\alpha^{n+3} - \beta^{n+3} > 0.94\alpha^{n+3}.$$

Recall that Binet's formula says that

$$f_n = \frac{1}{\sqrt{5}}(\alpha^n - \beta^n),$$

and so

$$a \geq f_{n+3} = \frac{1}{\sqrt{5}}(\alpha^{n+3} - \beta^{n+3}) > \frac{0.94}{\sqrt{5}}\alpha^{n+3}.$$

Taking logs,

$$\log_\alpha a > \log_\alpha \left(\frac{0.94}{\sqrt{5}}\right) + (n+3) > -1.9 + n + 3.$$

Finally,

$$\log_\alpha a > n + 1.$$

$\square$

The above result tells us that the Euclidean algorithm is quite fast. For example, suppose that $a = 10^{100}$, a huge number. Using a computer, we find that

$$\log_\alpha 10^{100} \doteq 478.497.$$

Hence, the Euclidean algorithm would take no more than 478 steps given inputs a and b, with $b \leq a$.

Exercises

2.68 Prove Binet's formula by mathematical induction.

Nonunique factorization

How special is the unique factorization property of the integers? We learn this early on, and so it is perhaps hard to appreciate. In fact, there are many other sets of numbers in which prime factorization is not at all unique. Let **E** be the set of even numbers, and consider how factorization works within the set. What should the definition of a *prime* within this set be? Back in the set of all integers, prime numbers are those numbers that cannot be written as the product of two strictly smaller positive numbers. Let's take the same definition for a *prime* in **E**. That is, we will call an even number an **E**-prime if it cannot be written as the product of two strictly smaller positive even numbers. Thus, the number 6 is an **E**-prime while the number 4 is not. Now consider the factorizations of 36 in **E** below:

$$36 = 6 \cdot 6 = 2 \cdot 18.$$

The numbers 2, 6, and 18 are all **E**-primes, and so we have two different **E**-prime factorizations of the number 36. (In **E**, the primes are the numbers of the form $4n + 2$ and the composites are the numbers of the form $4n$.)

The failure of unique factorization in general was the key stumbling block in an incorrect proof of Fermat's Last Theorem presented before the Paris Academy in March of 1847. Several correct proofs of special cases had made use of factorizations of $z^n - y^n$ into linear factors. For $n > 2$ these factorizations required the use of complex numbers. Thus, these factorizations were taking place in an enlargement of the set of integers. In a talk at the meeting of the Paris Academy, the French mathematician Gabriel Lamé (1795–1870) claimed he could prove Fermat's Last Theorem making use of the factorization

$$x^n = z^n - y^n = (z - y)(z - \omega y)(z - \omega^2 y) \ldots (z - \omega^{n-1} y),$$

where ω is a primitive nth root of 1, and generalizing the previous methods. In order for his proof to be correct he would need these enlarged sets of integers to possess the unique factorization property. Joseph Liouville (1809–1882) immediately pointed out this gap in Lamé's "proof." Augustin Cauchy (1789–1857), on the other hand, believed that Lamé's arguments could be made to work.

For the next two months, debate continued on the subject with Cauchy and Lamé standing firm in their beliefs that a proof was close at hand. Finally in late May of that year, Liouville received a letter from the German mathematician Eduard Kummer (1810–1893). In the letter Kummer informed Liouville that his objections were well-founded, and Kummer included a copy of a paper of his published several years earlier in which he proved that unique factorization does fail in some cases. For example, unique factorization fails in the case $n = 23$. In fact, unique factorization holds in only a finite number of cases. In a second paper published in 1846, Kummer introduced the term *ideal number* (which is actually a set of numbers), and showed that a sort of unique factorization holds for these ideal numbers. This result allows one to prove Fermat's last theorem whenever the exponent n is a *regular* prime number. For a more detailed account of this story and others related to Fermat's Last Theorem, see Sections 9.5 and 11.8, and [9].

Exercises

†⋆2.69 Let R be the set of numbers of the form $a + b\sqrt{-5}$, where a and b are any integers. The goal of this set of exercises is to prove that the unique factorization property fails in R.

The *norm* of an element of R is defined to be

$$N(a+b\sqrt{-5}) = a^2 + 5b^2.$$

(a) Show that if α and β are any two elements of R, then $N(\alpha\beta) = N(\alpha)N(\beta)$.

(b) Show that if $N(\alpha) = 1$, then $\alpha = \pm 1$.

A nonzero element $\alpha \neq \pm 1$ of R is called *irreducible* if $\pm\alpha$ and ± 1 are its only divisors. In other words, α is irreducible if whenever $\alpha = \beta\gamma$, either $\beta = \pm 1$ or $\gamma = \pm 1$.

(c) Show that any nonzero element $\alpha \neq \pm 1$ is irreducible or can be written as a product of irreducible elements of R.
Hint: Use induction on $N(\alpha)$.

(d) Show that 2, 3, $1-\sqrt{-5}$, and $1+\sqrt{-5}$ are irreducible and conclude that 6 can be factored as a product of irreducible elements in two different ways.

We say that a nonzero element α divides β if there exists γ in R such that $\beta = \alpha\gamma$. A nonzero element $\alpha \neq \pm 1$ of R is called "prime" if whenever α divides $\beta\gamma$, α divides either β or γ.

(e) Show that if π and π' are prime and π divides π', then $\pi' = \pm\pi$.

(f) Show that an element of R can be written as a product of prime elements of R in at most one way (up to order and choice of signs of the prime factors).

(g) Show that 6 cannot be written as a product of prime elements of R.

Chapter 3

Congruences

> Mathematicians have tried in vain to this day to discover some order in the sequence of prime numbers, and we have reason to believe that it is a mystery into which the human mind will never penetrate.
>
> LEONHARD EULER (1707–1783)

Congruences are a natural and convenient notation for formulating many interesting assertions in number theory. In this chapter, we introduce and work with congruences, we show how to solve linear congruences, and we use congruences to present three key theorems: Fermat's theorem, Euler's theorem, and Wilson's theorem. We also discuss applications of congruences to parity check codes, e.g., the ones used in the ISBN (International Standard Book Number) system.

3.1 Residue classes

DEFINITION 3.1 *Given an integer $n \geq 2$, two integers a and b are* congruent modulo n *if n divides $a - b$. We write*

$$a \equiv b \pmod{n}.$$

Notice that $a \equiv b \pmod{n}$ if and only if $a = b + kn$, for some integer k.

Example 3.2 We have

$$41 \equiv 21 \pmod{10},$$

since $41 - 21 = 20$ and $10 \mid 20$. □

We can perform algebra on congruences in many of the same ways we do with equalities.

PROPOSITION 3.3
If $a \equiv a' \pmod{n}$ *and* $b \equiv b' \pmod{n}$, *then*

$$a + b \equiv a' + b' \pmod{n}$$

and

$$ab \equiv a'b' \pmod{n}.$$

PROOF By assumption,

$$a' = a + kn \quad \text{and} \quad b' = b + ln,$$

for some integers k and l. It follows that

$$a' + b' = a + b + (k + l)n$$

and

$$a'b' = ab + (kb + la + kln)n,$$

so that $a' + b' \equiv a + b \pmod{n}$ and $a'b' \equiv ab \pmod{n}$. ∎

Example 3.4 Using the proposition, we can "solve" the congruence

$$3x + 2 \equiv 4 \pmod{5}.$$

First, we subtract 2 from both sides of the congruence and obtain

$$3x \equiv 2 \pmod{5}.$$

Multiplying both sides by 2, the congruence becomes

$$6x \equiv 4 \pmod{5}.$$

Since $6 \equiv 1 \pmod{5}$, we conclude that

$$x \equiv 4 \pmod{5}.$$

∎

PROPOSITION 3.5
The congruence relation $\equiv$ *(modulo* n*) is an equivalence relation on* $\mathbf{Z}$. *That is, the congruence relation satisfies the following.*

(i) *(Reflexivity)* $a \equiv a \pmod{n}$ *for all* a.

(ii) *(Symmetry) If* $a \equiv b \pmod{n}$, *then* $b \equiv a \pmod{n}$.

(iii) *(Transitivity) If* $a \equiv b \pmod{n}$ *and* $b \equiv c \pmod{n}$, *then* $a \equiv c \pmod{n}$.

PROOF (Reflexivity) Since n divides $a-a=0$, we see that $a \equiv a \pmod{n}$.

(Symmetry) Suppose that $a \equiv b \pmod{n}$. We can write $a - b = nk$, for some integer k. Hence,

$$b - a = n(-k),$$

and so $b \equiv a \pmod{n}$.

(Transitivity) Suppose that $a \equiv b \pmod{n}$ and $b \equiv c \pmod{n}$. We can write $a - b = nk$ and $b - c = nl$, for some integers k and l. Adding, we obtain

$$(a - b) + (b - c) = nk + nl.$$

Thus $a - c = n(k + l)$, and so $a \equiv c \pmod{n}$. ◻

An equivalence relation on a set partitions the set into equivalence classes. Every element of the set belongs to a single equivalence class, and the equivalence class of an element consists of all elements of the set equivalent to the given element.

DEFINITION 3.6 *The equivalence classes of the $\equiv$ modulo n equivalence relation are* residue classes *(modulo n).*

We denote the residue class of a as $[a]$, or simply as a. We denote the set of residue classes modulo n as $\mathbf{Z}_n$. Note that all members of $[a]$ are of the form $a + kn$, where $k \in \mathbf{Z}$.

Example 3.7 Let $n = 3$. The equivalence class of 0 consists of numbers a satisfying $a \equiv 0 \pmod{3}$. Thus the equivalence class of 0 is the set of numbers divisible by 3. So

$$[0] = \{\ldots, -6, -3, 0, 3, 6, \ldots\}.$$

The equivalence class of 1 consists of numbers a satisfying $a \equiv 1 \pmod{3}$. Thus the equivalence class of 1 is the set of numbers of the form $1 + 3k$. We see that

$$[1] = \{\ldots, -5, -2, 1, 4, 7, \ldots\}.$$

Likewise, the equivalence class of 2 is the set of numbers of the form $2 + 3k$. We conclude that

$$[2] = \{\ldots, -4, -1, 2, 5, 8, \ldots\}.$$

Every integer shows up in one of these classes, and so $\mathbf{Z}_3 = \{[0], [1], [2]\}$. ◻

We can generalize this example. By the division algorithm, for any integer a, we can find an integer r such that $0 \leq r < n$ and $a \equiv r \pmod{n}$. Moreover, if $0 \leq a < b < n$, then $b - a$ is not divisible by n, and so $[a] \neq [b]$. Putting these two observations together, we see that $\mathbf{Z}_n = \{[0], [1], \ldots, [n-1]\}$.

Any element of an equivalence class is called a "representative" of the class. We have just seen that every residue class modulo n has a unique nonnegative representative less than n. Recall that the *Mathematica* command `Mod[a,n]` and the *Maple* command `modp(a,n)` compute the remainder in the division algorithm when a is divided by n. Thus, these commands compute the least nonnegative element in the residue class of a modulo n. In performing calculations with residue classes, it will often be convenient to work with this least nonnegative residue. We will say that we are "reducing a mod n" when we calculate this residue. For example, when we reduce 17 mod 3 we obtain 2.

DEFINITION 3.8 *A* complete residue system *modulo n is a collection of integers containing exactly one representative of each residue class.*

Example 3.9 Two complete residue systems modulo 5 are $\{0, 1, 2, 3, 4\}$ and $\{10, 6, -3, 23, -1\}$. □

In solving congruences as we did in Example 3.4, we may regard the variable as taking on integer values, but we may also regard it as a variable representing a residue class. So in the example, we would say the solution to the congruence is the residue class of 4, and thus, from this perspective, it makes sense to say that the congruence has exactly one solution.

Exercises

3.1 Which of the following numbers are congruent modulo 11 to 3?

$$-8, \ -3, \ 8, \ 33, \ 36, \ 124$$

3.2 Reduce the following numbers mod 8 (i.e., compute the least nonnegative element in the residue class modulo 8 of each of the following numbers).

(a) 5

(b) -5

(c) 27

(d) 83

(e) -83

(f) 152

3.3 Suppose that $a \equiv a' \pmod{n}$ and $b \equiv b' \pmod{n}$. Show that $a - b \equiv a' - b' \pmod{n}$.

3.4 Find an example of numbers a, b and n such that $2a \equiv 2b \pmod{n}$ but $a \not\equiv b \pmod{n}$.

3.5 Which of the following are complete residue systems modulo 6?

$$\{4, 5, 6, 7, 8, 9\}$$
$$\{1, 5, 9, 13, 17, 21\}$$
$$\{1, 6, 11, 16, 21, 26\}$$
$$\{-2, -1, 0, 1, 2, 3\}$$
$$\{-14, -3, 1, 11, 36, 62\}$$

3.6 Show that the set $\{1, 2, \ldots, n\}$ is a complete residue system modulo n.

3.7 Show that every residue class modulo n contains a number r satisfying $|r| \leq n/2$.

3.8 Suppose that $m \mid n$. Show that $a \equiv b \pmod{n}$ implies that $a \equiv b \pmod{m}$.

†**3.9** Suppose that $n = n_1 n_2$, where $\gcd(n_1, n_2) = 1$. Show that $a \equiv b \pmod{n}$ if and only if $a \equiv b \pmod{n_1}$ and $a \equiv b \pmod{n_2}$.

3.2 Linear congruences

For which elements of $\mathbf{Z}_9$ does the congruence

$$3x \equiv 6 \pmod{9}$$

hold? We might expect that a *linear congruence* of this type should have one solution, but checking the nine elements of $\mathbf{Z}_9$, we find that there are actually three solutions: $x \equiv 2$, 5, and 8 (mod 9). We can also solve the congruence by using the `Solve` command in *Mathematica*.

Mathematica

```
Solve[{3x==6,Modulus==9},x]

{{Modulus → 9, x → 2}, {Modulus → 9, x → 5}, {Modulus →
9, x → 8}}
```

The corresponding *Maple* command is `msolve`.

Maple

```
msolve(3*x=6,9);

{x = 2}, {x = 5}, {x = 8};
```

Hence, we have verified that the congruence has three solutions: $x \equiv 2 \pmod 9$, $x \equiv 5 \pmod 9$, and $x \equiv 8 \pmod 9$.

Some linear congruences have no solutions. For example, the congruence

$$3x \equiv 5 \pmod 9$$

does not have a solution. If x_0 were a solution, we could find an integer k such that $3x_0 - 5 = 9k$, or

$$3x_0 - 9k = 5.$$

Since the left side is divisible by 3 and the right side is not, we have a contradiction.

When does a congruence of the form

$$ax \equiv b \pmod n$$

have a solution in $\mathbf{Z}_n$? In light of our example above, we see that a necessary condition for a solution to exist is that any common divisor of a and n must also divide b. This turns out to be a sufficient condition as well.

THEOREM 3.10

The congruence

$$ax \equiv b \pmod n \tag{3.1}$$

has a solution in $\mathbf{Z}_n$ *if and only if* $\gcd(a, n)$ *divides* b. *In this case, the number of solutions in* $\mathbf{Z}_n$ *is exactly* $\gcd(a, n)$.

PROOF The congruence in (3.1) has a solution if and only if we can find integers x and k such that $ax = b + nk$. In other words, the congruence has a solution if and only if the linear Diophantine equation

$$ax + (-n)k = b \tag{3.2}$$

has a solution. According to Theorem 2.26, this happens exactly when $\gcd(a, n)$ divides b.

Now we count the solutions to the congruence. Let $g = \gcd(a, n)$, and suppose that (x_0, k_0) is a solution to (3.2). Theorem 2.30 tells us that the solution, x, to (3.2) has the form $x = x_0 + mn/g$. Furthermore, for every integer m there is a solution whose x value has this form. Thus, every solution to the congruence has the form

$$x_0 + m(n/g),$$

for some integer m, and every number of this form represents a solution. We must determine the number of distinct congruence classes represented by these solutions. We will prove that the set of residue classes

$$\{x_0 + r(n/g) \colon 0 \le r < g\}$$

has exactly g distinct elements and is the complete set of solutions in $\mathbf{Z}_n$. Given any solution $x_0 + m(n/g)$ to the congruence, write

$$m = gq + r,$$

where $0 \le r < g$. Then

$$\begin{aligned} x_0 + m(n/g) &= x_0 + qn + r(n/g) \\ &\equiv x_0 + r(n/g) \pmod{n}. \end{aligned}$$

Thus, every solution to the congruence belongs to a residue class from our set above. Now suppose that $0 \le r_1 < r_2 < g$. Then

$$0 < r_2 - r_1 < g,$$

and so

$$0 < r_2(n/g) - r_1(n/g) < n.$$

Hence, the quantity $r_2(n/g) - r_1(n/g)$ is not divisible by n. As a result, the congruence classes represented by $x_0 + r_1(n/g)$ and $x_0 + r_2(n/g)$ are not the same, and our set above does, in fact, contain g distinct elements of $\mathbf{Z}_n$. ▯

Example 3.11 Let's find all solutions to the linear congruence

$$12345x \equiv 99 \pmod{54321}.$$

In Example 2.28, we found a solution to the equation $12345x + 54321y = 99$ in which $x = 119361$. Thus $x = 119361$ is one solution to the congruence. Since $119361 \equiv 10719 \pmod{54321}$, we can work with the value $x = 10719$ as our first solution to the congruence. In Example 2.28, we also found that $\gcd(12345, 54321) = 3$, and so there are exactly three solutions to the congruence. According to the theorem, they are

$$\begin{aligned} x_1 &= 10719 \\ x_2 &= 10719 + 1 \cdot 54321/3 = 28826 \\ x_3 &= 10719 + 2 \cdot 54321/3 = 46933. \end{aligned}$$

▯

COROLLARY 3.12

If p is a prime number, and a is not divisible by p, then the congruence

$$ax \equiv b \pmod{p}$$

has exactly one solution.

This corollary tells us that a linear congruence has exactly one solution when we are working in $\mathbf{Z}_p$ for a prime p.

COROLLARY 3.13
If $\gcd(a, n) = 1$ *and*

$$ax \equiv ab \pmod{n},$$

then $x \equiv b \pmod{n}$.

PROOF As a result of the theorem, there is exactly one solution to this congruence, and since $x \equiv b \pmod{n}$ is a solution, it must be the only one. ▯

This corollary says that we may cancel like terms relatively prime to n from both sides of a congruence in $\mathbf{Z}_n$. If n is a prime number, then the condition $\gcd(a, n) = 1$ is equivalent to $a \not\equiv 0 \pmod{n}$. If n is not prime, it is possible to have $ax \equiv ab \pmod{n}$, but $x \not\equiv b \pmod{n}$ even when $a \not\equiv 0 \pmod{n}$. For instance,

$$4 \cdot 2 \equiv 4 \cdot 5 \pmod{6}.$$

This does not contradict the corollary since $\gcd(4, 6) \neq 1$.

Let $\mathbf{Z}_n^*$ be the subset of $\mathbf{Z}_n$ consisting of residue classes containing numbers relatively prime to n. This set is closed under multiplication, as a product of two numbers relatively prime to n is also relatively prime to n. The theorem tells us that the congruence

$$ax \equiv 1 \pmod{n}$$

has a solution if and only if a is in $\mathbf{Z}_n^*$. Furthermore, when a is in $\mathbf{Z}_n^*$, the solution to the congruence is unique modulo n.

DEFINITION 3.14 *If a is an element of $\mathbf{Z}_n^*$, the inverse of a (modulo n) is the element b in $\mathbf{Z}_n^*$ satisfying*

$$ab \equiv 1 \pmod{n}.$$

Example 3.15 The inverse of 7 modulo 10 is 3 since

$$7 \cdot 3 \equiv 1 \pmod{10}.$$

Notice that 9 is its own inverse (modulo 10) since

$$9 \cdot 9 \equiv 1 \pmod{10}.$$

▯

We will use the notation a^{-1} to represent the residue class of the inverse of a modulo n. Furthermore, if $n > 0$, we let $a^{-n} = (a^{-1})^n$, and we also let $a^0 = 1$. With this notation, the usual exponent rules hold for multiplication in $\mathbf{Z}_n^*$, i.e., if a is in $\mathbf{Z}_n^*$ then $(a^m)^n = a^{mn}$ and $a^m a^n = a^{m+n}$ for all integers m and n.

Exercises

3.10 What is the value of x modulo 47 that satisfies

$$7x \equiv 20 \pmod{47}\,?$$

3.11 Find all solutions in $\mathbf{Z}_{68}$ to the congruence

$$12x \equiv 20 \pmod{68}.$$

3.12 Determine the number of distinct solutions in $\mathbf{Z}_{56}$ to the following congruences.

(a) $21x \equiv 42 \pmod{56}$

(b) $24x \equiv 42 \pmod{56}$

(c) $27x \equiv 42 \pmod{56}$

3.13 Find a solution (by hand) to the congruence

$$1234x \equiv 123 \pmod{4321}.$$

3.14 (a) Considering $a = 5$ as an element of $\mathbf{Z}_{11}^*$, determine a^{-1}.

(b) Considering $a = 5$ as an element of $\mathbf{Z}_{12}^*$, determine a^{-1}.

$\diamond$**3.15** (a) Determine the inverse of 21 modulo 100.

(b) Use a computer to find the inverse of 21 modulo 1999.

3.16 Let p be a prime number and a a number not divisible by p. Show that the congruence

$$ax \equiv b \pmod{p^n}$$

has exactly one solution in $\mathbf{Z}_{p^n}$.

3.17 (a) Prove that a positive integer is divisible by 9 if and only if the sum of its digits is divisible by 9.

(b) Prove that a positive integer is divisible by 11 if and only if the alternating sum of its digits is divisible by 11.

3.18 Prove that if $d \mid a$, $d \mid b$, and $a \equiv b \pmod{m}$, then $a/d \equiv b/d \pmod{m/\gcd(m,d)}$.

3.19 This exercise provides a method for constructing a round-robin tournament—that is, a tournament in which every team plays every other team exactly once, playing only one game per round. Obviously, if the number of teams is odd, not every team can play in a round. In this case we add a team called 'bye' to the list, and the team that is matched with 'bye' in a given round does not play in that round. Number the teams 1, 2, ..., n (and we now assume that n is even). For $a, b < n$, schedule teams a and b to play in round r if

$$a + b \equiv r \pmod{n-1}.$$

Team a plays team n in round r if

$$2a \equiv r \pmod{n-1}.$$

(a) Prove that, under this assignment, each team plays exactly once a round and that each team plays every other team at some point.

(b) Write out a round-robin tournament schedule for five teams.

3.3 Application: Check digits and the ISBN system

Congruences are frequently used to provide an efficient way to detect errors in data transmission. Suppose that we have a sequence of nine-digit numbers that we need to enter into a computer. It is important that the data be entered correctly, but the quantity of numbers to be entered is large enough that we prefer not to double-check them. Instead, we add a tenth digit, called a "check digit," to each number that will detect some of our errors. If our nine-digit number is $x_1x_2\ldots x_9$, we define the tenth digit x_{10} to be the number that satisfies

$$x_{10} \equiv (x_1 + x_2 + \cdots + x_9) \pmod{10}. \tag{3.3}$$

We ask the computer to alert us any time we enter a ten-digit number that does not satisfy the above congruence. Notice that if we take a ten-digit number $x_1x_2\ldots x_{10}$ that satisfies the congruence and replace exactly one digit with a different digit, then the resulting number no longer satisfies the congruence. Thus, our tenth digit detects an error when we have typed exactly one digit incorrectly. Obviously, this will not catch all of our errors, but we only needed to enter one extra digit rather than retype all nine digits.

The ISBN (International Standard Book Number) scheme employs a slightly more sophisticated check digit. The ISBN of a book is a 10-digit number grouped into four blocks of numbers. For example, the ISBN for the fourth edition of *The Mathematica Book* [34] is 0-521-64314-7. The first block is determined by the country of publication. For books published in the U.S., U.K., Australia, New Zealand, or Canada, this number is 0. The second block indicates the publisher. Any book that has a 521 as its second block is published by Cambridge University Press. The third block of the ISBN identifies the title and edition of the book. The final block is the check digit. If x_1, x_2, ..., x_9 are the first nine digits of an ISBN, then the check digit is the number

x_{10} satisfying

$$x_{10} \equiv \sum_{i=1}^{9} i x_i \pmod{11}. \tag{3.4}$$

(In the case $x_{10} = 10$, the character X is used as the tenth digit.) Let's perform this calculation for the check digit of *The Mathematica Book.*

$$1 \cdot 0 + 2 \cdot 5 + 3 \cdot 2 + 4 \cdot 1 + 5 \cdot 6 + 6 \cdot 4 + 7 \cdot 3 + 8 \cdot 1 + 9 \cdot 4 \equiv 7 \pmod{11}$$

The check digit of an ISBN detects not only errors in which one digit has been incorrectly entered, but also errors in which two digits have been interchanged. Certainly, these would be very typical kinds of errors if the numbers are entered by hand. Let's see what happens when we make errors of these two types. First, consider the number 0-521-64714-7 obtained from the ISBN of *The Mathematica Book* by changing the seventh digit from a 3 to 7.

$$1 \cdot 0 + 2 \cdot 5 + 3 \cdot 2 + 4 \cdot 1 + 5 \cdot 6 + 6 \cdot 4 + 7 \cdot 7 + 8 \cdot 1 + 9 \cdot 4 \equiv 2 \pmod{11}$$

The check digit formula produces a 2 instead of 7. If we swap the third and fourth digits (in the original ISBN) we obtain the number 0-512-64314-7.

$$1 \cdot 0 + 2 \cdot 5 + 3 \cdot 1 + 4 \cdot 2 + 5 \cdot 6 + 6 \cdot 4 + 7 \cdot 3 + 8 \cdot 1 + 9 \cdot 4 \equiv 8 \pmod{11}$$

Again, the congruence defining the check digit produces a number other than 7. This will always happen for these types of errors.

PROPOSITION 3.16

If $x_1 x_2 \ldots x_{10}$ is a valid ISBN and the number $x'_1 x'_2 \ldots x'_{10}$ is obtained from that number by either altering exactly one digit or by interchanging two unequal digits, then

$$x'_{10} \not\equiv \sum_{i=1}^{9} i x'_i \pmod{11}.$$

PROOF In a valid ISBN,

$$\sum_{i=1}^{10} i x_i \equiv 0 \pmod{11}.$$

We will prove that

$$\sum_{i=1}^{10} i x'_i \not\equiv 0 \pmod{11}.$$

Suppose that we have an error of the first type; say, $x_j \neq x'_j$ for some j. Then

$$\begin{aligned} \sum_{i=1}^{10} i x'_i &\equiv \sum_{i=1}^{10} i x_i - j x_j + j x'_j \pmod{11} \\ &\equiv j(x'_j - x_j) \pmod{11}. \end{aligned}$$

Since $j(x'_j - x_j)$ is not divisible by 11, we have proved that the required congruence fails.

Now suppose that we have an error of the second type, i.e., $x'_j = x_k$ and $x'_k = x_j$, for some $k \neq j$ and $x_j \neq x_k$. Then

$$\begin{aligned}\sum_{i=1}^{10} ix'_i &\equiv \left(\sum_{i=1}^{10} ix_i\right) + jx_k - jx_j + kx_j - kx_k \pmod{11}\\ &\equiv (k-j)(x_j - x_k) \pmod{11}.\end{aligned}$$

The integers $k-j$ and $x_j - x_k$ are nonzero and have absolute value less than 10. In particular, neither is divisible by 11, and so their product is not divisible by 11. □

Exercises

3.20 Which of the following numbers are valid ISBN's?

(a) 0-374-46868-0
(b) 0-124-98040-7
(c) 0-395-33957-X

3.21 Compute the check digit for the ISBN 3-564-56764-.

3.22 The following code number was obtained from a valid ISBN by changing the fourth digit: 5-382-14572-2. Determine the original ISBN.

3.23 The following code number was obtained from a valid ISBN by transposing two adjacent digits: 5-382-14572-2. Determine the original ISBN.

3.24 Give an example of a ten-digit number that does not satisfy the ISBN congruence, but can be modified by interchanging a single pair of adjacent digits to obtain a valid ISBN in several different ways.

Exercises 3.25–3.30 refer to the Universal Product Code, which we now describe. The Universal Product Code (UPC) is another product numbering scheme that makes use of a check digit. The code consists of a 12-digit number found underneath the familiar bar code. The first six digits identify the manufacturer, the next five identify the product, and the last digit serves as a check digit. The digits of a valid UPC number $x_1x_2 \dots x_{12}$ satisfy

$$3x_1 + x_2 + 3x_3 + x_4 + \cdots + 3x_{11} + x_{12} \equiv 0 \pmod{10}.$$

3.25 Which of the following are valid UPC numbers?

(a) 6-97363-76105-3
(b) 3-58520-98817-1

(c) 9-84774-13191-0

3.26 Compute the check digit for the UPC number whose first 11 digits are 7-40755-87915-.

3.27 If $x_1x_2\ldots x_{12}$ is a valid UPC number, show that

$$x_{12} \equiv 7x_1 + 9x_2 + 7x_3 + 9x_4 + \cdots + 9x_{10} + 7x_{11} \pmod{10}.$$

3.28 Show that the UPC scheme detects errors in which exactly one digit is incorrect. That is, show that if $x'_1x'_2\ldots x'_{12}$ is obtained from a valid UPC number by changing exactly one digit, then

$$3x'_1 + x'_2 + 3x'_3 + x'_4 + \cdots + 3x'_{11} + x'_{12} \not\equiv 0 \pmod{10}.$$

3.29 Under what conditions will the UPC scheme detect an error in which two consecutive digits are interchanged?

3.30 The 12-digit number 2-38074-90195-2 is not a valid UPC number. Determine the possible correct UPC numbers under each of the following assumptions.

(a) The third number is incorrect.

(b) The fourth number is incorrect.

(c) Two consecutive digits are interchanged.

3.4 Fermat's theorem and Euler's theorem

High school math contests often have problems that ask the students to compute the ones digit of a number such as 3^{2005}. The trick is to observe a pattern in the ones digit as successive powers of 3 are computed.

k	3^k	Ones digit of 3^k
1	3	3
2	9	9
3	27	7
4	81	1
5	243	3
6	729	9
7	2187	7
8	6561	1

The last column will continue to cycle through the numbers 3, 9, 7, 1. In particular, if k is a multiple of 4, then the ones digit of 3^k is 1. Since 2005 is

1 more than a multiple of 4 (i.e., $2005 = 4 \cdot 501 + 1$), the ones digit of 3^{2005} is the number that comes after 1 in our cycle, which is 3.

What we have done is a computation mod 10. The numbers in the second and third columns are congruent mod 10. If k is a multiple of 4, then

$$3^k \equiv 1 \pmod{10}.$$

If we replace the base with numbers that are not divisible by 2 or 5, we get a similar result. That is,

$$a^4 \equiv 1 \pmod{10}$$

for all numbers a not divisible by 2 or 5. (Why could this statement never be true for numbers divisible by 2 or 5?) In this section, we will show that, given $n > 0$, there exists a number $k > 0$ such that

$$a^k \equiv 1 \pmod{n}$$

whenever $\gcd(a, n) = 1$, and we will explain how to find such a k.

Example 3.17 Let's find a number k such that

$$a^k \equiv 1 \pmod{5} \tag{3.5}$$

for all numbers a relatively prime to 5. Note, for example, if we find a k that works for $a = 2$, then it will also work for all members of the equivalence class of 2 modulo 5. So we just need to find a k that satisfies (3.5) when $a = 1$, 2, 3, and 4. Based on the following table we see that $k = 4$ suffices.

a	$a^2 \pmod{5}$	$a^3 \pmod{5}$	$a^4 \pmod{5}$
1	1	1	1
2	4	3	1
3	4	2	1
4	1	4	1

□

THEOREM 3.18 (Fermat's (little) theorem)
If p is prime and $a \not\equiv 0 \pmod{p}$, then

$$a^{p-1} \equiv 1 \pmod{p}.$$

PROOF Consider the residues mod p of the following list of numbers:

$$a,\ 2a,\ 3a,\ \ldots,\ (p-1)a. \tag{3.6}$$

We claim that these residues are distinct. If $xa \equiv ya \pmod{p}$, then Corollary 3.13 implies that $x \equiv y \pmod{p}$. If xa and ya are in the list in (3.6), then $1 \le x, y \le p - 1$, and so $x = y$.

Since there are $p-1$ distinct residues in (3.6) and they are all nonzero, this list must be the list 1, 2, 3, ..., $p-1$, in some order. Hence,

$$\begin{aligned}\prod_{x=1}^{p-1} x &\equiv \prod_{x=1}^{p-1} ax \pmod{p} \\ &\equiv a^{p-1} \prod_{x=1}^{p-1} x \pmod{p}.\end{aligned}$$

Since $\prod_{x=1}^{p-1} x$ is not divisible by p, we may apply Corollary 3.13 again to cancel this factor and obtain $1 \equiv a^{p-1} \pmod{p}$. □

Example 3.19 Since 7 is a prime number, we have

$$2^6 \equiv 1 \pmod{7}.$$

We check: $2^6 = 64$ and $7 \mid (64-1)$. □

A version of Fermat's little theorem holds for all integers a by including the trivial congruence $0^p \equiv 0 \pmod{p}$.

COROLLARY 3.20
For any prime p and any a, we have

$$a^p \equiv a \pmod{p}.$$

We can use the contrapositive of Fermat's theorem to try to test whether a given number is prime. For instance, consider the following computations.

Mathematica

```
Mod[2^320,321]

4
```

Maple

```
modp(2^320,321);

4
```

If 321 were prime, Fermat's theorem would imply that

$$2^{320} \equiv 1 \pmod{321}.$$

Since our computations indicate that this is not true, then we must conclude that 321 is not prime.

It is important to note that the converse of Fermat's theorem is not true in general. That is, there exist numbers n and a, with n composite, for which

$$a^{n-1} \equiv 1 \pmod{n}.$$

For example, let $a = 2$ and $n = 341 = 11 \cdot 31$. Since 11 and 31 are prime numbers, we can calculate using Fermat's little theorem: $2^{340} \equiv (2^{10})^{34} \equiv 1^{34} \equiv 1 \pmod{11}$, and $2^{340} \equiv (2^{30})^{11} \cdot 2^{10} \equiv 1^{11} \cdot (1024) \equiv 1 \pmod{31}$. It follows by Exercise 2.22 that $2^{340} \equiv 1 \pmod{341}$.

DEFINITION 3.21 *If $a > 1$ and $a^{n-1} \equiv 1 \pmod{n}$, for n composite, then n is a* pseudoprime *to the* base a.

Note that we have shown that 341 is pseudoprime to the base 2.

Let $P_a(x)$ be the number of pseudoprimes to the base a less than or equal to x. Erdős showed that

$$\lim_{x \to \infty} \frac{P_a(x)}{x} = 0.$$

That is, pseudoprimes to the base a are quite rare. In fact, among the first $25 \cdot 10^9$ positive integers, the number of pseudoprimes to the bases 2 and 3 simultaneously is only 4704. The rarity of pseudoprimes yields a fairly reliable test of whether a given number is prime. If there are several bases, a, such that $a^{n-1} \equiv 1 \pmod{n}$, then n is probably a prime. See Chapter 7 for details.

We now consider the general problem of finding, for any positive number n, a positive number k such that

$$a^k \equiv 1 \pmod{n},$$

for $\gcd(a, n) = 1$. Recall from the discussion at the beginning of the section that when $n = 10$, we may take $k = 4$. It is not an accident that 4 is also the number of residue classes modulo 10 containing numbers relatively prime to 10.

DEFINITION 3.22 *Euler's ϕ-function (totient function), defined for all natural numbers, is given by*

$$\phi(n) = |\{x : 1 \leq x \leq n \text{ and } \gcd(x, n) = 1\}|. \tag{3.7}$$

Example 3.23 Let $n = 10$. There are four numbers in the set $\{1, \ldots, 10\}$ that are relatively prime to 10, namely, the integers 1, 3, 7, and 9. Therefore $\phi(n) = 4$. □

Example 3.24 If p is a prime number, then the numbers 1, 2, ..., $p-1$ are all relatively prime to p, and so $\phi(p) = p - 1$. ∎

THEOREM 3.25 (Euler's theorem)
If $\gcd(a, n) = 1$, *then*

$$a^{\phi(n)} \equiv 1 \pmod{n}.$$

This result generalizes Fermat's little theorem. In fact, the proof we will give for Euler's theorem is nearly identical to the one we gave for Fermat's little theorem. The key to the latter proof was a bijection from the set $\mathbf{Z}_p \setminus \{0\}$ to itself. Recall that $\mathbf{Z}_n^*$ is the set of classes in $\mathbf{Z}_n$ containing numbers relatively prime to n. The number of elements in $\mathbf{Z}_n^*$ is $\phi(n)$. To prove Euler's theorem, we will exhibit a bijection from this set to itself.

PROOF Define a function $f : \mathbf{Z}_n^* \to \mathbf{Z}_n^*$ by the rule $f(x) = ax$. Corollary 3.13 implies that this function is one-to-one. Since f maps a finite set to itself, it is a bijection. It follows that the product of all the domain elements of the function is congruent to the product of all the range elements (the two sets are the same):

$$\prod_{x \in \mathbf{Z}_n^*} ax \equiv \prod_{x \in \mathbf{Z}_n^*} x \pmod{n}.$$

By Corollary 3.13, we may cancel the x's on both sides and obtain

$$a^{\phi(n)} \equiv 1 \pmod{n}.$$

∎

Example 3.26 Since $\phi(10) = 4$ and $\gcd(7, 10) = 1$, Euler's theorem tells us that

$$7^4 \equiv 1 \pmod{10}.$$

∎

Euler's theorem provides another way to solve linear congruences. The *general linear congruence*

$$ax + b \equiv 0 \pmod{n}, \tag{3.8}$$

where $\gcd(a, n) = 1$, has the solution

$$x \equiv -a^{\phi(n)-1}b \pmod{n}. \tag{3.9}$$

However, the method shown in Chapter 2, using the Euclidean algorithm, is much faster for large values of n.

Exercises

3.31 Reduce the following numbers modulo 7 (without using a calculator/computer

(a) 3^4

(b) 3^6

(c) 3^7

(d) 3^{18}

(e) 3^{19}

(f) 3^{74}

3.32 Let p be a prime number, and suppose that $s \equiv t \pmod{p-1}$ for some positive integers s and t. Show that

$$a^s \equiv a^t \pmod{p}$$

for all integers a.

3.33 Reduce the following numbers modulo 15 (without using a calculator/computer).

(a) 2^3

(b) 2^8

(c) 2^9

(d) 2^{32}

(e) 2^{83}

3.34 Let n be a positive integer, and suppose that $s \equiv t \pmod{\phi(n)}$ for some positive numbers s and t. Show that

$$a^s \equiv a^t \pmod{n}$$

for all numbers a satisfying $\gcd(a, n) = 1$.

†3.35 Let p be a prime. Prove by mathematical induction that $a^p \equiv a \pmod{p}$ for all integers $a \geq 0$.

⋆3.36 Let q_n be the probability that an integer m chosen at random from the first n integers will have the property that $m^m - 1$ is divisible by 7. Find $\lim_{n\to\infty} q_n$.

⋄3.37 Find an even number n such that there exists no odd number m with $\phi(m) = \phi(n)$.

◇**3.38** Execute the *Mathematica* command

```
Table[GCD[10^{k} + 234234371, 10^500 + 234523452532457],
{k, 1, 1000}]
```

or the equivalent *Maple* command

```
seq(gcd(10^k + 234234371, 10^500 + 234523452532457),
k=1..1000);
```

and explain the output.

†**3.39** Show that if n is a pseudoprime to the base 2, then $2^n - 1$ is a greater pseudoprime to the base 2. Since 341 is a pseudoprime to the base 2, this result shows that there are infinitely many pseudoprimes.

3.5 The Chinese remainder theorem

In light of Euler's theorem, we will want to be able to compute $\phi(n)$. We have already observed that $\phi(p) = p - 1$ when p is a prime number. If n is a large composite number, then it is impractical to count the number of integers between 1 and n that are relatively prime to n. However if we know the factorization of n, then we will be able to break the problem down. We will show that if $n = n_1 n_2$, where n_1 and n_2 are relatively prime, then $\phi(n) = \phi(n_1)\phi(n_2)$. To prove this, we will show that there is a one-to-one correspondence between elements of $\mathbf{Z}_n^*$ and ordered pairs (r_1, r_2) in $\mathbf{Z}_{n_1}^* \times \mathbf{Z}_{n_2}^*$.

Consider the system of congruences

$$\begin{aligned} x &\equiv r_1 \pmod{n_1} \\ x &\equiv r_2 \pmod{n_2}. \end{aligned} \tag{3.10}$$

All numbers of the form $x = r_1 + n_1 k$ are solutions to the first congruence. If we substitute this into the second congruence and rearrange, we obtain

$$n_1 k = r_2 - r_1 \pmod{n_2}. \tag{3.11}$$

So a simultaneous solution to (3.10) exists if a value of k exists such that (3.11) holds. If n_1 and n_2 are relatively prime, then such a k is guaranteed to exist.

We further claim, under the assumption that $\gcd(n_1, n_2) = 1$, that the solution to (3.10) is unique modulo $n_1 n_2$. If x_1 and x_2 are solutions, then $x_1 \equiv x_2 \pmod{n_1}$ and $x_1 \equiv x_2 \pmod{n_2}$. It follows that n_1 and n_2 both divide $x_1 - x_2$, and since n_1 and n_2 are relatively prime, their product will also divide $x_1 - x_2$. Hence, $x_1 \equiv x_2 \pmod{n_1 n_2}$.

We now generalize this to a system of k congruences.

THEOREM 3.27 (Chinese remainder theorem)
If $n_1, n_2, \ldots, n_k$ are pairwise relatively prime numbers, and $r_1, r_2, \ldots, r_k$ are any numbers, then there exists a value of x satisfying the simultaneous congruences

$$\begin{aligned} x &\equiv r_1 \pmod{n_1} \\ x &\equiv r_2 \pmod{n_2} \\ &\vdots \\ x &\equiv r_k \pmod{n_k}. \end{aligned}$$

Furthermore, x is unique modulo $n_1 n_2 \ldots n_k$.

PROOF Let $n = n_1 n_2 \ldots n_k$. By Theorem 3.10, there exists, for $1 \leq i \leq k$, a number s_i such that $(n/n_i)s_i \equiv r_i \pmod{n_i}$. Let

$$x = \sum_{i=1}^{k} \frac{n}{n_i} s_i.$$

Notice that n_i divides n/n_j, for $i \neq j$. Hence

$$x \equiv \frac{n}{n_i} s_i \equiv r_i \pmod{n_i},$$

and so x does indeed satisfy the above congruences.

To check the uniqueness, we proceed by induction. Clearly, there is a unique value of x modulo n_1 satisfying the first congruence. Now assume that the solution to the first $k-1$ congruences is unique modulo $n_1 n_2 \ldots n_{k-1}$. Suppose that x_1 and x_2 are both solutions to all k congruences. Then

$$\begin{aligned} x_1 &\equiv x_2 \pmod{n_1 n_2 \ldots n_{k-1}} \\ x_1 &\equiv x_2 \pmod{n_k}. \end{aligned}$$

It follows that $x_1 - x_2$ is divisible by both $n_1 n_2 \ldots n_{k-1}$ and n_k. Since $n_1 n_2 \ldots n_{k-1}$ and n_k are relatively prime, their product divides $x_1 - x_2$, and so $x_1 \equiv x_2 \pmod{n_1 n_2 \ldots n_k}$. ∎

Example 3.28 Let's apply the Chinese remainder theorem to the system of three congruences $x \equiv 2 \pmod 4$, $x \equiv 3 \pmod 5$, and $x \equiv 7 \pmod 9$. The proof of the theorem illustrates the method for obtaining a simultaneous solution to these congruences. Let $n = 4 \cdot 5 \cdot 9 = 180$. Then we obtain the three auxiliary congruences $45 s_1 \equiv 2 \pmod 4$, $36 s_2 \equiv 3 \pmod 5$, and $20 s_3 \equiv 7 \pmod 9$, whose solutions are $s_1 = 2$, $s_2 = 3$, and $s_3 = 8$. Thus, a simultaneous solution to the given system of congruences is $x = 45 s_1 + 36 s_2 + 20 s_3 = 358 \equiv 178 \pmod{180}$. ∎

We now apply the Chinese remainder theorem to deduce the previously stated multiplicativity property of ϕ.

PROPOSITION 3.29
If n_1 and n_2 are relatively prime positive integers, then

$$\phi(n_1 n_2) = \phi(n_1)\phi(n_2).$$

PROOF As mentioned above, we will establish a bijection between $\mathbf{Z}_{n_1 n_2}^*$ and $\mathbf{Z}_{n_1}^* \times \mathbf{Z}_{n_2}^*$. Suppose that $z \in \mathbf{Z}_{n_1 n_2}^*$. Reducing z modulo n_1, there is a unique $x \in \mathbf{Z}_{n_1}^*$ such that $z \equiv x \pmod{n_1}$. Similarly, there is a unique $y \in \mathbf{Z}_{n_2}^*$ such that $z \equiv y \pmod{n_2}$. Hence, each $z \in \mathbf{Z}_{n_1 n_2}^*$ is associated with a unique pair $(x, y) \in \mathbf{Z}_{n_1}^* \times \mathbf{Z}_{n_2}^*$. Given an ordered pair $(x, y) \in \mathbf{Z}_{n_1}^* \times \mathbf{Z}_{n_2}^*$, the Chinese remainder theorem guarantees the existence of a unique value of z (modulo n) such that $z \equiv x \pmod{n_1}$ and $z \equiv y \pmod{n_2}$. (Note that z must be relatively prime to $n_1 n_2$, since x is relatively prime to n_1 and y is relatively prime to n_2.) This establishes the bijection between $\mathbf{Z}_{n_1}^* \times \mathbf{Z}_{n_2}^*$ and $\mathbf{Z}_{n_1 n_2}^*$. □

Example 3.30 We have

$$\phi(77) = \phi(7)\phi(11) = 6 \cdot 10 = 60.$$

□

More generally, if p and q are any pair of distinct prime numbers, then

$$\phi(pq) = (p-1)(q-1).$$

THEOREM 3.31
Suppose that the canonical factorization of n is

$$n = \prod_{i=1}^{k} p_i^{\alpha_i}.$$

Then

$$\phi(n) = n \prod_{i=1}^{k} \left(1 - \frac{1}{p_i}\right).$$

PROOF First, we compute $\phi(p^k)$, where p is a prime number. Of the numbers $1, \ldots, p^k$, the ones *not* relatively prime to p are $p, 2p, 3p, \ldots, p^k$.

Therefore, $\phi(p^k) = p^k - p^{k-1}$. Now, using Proposition 3.29,

$$\begin{aligned}\phi(n) &= \prod_{i=1}^{k} \phi(p_i^{\alpha_i}) \\ &= \prod_{i=1}^{k} (p_i^{\alpha_i} - p_i^{\alpha_i - 1}) \\ &= \prod_{i=1}^{k} p_i^{\alpha_i} \left(1 - \frac{1}{p_i}\right) \\ &= n \prod_{i=1}^{k} \left(1 - \frac{1}{p_i}\right).\end{aligned}$$

□

Example 3.32

$$\phi(360) = \phi(2^3 \cdot 3^2 \cdot 5) = 360 \left(1 - \frac{1}{2}\right)\left(1 - \frac{1}{3}\right)\left(1 - \frac{1}{5}\right) = 96$$

□

Example 3.33 We compute

$$\phi(561) = \phi(3 \cdot 11 \cdot 17) = 561 \left(1 - \frac{1}{3}\right)\left(1 - \frac{1}{11}\right)\left(1 - \frac{1}{17}\right) = 320.$$

Hence, $a^{320} \equiv 1 \pmod{561}$ whenever $\gcd(a, 561) = 1$. The number 561 has the curious property that if a prime p divides 561, then $p - 1$ divides $561 - 1$ (i.e., 2, 10, and 16 divide 560). Applying Fermat's theorem, we have

$$\begin{aligned}a^{560} &\equiv (a^2)^{280} \equiv 1 \pmod{3} \\ a^{560} &\equiv (a^{10})^{56} \equiv 1 \pmod{11} \\ a^{560} &\equiv (a^{16})^{35} \equiv 1 \pmod{17},\end{aligned}$$

whenever $\gcd(a, 561) = 1$. For such a, the Chinese remainder theorem implies that

$$a^{560} \equiv 1 \pmod{561}.$$

That is, 561 is a pseudoprime to every base a relatively prime to 561. □

DEFINITION 3.34 *A* Carmichael number *is a composite number n that satisfies $a^{n-1} \equiv 1 \pmod{n}$ for every base a with $\gcd(a, n) = 1$. Carmichael numbers are also called* absolute pseudoprimes.

THEOREM
There are infinitely many Carmichael numbers.

See [1] for a proof of this assertion. Carmichael numbers are named after Robert Carmichael (1879–1967).

Exercises

3.40 Find an integer x that satisfies

$$\begin{aligned} x &\equiv 6 \pmod{15} \\ x &\equiv 9 \pmod{14}. \end{aligned}$$

3.41 Find an integer x that satisfies

$$\begin{aligned} x &\equiv 3 \pmod{4} \\ x &\equiv 2 \pmod{9} \\ x &\equiv 1 \pmod{25}. \end{aligned}$$

3.42 Find a simultaneous solution modulo 455 to the system of congruences $x \equiv 1 \pmod{5}$, $x \equiv 2 \pmod{7}$, and $x \equiv 3 \pmod{13}$.

3.43 Show that there does not exist an integer x such that

$$\begin{aligned} x &\equiv 5 \pmod{6} \\ x &\equiv 4 \pmod{10}. \end{aligned}$$

3.44 Show that the system of congruences

$$\begin{aligned} x &\equiv 5 \pmod{6} \\ x &\equiv 7 \pmod{10} \end{aligned}$$

has more than one solution modulo 60.

3.45 Show that the system of congruences

$$\begin{aligned} x &\equiv a \pmod{n_1} \\ x &\equiv b \pmod{n_2} \end{aligned}$$

has a solution if and only if $\gcd(n_1, n_2) \mid b - a$.

3.46 Compute the following.

(a) $\phi(31)$

(b) $\phi(35)$

(c) $\phi(81)$

(d) $\phi(100)$

(e) $\phi(144)$

3.47 Show that $\phi(n)$ is even for $n > 2$.

†⋆3.48 The inclusion–exclusion principle is a counting technique that generalizes the familiar Venn diagram rule $|A \cup B| = |A| + |B| - |A \cap B|$ (where A and B are finite sets).

Inclusion–exclusion principle: If $A_1, \ldots, A_n$ are subsets of a finite set S, then

$$|A_1 \cup \cdots \cup A_n| = \sum_{i=1}^{n} (-1)^{i+1} \sum |A_{k_1} \cap \cdots \cap A_{k_i}|,$$

where the second sum is over all i-tuples $(k_1, \ldots, k_i)$ with $1 \leq k_1 < \cdots < k_i \leq n$.

(a) Use induction to prove the inclusion–exclusion principle.

(b) Use the inclusion–exclusion principle to derive the formula of Theorem 3.31 for Euler's phi function $\phi(n)$.

◇3.49 Find a Carmichael number greater than 561.

⋆3.50 Let $a, b, c > 1$ and $\gcd(a, b) = 1$. Show that there exists a nonnegative integer n such that $\gcd(a + bn, c) = 1$.

†⋆3.51 The goal of this exercise is to show that

$$\sum_{d|n} \phi(d) = n.$$

Let $f(n) = \sum_{d|n} \phi(d)$.

(a) Show that if p is a prime, then $f(p^k) = p^k$.

(b) Show that if $\gcd(a, b) = 1$, then $f(ab) = f(a)f(b)$.

(c) Show that $f(n) = n$ for all positive numbers n.

3.6 Wilson's theorem

Wilson's theorem was first proved by Joseph-Luis Lagrange (1736–1813) and later by Carl Friedrich Gauss (1777–1855). John Wilson (1741–1793) is credited with conjecturing the result but not proving it.

THEOREM 3.35 (Wilson's theorem)
The congruence

$$(n-1)! \equiv -1 \pmod{n}$$

holds if and only if n is prime.

Although Wilson's theorem provides an "if and only if" criterion for whether a number n is prime, it is impractical when n is large.

PROOF First assume that n is composite: $n = ab$, for some $a > 1$, $b > 1$. Then a appears in the product $(n-1)!$, so $(n-1)! + 1 = nk$ cannot hold, for a would have to divide 1, which is impossible.

Now assume that n is prime. Then, for each $x \in \mathbf{Z}_n^*$, there exists a unique $y \in \mathbf{Z}_n^*$ with $xy \equiv 1 \pmod{n}$. The only values of x for which x is equal to y are solutions to the congruence $x^2 \equiv 1 \pmod{n}$. For these values of x, $n \mid (x^2-1) = (x-1)(x+1)$, i.e., $x \equiv \pm 1 \pmod{n}$, since n is prime. As the product of all the elements of $\mathbf{Z}_n^*$ is $(n-1)!$ and the elements other than 1 and -1 are paired to multiply to 1, we have

$$(n-1)! \equiv 1^{(n-3)/2} \cdot 1 \cdot (-1) \equiv -1 \pmod{n}.$$

∎

Looking further at Wilson's theorem, we notice that, for p an odd prime,

$$\begin{aligned}(p-1)! &= \left(1 \cdot 2 \cdot \cdots \cdot \left(\frac{p-1}{2}\right)\right)\left(\left(\frac{p+1}{2}\right) \cdot \cdots \cdot (p-1)\right) \\ &\equiv \left(1 \cdot 2 \cdot \cdots \cdot \left(\frac{p-1}{2}\right)\right)\left(\left(\frac{-p+1}{2}\right) \cdot \cdots \cdot (-1)\right) \pmod{p} \\ &\equiv \left(1 \cdot 2 \cdot \cdots \cdot \left(\frac{p-1}{2}\right)\right)^2 (-1)^{(p-1)/2} \pmod{p}.\end{aligned}$$

What is the value of $(-1)^{(p-1)/2}$? Since p is odd, $p = 4k+1$ or $p = 4k+3$, and $(-1)^{(p-1)/2} = +1$ or -1, respectively. Therefore, if $p = 4k+1$, then $x = 1 \cdot 2 \cdot \cdots \cdot (\frac{p-1}{2})$ is a solution to the congruence

$$x^2 \equiv -1 \pmod{p}.$$

Example 3.36 If we take $x = 1 \cdot 2 \cdot 3 \cdot 4 \cdot 5 \cdot 6 \equiv 5 \pmod{13}$, then $x^2 \equiv -1 \pmod{13}$. ∎

THEOREM 3.37
Let p be a prime. Then the congruence $x^2 \equiv -1 \pmod{p}$ has solutions if and only if $p = 2$ or $p \equiv 1 \pmod{4}$.

PROOF For $p = 2$, we have the solution $x = 1$. Suppose that $p = 4k + 1$. By the previous expansion of $(p-1)!$, we obtain a solution $x = \prod_{j=1}^{2k} j$.

Conversely, assume that $p \equiv 3 \pmod 4$, say $p = 4k + 3$, and suppose that there exists an integer x with $x^2 \equiv -1 \pmod p$. Then

$$x^{p-1} = (x^2)^{2k+1} \equiv (-1)^{2k+1} \equiv -1 \pmod p.$$

However, since $p \nmid x$, this contradicts Fermat's little theorem, which asserts that $x^{p-1} \equiv 1 \pmod p$. ▯

Recall that we have shown that there are infinitely many primes of the form $4n + 3$ (Proposition 2.38). The same is true for primes of the form $4n + 1$.

PROPOSITION 3.38

There exist infinitely many primes of the form $4n + 1$.

PROOF Suppose to the contrary that only a finite number of such primes exist and let N denote their product. Any prime factor p of $4N^2 + 1$ is then congruent to 3 modulo 4, but for $x = 2N$, we have $x^2 \equiv -1 \pmod p$, which contradicts the result of the previous theorem. ▯

Exercises

3.52 Show that if n is a composite number greater than 4, then $(n-1)! \equiv 0 \pmod n$.

◇**3.53** For a positive integer n, let

$$W(n) = \frac{(n-1)! + 1}{n}.$$

(a) Show that $W(n)$ is an integer if and only if $n = 1$ or n is prime.

(b) A *Wilson prime* is a prime number p such that $W(p)$ is divisible by p. There are only three known Wilson primes. Can you find them?

3.54 Find a solution to the congruence $x^2 \equiv -1 \pmod{17}$. How many solutions modulo 17 are there?

3.55 Find a solution to the congruence $x^2 \equiv -1 \pmod{65}$. How many solutions modulo 65 are there?

†**3.56** Suppose that p and q are distinct primes satisfying $p, q \equiv 1 \pmod 4$. Show that the congruence $x^2 \equiv -1 \pmod{pq}$ has a solution.

†**3.57** Suppose that n has a prime divisor p such that $p \equiv 3 \pmod 4$. Show that the congruence $x^2 \equiv -1 \pmod n$ does not have a solution.

$\star$**3.58** (For those who have studied field theory) Assume that F is a finite field with n elements. Show that F contains a solution to the equation $x^2 = -1$ only if n is even or $n = 4k + 1$.

Hint: The nonzero elements of a field form a cyclic group under multiplication. Consider a generator of the group.

3.7 Order of an element mod n

Euler's theorem tells us that if a is in $\mathbf{Z}_n^*$, then $a^{\phi(n)} \equiv 1 \pmod{n}$. For instance,

$$a^4 \equiv 1 \pmod{8},$$

for $a = 1, 3, 5, 7$. Notice that we can replace 4 with the smaller exponent 2. Thus

$$a^2 \equiv 1 \pmod{8},$$

for $a = 1, 3, 5, 7$.

DEFINITION 3.39 *Let a be an element of $\mathbf{Z}_n^*$. The* order of a modulo n *is the smallest positive number d such that*

$$a^d \equiv 1 \pmod{n}.$$

We may also say that d is the "order of a in $\mathbf{Z}_n^*$," or simply the "order of a" if the context is clear.

Example 3.40 The order of 1 modulo any n is always 1. In $\mathbf{Z}_8^*$, the orders of the remaining elements are all 2. ▯

Example 3.41 In $\mathbf{Z}_5^*$, the element 4 has order 2, while the elements 2 and 3 have order 4. ▯

Euler's theorem tells us that the order of an element in $\mathbf{Z}_n^*$ is at most $\phi(n)$. In fact, the order of a must divide $\phi(n)$. We will obtain this as a consequence of the following proposition.

PROPOSITION 3.42
Let a be an element of order d in $\mathbf{Z}_n^*$. Then

$$a^i \equiv a^j \pmod{n}$$

if and only if d divides $i - j$. In particular, $a^i \equiv 1 \pmod{n}$ if and only if d divides i.

PROOF Using the division algorithm, write

$$i - j = dq + r,$$

where $0 \le r < d$. We observe that

$$\begin{aligned} a^i &= a^{j+dq+r} \\ &= (a^j)(a^d)^q(x^r) \\ &\equiv (a^j)(a^r) \pmod{n}. \end{aligned}$$

If d divides $i - j$, then $r = 0$, and from the above calculation we see that $a^i \equiv a^j \pmod{n}$. Conversely, if $a^i \equiv a^j \pmod{n}$, then we may cancel them from both sides of the congruence above and obtain $a^r \equiv 1 \pmod{n}$. Since $0 \le r < d$, this implies that $r = 0$. ∎

COROLLARY 3.43

If the order of a modulo n is d, then $d \mid \phi(n)$

PROOF This follows at once from Euler's theorem and Proposition 3.42. ∎

Notice that if a is an element of order d modulo n, then

$$(a^k)^d \equiv a^{dk} \equiv (a^d)^k \equiv 1 \pmod{n}$$

for all integers k. Thus, by the theorem, a power of an element of order d has order dividing d. We can say precisely which divisor of d the order is in terms of the greatest common divisor of d and k.

PROPOSITION 3.44

Let a be any element of $\mathbf{Z}_n^*$ and d be the order a in $\mathbf{Z}_n^*$. Then the order of a^k in $\mathbf{Z}_n^*$ is $d/\gcd(d,k)$.

PROOF Let d_k be the order of a^k in $\mathbf{Z}_n^*$. On the one hand, we have

$$a^{kd_k} = (a^k)^{d_k} \equiv 1 \pmod{n}.$$

From Proposition 3.42, it follows that d divides kd_k and so $d/\gcd(d,k)$ divides $(k/\gcd(d,k))d_k$. Using Exercise 2.21 we conclude that $d/\gcd(d,k)$ divides d_k.

On the other hand,

$$\begin{aligned} (a^k)^{d/\gcd(d,k)} &\equiv (a^d)^{k/\gcd(d,k)} \pmod{n} \\ &\equiv 1^{k/\gcd(d,k)} \pmod{n} \\ &\equiv 1 \pmod{n}. \end{aligned}$$

Hence, by Proposition 3.42, d_k divides $d/\gcd(d,k)$, and so $d_k = d/\gcd(d,k)$. □

Example 3.45 Let's compute the orders of elements in $\mathbf{Z}_{11}^*$. We begin by computing powers of 2 modulo 11.

k	1	2	3	4	5	6	7	8	9	10
2^k mod 11	2	4	8	5	10	9	7	3	6	1

Thus, we see that the order of 2 modulo 11 is 10. Using Proposition 3.44, we can now quickly determine the orders of other elements. For instance, $5 \equiv 2^4 \pmod{11}$, and so the proposition tells us that 5 has order $10/\gcd(10,4) = 5$. As every element of $\mathbf{Z}_{11}^*$ is a power of 2, we can perform similar calculations to determine the orders of the remaining elements.

a	1	2	3	4	5	6	7	8	9	10
k (where $a \equiv 2^k \pmod{11}$)	0	1	8	2	4	9	7	3	6	5
$\gcd(10,k)$	10	1	2	2	2	1	1	1	2	5
Order of a modulo 11	1	10	5	5	5	10	10	10	5	2

□

Exercises

3.59 Determine the orders of elements in $\mathbf{Z}_{13}^*$.

3.60 Determine the orders of elements in $\mathbf{Z}_{14}^*$.

3.61 Determine the orders of elements in $\mathbf{Z}_{20}^*$.

3.62 Show that $n-1$ has order 2 modulo n.

3.63 Suppose that $\mathbf{Z}_n^*$ has an element of order 15. Show that it also has an element of order 3.

3.64 Show that if $\mathbf{Z}_n^*$ has an element of order d, then it has at least $\phi(d)$ elements of order d.

3.65 Let $a \in \mathbf{Z}_n^*$, and suppose that $a^k \not\equiv 1 \pmod{n}$, for $1 \le k \le \phi(n)/2$. Show that the order of a modulo n is $\phi(n)$.

3.66 Let a be an element of $\mathbf{Z}_n^*$ of order k. Show that if m divides n, then the order of a modulo m divides k.

3.67 Let a be an element of $\mathbf{Z}_n^*$. Suppose that $n = n_1 n_2$ with $\gcd(n_1, n_2) = 1$. Show that if k is the order of a modulo n_1 and modulo n_2, then k is the order of a modulo n.

3.68 Show that the fraction a/b, where $\gcd(a, b) = 1$ and neither 2 nor 5 divides b, when written in decimal form has period equal to the order of 10 in $\mathbf{Z}_b^*$.

3.8 Existence of primitive roots

For some values of n, there exist elements of order $\phi(n)$ in $\mathbf{Z}_n^*$ (for example, $n = 5$), while for others there do not (for example, $n = 8$). The existence of such an element gives rise to a special kind of structure in $\mathbf{Z}_n^*$.

PROPOSITION 3.46
The order of an element a in $\mathbf{Z}_n^*$ is $\phi(n)$ if and only if every element b in $\mathbf{Z}_n^*$ satisfies a congruence

$$b \equiv a^i \pmod{n}$$

for some integer i.

PROOF First, suppose that a is an element of order $\phi(n)$. Proposition 3.42 says that the set

$$\left\{1, a, a^2, a^3, \ldots, a^{\phi(n)-1}\right\}$$

contains $\phi(n)$ distinct elements of $\mathbf{Z}_n^*$. Thus, this set consists of every element of $\mathbf{Z}_n^*$.

Conversely, suppose that every element of $\mathbf{Z}_n^*$ can be expressed as a power of a. Let d be the order of a modulo n. Given an integer i, write

$$i = dq + r,$$

where $0 \le r < d$. Then

$$\begin{aligned} a^i &= a^{dq+r} \\ &= (a^d)^q a^r \\ &\equiv a^r \pmod{n}. \end{aligned}$$

Thus, every element of $\mathbf{Z}_n^*$ can be written in the form a^r, where $0 \le r < d$. It now follows that $d = \phi(n)$. ∎

DEFINITION 3.47 *An element g of $\mathbf{Z}_n^*$ is a* primitive root modulo n *if every element of $\mathbf{Z}_n^*$ is congruent modulo n to g^k for some integer k. Equivalently, g is a primitive root in $\mathbf{Z}_n^*$ if g has order $\phi(n)$.*

Note: A primitive root modulo n is also called a "generator" of $\mathbf{Z}_n^*$.

Example 3.48 We observe that 3 is a primitive root modulo 10, for

$$\mathbf{Z}_{10}^* = \{1, 3, 3^2, 3^3\} = \{1, 3, 9, 27\} = \{1, 3, 7, 9\}.$$

Notice that there are $\phi(4) = 2$ primitive roots modulo 10: 3^1 and 3^3. These are powers of 3 in which the exponents are relatively prime to 4. □

As we have observed, primitive roots modulo n do not exist for all n. For example, $\mathbf{Z}_8^*$ does not have a primitive root. Later, we will describe the set of integers n for which $\mathbf{Z}_n^*$ does contain a primitive root. For now, we observe that if a primitive root exists, then all primitive roots may be obtained as in the above example.

PROPOSITION 3.49
Suppose that $\mathbf{Z}_n^*$ contains a primitive root g. Then an element g^k in $\mathbf{Z}_n^*$ is another primitive root if and only if k is relatively prime to $\phi(n)$. Hence, the total number of primitive roots for such an n is $\phi(\phi(n))$.

PROOF By Proposition 3.44, the order of g^k is

$$\frac{\phi(n)}{\gcd(k, \phi(n))}.$$

This is equal to $\phi(n)$ if and only if k and $\phi(n)$ are relatively prime. □

Before we begin to investigate the existence of primitive roots, we will need to record some facts about the behavior of polynomials mod p for a prime p. We denote the set of polynomials with coefficients in $\mathbf{Z}_p$ as $\mathbf{Z}_p[x]$. We can think of elements of $\mathbf{Z}_p[x]$ as polynomials with integer coefficients where we identify two polynomials whose coefficients are the same mod p. We can add and multiply two polynomials in $\mathbf{Z}_p[x]$ in the usual way, reducing coefficients mod p as we go. For example, in $\mathbf{Z}_5[x]$, if

$$f(x) = 3x^2 + 2x + 1 \text{ and } g(x) = 3x + 1,$$

then

$$f(x) + g(x) = 3x^2 + 2 \text{ and } f(x)g(x) = 4x^3 + 4x^2 + 1.$$

Many theorems that are true for polynomials with coefficients in $\mathbf{Q}$, $\mathbf{R}$, etc. continue to be true for $\mathbf{Z}_p[x]$. These theorems hinge on the existence of a division algorithm similar to the one satisfied by $\mathbf{Z}$.

THEOREM 3.50 (Division algorithm for polynomials)
Let $f(x)$ and $g(x)$ be polynomials in $\mathbf{Z}_p[x]$ with $g(x) \neq 0$. There exist unique polynomials $q(x)$ and $r(x)$ in $\mathbf{Z}_p[x]$ such that

$$f(x) = g(x)q(x) + r(x),$$

where $r(x) = 0$, or the degree of $r(x)$ is less than the degree of $g(x)$.

The proof of this theorem is similar to that of Theorem 2.9. The key step is to show that if $f_1(x)$ and $g_1(x)$ are polynomials such that $f_1(x)$ has degree greater than or equal to that of $g_1(x)$, then there exists an element a of $\mathbf{Z}_p$ such that $f_1(x) - ax^k g_1(x)$ has smaller degree than $f_1(x)$, for some k. Such a step cannot in general be accomplished in $\mathbf{Z}_n[x]$ if n is composite. We leave the details of the proof as an exercise.

As a result of Proposition 3.3, if $f(x)$ and $g(x)$ represent the same polynomial in $\mathbf{Z}_p[x]$, then for any integer a, $f(a)$ and $g(a)$ will represent the same congruence class in $\mathbf{Z}_p$. Thus, it makes sense to evaluate elements of $\mathbf{Z}_p[x]$ at elements of $\mathbf{Z}_p$. We say that an element a of $\mathbf{Z}_p$ is a "root" of $f(x)$ in $\mathbf{Z}_p[x]$ if $f(a) \equiv 0 \pmod{p}$.

COROLLARY 3.51
Let $f(x)$ be a polynomial in $\mathbf{Z}_p[x]$ and a an element of $\mathbf{Z}_p$. Then a is a root of $f(x)$ if and only if $f(x) = (x - a)q(x)$ for some $q(x)$ in $\mathbf{Z}_p[x]$.

PROOF Let a be any element of $\mathbf{Z}_p$. By the previous theorem, we have

$$f(x) = (x - a)q(x) + r(x)$$

from some $q(x)$ and $r(x)$ such that the degree of $r(x)$ is less than the degree of $x - a$. Thus, $r(x)$ is a constant. In fact, we see that $r(x) = f(a)$, and so $f(a) \equiv 0 \pmod{p}$ if and only if $r(x) \equiv 0 \pmod{p}$. □

The main result we will need is the following theorem, which probably sounds familiar.

THEOREM 3.52
If $f(x)$ is a polynomial of degree n in $\mathbf{Z}_p[x]$, then $f(x)$ has at most n distinct roots in $\mathbf{Z}_p$.

PROOF We observe that Corollary 3.12 implies that a polynomial of degree 1 has exactly one root. Now suppose, by way of induction, that the result holds for polynomials of degree $n-1$. Let a be a root of $f(x)$. By the corollary above, we obtain a factorization

$$f(x) = (x - a)g(x),$$

where $g(x)$ has degree $n-1$. Suppose that a' is a second root of $f(x)$, different from a. Then

$$f(a') = (a - a')g(a') \equiv 0 \pmod{p}.$$

Since a and a' represent different elements of $\mathbf{Z}_p$, it must be the case that $g(a') \equiv 0 \pmod{p}$. Thus, all roots of $f(x)$ other than a are roots of $g(x)$. By our inductive hypothesis, there are at most $n-1$ of them. ∎

This theorem is not true in general in $\mathbf{Z}_n[x]$ when n is not prime. For instance, the polynomial $x^2 - 1$ in $\mathbf{Z}_8[x]$ has four roots in $\mathbf{Z}_8$.

We now return to our discussion of primitive roots.

THEOREM 3.53

If p is a prime, then $\mathbf{Z}_p^*$ has a primitive root.

PROOF For each d dividing $p-1$, let N_d be the number of elements in $\mathbf{Z}_p^*$ of order d. From Proposition 3.42 and Fermat's little theorem, we know that the order of every element divides $p-1$. As a result,

$$\sum_{d|(p-1)} N_d = p - 1.$$

Suppose that a is an element of $\mathbf{Z}_p^*$ of order d. By Proposition 3.42, the set $\{1, a, a^2, \ldots, a^{d-1}\}$ contains d distinct elements of $\mathbf{Z}_p^*$. Furthermore,

$$(a^k)^d = (a^d)^k \equiv 1 \pmod{p}.$$

Theorem 3.52 tells us that the congruence $x^d \equiv 1 \pmod{p}$ has at most d solutions in $\mathbf{Z}_p^*$, and so our set of powers of a must be the complete list of solutions. By Proposition 3.44, a^k will have order d if and only if d and k are relatively prime. We have shown that if $N_d > 0$, then $N_d = \phi(d)$. In particular, $N_d \leq \phi(d)$ for all divisors d of $p-1$. Using Exercise 3.51, we have

$$p - 1 = \sum_{d|(p-1)} \phi(d) \geq \sum_{d|(p-1)} N_d = p - 1.$$

Hence, $N_d = \phi(d)$ for all d dividing $p-1$, and in particular, there are elements ($\phi(p-1)$ of them) of order $p-1$. ∎

In the discussion that follows, let p be an odd prime. We can now use the above theorem to inductively obtain primitive roots in $\mathbf{Z}_{p^m}^*$ for all m. If g is an integer that is a primitive root modulo p, then by modifying g a bit we will be able to obtain a primitive root modulo p^2. Subsequently, we will be able to show that if g is an integer that is a primitive root modulo p^2, then g will also be a primitive root modulo p^m for all m.

THEOREM 3.54
If p is an odd prime, then $\mathbf{Z}_{p^2}^*$ has a primitive root.

PROOF Let g be an integer that is a primitive root modulo p. We set

$$g' = g + px$$

and will show that g' is a primitive root modulo p^2, for some choice of x. That is, we will show that an x exists so that the order of g' modulo p^2 is exactly $p(p-1)$.

By Corollary 3.43, we know that the order of g' in $\mathbf{Z}_{p^2}^*$ divides $p(p-1)$. Hence, the order of g' in $\mathbf{Z}_{p^2}^*$ is either k or kp for some integer k dividing $p-1$. We first show that we can choose x so that the order is not k for any k dividing $p-1$. To accomplish this, it is enough to show that an x exists such that $(g')^{p-1} \not\equiv 1 \pmod{p^2}$. Now

$$(g')^{p-1} \equiv g^{p-1} + p(p-1)g^{p-2}x \pmod{p^2}.$$

Since $g^{p-1} \equiv 1 \pmod{p}$, we can write $g^{p-1} = 1 + pr$ for some r. Factoring out g^{p-1} and replacing it with $1 + pr$, the right side from above is congruent to

$$(1 + pr)(1 + p(p-1)g^{-1}x) \pmod{p^2}.$$

(Note that g^{-1} is an integer from the residue class of the inverse of g modulo p.) Upon reducing mod p^2 we obtain

$$1 + p(r + (p-1)g^{-1}x).$$

Corollary 3.12 says that we can solve the congruence

$$r + (p-1)g^{-1}x \equiv c \pmod{p}$$

for any c that we like. In particular, we can find an x such that $r + (p-1)g^{-1}x$ is not divisible by p. For such an x, $(g')^{p-1} \not\equiv 1 \pmod{p^2}$, and so the order of g' in $\mathbf{Z}_{p^2}^*$ does not divide $p-1$.

With our choice of x, we now know that the order of g' in $\mathbf{Z}_{p^2}^*$ must be of the form kp for some k dividing $p-1$. Observe that

$$(g')^{kp} \equiv g^{kp} \pmod{p^2}.$$

Thus, if kp is the order of g' in $\mathbf{Z}_{p^2}^*$, then $g^{kp} \equiv 1 \pmod{p^2}$. Since $g^p \equiv g \pmod{p}$, we see that g^p is a primitive root modulo p, and so $g^{kp} \equiv 1 \pmod{p}$ if and only if $p-1$ divides k. We conclude that $k = p-1$, and so the order of g' in $\mathbf{Z}_{p^2}^*$ is in fact $p(p-1)$, as desired. $\square$

THEOREM 3.55
Let p be an odd prime. If g is an integer that is a primitive root in $\mathbf{Z}_{p^2}^*$, then g is also a primitive root in $\mathbf{Z}_{p^n}^*$ for all n.

PROOF By Corollary 3.43, the order of g in $\mathbf{Z}_{p^n}^*$ is kp^i for some k dividing $p-1$, and some integer i satisfying $0 \leq i \leq n-1$. Since g is a primitive root modulo p^2, it is also a primitive root modulo p. As in the above proof, g^{p^i} is also a primitive root modulo p, and so $g^{kp^i} \equiv 1 \pmod{p}$ forces k to be divisible by $p-1$. We conclude that the order of g in $\mathbf{Z}_{p^n}^*$ must actually be $(p-1)p^i$ for some i satisfying $0 \leq i \leq n-1$.

We now show that g is a primitive root in $\mathbf{Z}_{p^3}^*$. Since g is a primitive root modulo p^2, we know that $g^{p-1} \equiv 1 \pmod{p}$, but $g^{p-1} \not\equiv 1 \pmod{p^2}$. Thus, we can write

$$g^{p-1} = 1 + rp,$$

where r is not divisible by p. Raising this element to the pth power, we get

$$(g^{p-1})^p \equiv 1 + rp^2 \pmod{p^3}.$$

(This is where we are using the fact that $p \neq 2$.) We conclude that g has order $(p-1)p^2$ in $\mathbf{Z}_{p^3}^*$, and so is a primitive root. Using an inductive argument, we see that

$$g^{(p-1)p^m} \equiv 1 + rp^{m+1} \pmod{p^{m+2}}.$$

It follows that g has order $(p-1)p^{m+1}$ in $\mathbf{Z}_{p^{m+2}}^*$, and so is a primitive root modulo p^{m+2}. ∎

With a bit more effort, one can completely classify all integers n for which $\mathbf{Z}_n^*$ has a primitive root.

THEOREM 3.56

$\mathbf{Z}_n^*$ has a primitive root if and only if $n = 1, 2, 4$, $n = p^m$, or $n = 2p^m$, for some odd prime p.

An outline of the proof of this theorem is provided in Exercises 3.74–3.77.

We close this section by giving a second proof of Theorem 3.37.

PROOF (Second proof of Theorem 3.37) Let g be a primitive root modulo p. Then the elements of $\mathbf{Z}_p^*$ are

$$1,\ g,\ g^2,\ g^3,\ \ldots,\ g^{p-2}.$$

The element -1 occurs in this list. In fact, $-1 \equiv g^{(p-1)/2}$. This follows from the fact that $x^2 \equiv 1$ has two solutions modulo p: $x \equiv \pm 1$. Since $(g^{(p-1)/2})^2 \equiv 1$ and $g^{(p-1)/2} \not\equiv 1$ (because g is a primitive root), it follows that $-1 \equiv g^{(p-1)/2}$. If $p = 4k+1$, then $g^{(p-1)/4}$ and $g^{3(p-1)/4}$ are solutions to the congruence $x^2 \equiv -1 \pmod{p}$. If $p = 4k+3$, then there are no solutions as in the first proof of Theorem 3.37. ∎

Exercises

3.69 Show that 2 is a primitive root modulo 13. How many primitive roots are there modulo 13? List them.

3.70 Find all the primitive roots modulo 17.

3.71 (a) How many elements of order 54 are there in $\mathbf{Z}_{81}^*$?

(b) How many elements of order 53 are there in $\mathbf{Z}_{81}^*$?

3.72 How many elements of $\mathbf{Z}_{100}^*$ have order 40?

⋆**3.73** (For those who have studied group theory) What cyclic group or product of cyclic groups is isomorphic to $\mathbf{Z}_{18}^*$? What about $\mathbf{Z}_{98}^*$?

†**3.74** Let p be an odd prime, and let g be a primitive root modulo p^m. Let

$$g' = \begin{cases} g & \text{if } g \text{ is odd} \\ g + p^m & \text{if } g \text{ is even.} \end{cases}$$

Show that g' is a primitive root modulo $2p^m$.

†**3.75** Suppose that $m \geq 3$ and $\gcd(a, 2^m) = 1$. Show that

$$a^{2^{m-2}} \equiv 1 \pmod{2^m}.$$

†**3.76** Suppose that $n = n_1 n_2$ where $n_1, n_2 > 2$ and $\gcd(n_1, n_2) = 1$. Show that

$$a^{\phi(n)/2} \equiv 1 \pmod{n}$$

for all a satisfying $\gcd(a, n) = 1$.

†**3.77** Use Exercises 3.74–3.76 to complete the proof of Theorem 3.56.

3.9 Application: Construction of the regular 17-gon

In 1796 Gauss discovered that the regular 17-gon (heptadecagon) can be constructed using a straightedge and compass. This was the first new such construction since the time of the ancient Greeks. We will give the details that show that this construction is possible, using the concept of primitive roots. As a warm-up, we demonstrate the constructibility of the regular pentagon (five sides).

Let's assume that we have a coordinatized plane, so that it suffices to construct a complex number z, with $z \neq 1$, such that

$$z^5 = 1. \tag{3.12}$$

(The values of z that satisfy this relation are called "5th roots of unity.") To be specific, we take $z = e^{2\pi i/5}$, a number that lies in the first quadrant of the plane. Since $z \neq 1$, it follows from (3.12) that

$$z^4 + z^3 + z^2 + z + 1 = 0.$$

Noting that 2 is a primitive root modulo 5, we reorder the terms in this equation:

$$z^{2^0} + z^{2^1} + z^{2^2} + z^{2^3} + 1 = 0,$$

or

$$z^1 + z^2 + z^4 + z^3 = -1.$$

Grouping together alternate summands yields

$$(z^1 + z^4) + (z^2 + z^3) = -1.$$

Now we introduce two auxiliary variables: $\alpha = z^1 + z^4$ and $\beta = z^2 + z^3$. Since $\alpha + \beta = -1$ and $\alpha\beta = (z^1 + z^4)(z^2 + z^3) = z^3 + z^4 + z^6 + z^7 = z^3 + z^4 + z + z^2 = -1$, it follows that α and β are roots of the quadratic equation

$$x^2 + x - 1 = 0.$$

Using the quadratic formula, we find that

$$\alpha,\, \beta = \frac{-1 \pm \sqrt{1+4}}{2} = \frac{-1 \pm \sqrt{5}}{2}.$$

We can see from the Argand diagram of the numbers z, z^2, z^3, z^4 (Figure 3.1) that the correct assignment of values is $\alpha = (-1+\sqrt{5})/2$ and $\beta = (-1-\sqrt{5})/2$. Returning to the variable z, we obtain $z^1 + z^4 = (-1 + \sqrt{5})/2$ and $z^1 \cdot z^4 = z^5 = 1$. Hence, z is a root of the equation

$$z^2 - \left(\frac{-1 + \sqrt{5}}{2}\right) z + 1 = 0.$$

Therefore,

$$z = \frac{\frac{-1+\sqrt{5}}{2} \pm \sqrt{\left(\frac{-1+\sqrt{5}}{2}\right)^2 - 4}}{2} = \frac{-1 + \sqrt{5}}{4} \pm \frac{\sqrt{2\sqrt{5} + 10}}{4} i.$$

Again, checking the Argand diagram, we find that the positive sign applies, so that

$$z = \frac{-1 + \sqrt{5}}{4} + \frac{\sqrt{2\sqrt{5} + 10}}{4} i. \tag{3.13}$$

As the operations of addition, subtraction, multiplication, division, and square root extraction can all be carried out using straightedge and compass, the

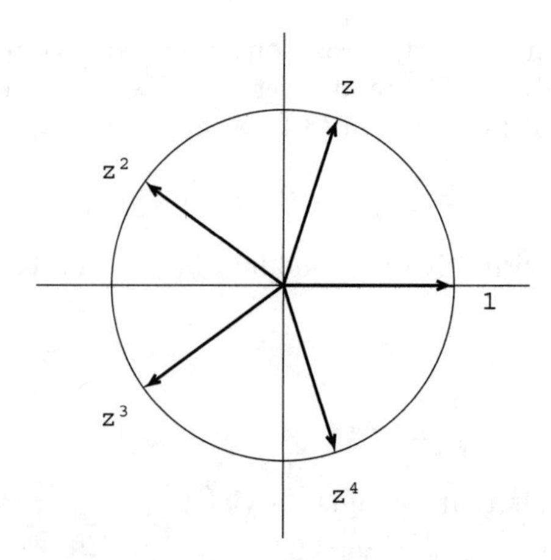

FIGURE 3.1: Argand diagram for the 5th roots of unity.

number z can be constructed and with it the regular pentagon. See Chapter Notes and [17] for more on straightedge and compass constructions.

Now we will show how to construct the regular 17-gon. Let $z = e^{2\pi i/17}$, a complex number in the first quadrant. Then

$$z^{17} = 1 \tag{3.14}$$

and

$$z^{16} + z^{15} + z^{14} + \cdots + z^3 + z^2 + z + 1 = 0.$$

By inspection, we find that 3 is a primitive root modulo 17, for the powers of 3 modulo 17 are

$$\{1, 3, 9, 10, 13, 5, 15, 11, 16, 14, 8, 7, 4, 12, 2, 6\}.$$

We rearrange the terms of the polynomial equation accordingly:

$$\begin{aligned} &z + z^3 + z^9 + z^{10} + z^{13} + z^5 + z^{15} + z^{11} + z^{16} \\ &+z^{14} + z^8 + z^7 + z^4 + z^{12} + z^2 + z^6 = -1. \end{aligned}$$

Now we introduce two auxiliary variables:

$$\begin{aligned} \alpha &= z + z^9 + z^{13} + z^{15} + z^{16} + z^8 + z^4 + z^2 \\ \beta &= z^3 + z^{10} + z^5 + z^{11} + z^{14} + z^7 + z^{12} + z^6. \end{aligned}$$

Direct calculation shows that

$$\begin{aligned} \alpha + \beta &= -1 \\ \alpha\beta &= 4\alpha + 4\beta = -4. \end{aligned}$$

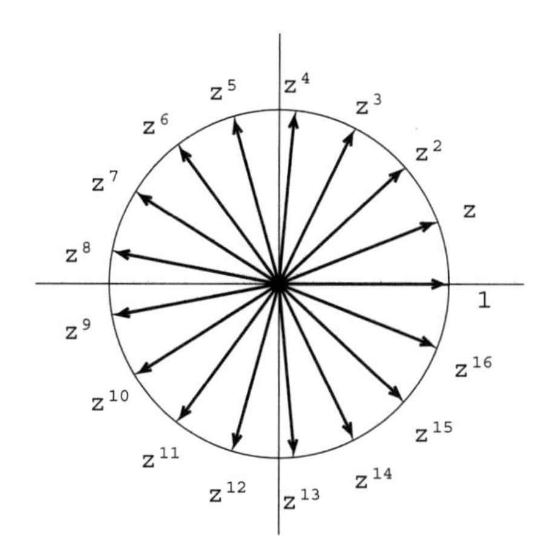

FIGURE 3.2: Argand diagram for the 17th roots of unity.

Thus, α and β are roots of the quadratic equation

$$x^2 + x - 4 = 0.$$

Hence,

$$\alpha,\ \beta = \frac{-1 \pm \sqrt{17}}{2}.$$

In fact, from the Argand diagram for the numbers $1,\ z,\ z^2,\ \ldots,\ z^{16}$ (Figure 3.2), we see that

$$\alpha = \frac{-1 + \sqrt{17}}{2} \quad \text{and} \quad \beta = \frac{-1 - \sqrt{17}}{2}.$$

We repeat this process with four new variables:

$$\begin{aligned}
\gamma &= z + z^{13} + z^{16} + z^4 \\
\delta &= z^9 + z^{15} + z^8 + z^2 \\
\epsilon &= z^3 + z^5 + z^{14} + z^{12} \\
\zeta &= z^{10} + z^{11} + z^7 + z^6.
\end{aligned}$$

Now we have

$$\begin{aligned}
&\gamma + \delta = \alpha \\
&\epsilon + \zeta = \beta \\
&\gamma\delta = \epsilon\zeta = \gamma + \delta + \epsilon + \zeta = -1.
\end{aligned}$$

We see that γ and δ satisfy the quadratic equation

$$x^2 - \alpha x - 1 = 0.$$

Therefore,

$$\gamma,\ \delta = \frac{\alpha \pm \sqrt{\alpha^2 + 4}}{2}.$$

Again, using the Argand diagram, we find that

$$\gamma = \frac{\alpha + \sqrt{\alpha^2 + 4}}{2} \quad \text{and} \quad \delta = \frac{\alpha - \sqrt{\alpha^2 + 4}}{2}.$$

Similarly, we compute ϵ and ζ:

$$\epsilon = \frac{\beta + \sqrt{\beta^2 + 4}}{2} \quad \text{and} \quad \zeta = \frac{\beta - \sqrt{\beta^2 + 4}}{2}.$$

We introduce two new variables:

$$\eta = z + z^{16}$$
$$\theta = z^{13} + z^4.$$

We obtain

$$\eta + \theta = \gamma \quad \text{and} \quad \eta\theta = \epsilon.$$

Hence, η and θ satisfy the quadratic equation

$$x^2 - \gamma x + \epsilon = 0.$$

It follows that

$$\eta,\, \theta = \frac{\gamma \pm \sqrt{\gamma^2 - 4\epsilon}}{2}.$$

By consulting the Argand diagram, we see that

$$\eta = \frac{\gamma + \sqrt{\gamma^2 - 4\epsilon}}{2}.$$

Now that we know η, we can compute z from the equation

$$z^{16} + z - \eta = 0.$$

Multiplying by z, we obtain

$$1 + z^2 - \eta z = 0,$$

or

$$z^2 - \eta z + 1 = 0.$$

Hence,

$$z = \frac{\eta \pm \sqrt{\eta^2 - 4}}{2}.$$

Consulting the Argand diagram, we find that the positive solution applies:

$$z = \frac{\eta + \sqrt{\eta^2 - 4}}{2}.$$

As each variable is defined in terms of the previous ones via the operations of addition, subtraction, multiplication, division, and square roots, the quantity z is constructible with straightedge and compass.

Gauss's work shows that a regular n-gon (with $n \geq 3$) is constructible if

$$n = 2^a p_1 \dots p_k,$$

where $a \geq 0$, $k \geq 0$, and the p_i are distinct primes of the form $2^l + 1$. This is a necessary condition as well (see [31]), and Gauss stated this but only provided a proof of the sufficiency.

If $2^l + 1$ is prime, then l is a power of 2 (see Exercises), so that the prime is of the form $2^{2^j} + 1$. Numbers of the form $2^{2^j} + 1$ are called "Fermat numbers." The first five Fermat numbers are primes: $3 = 2^1 + 1$, $5 = 2^2 + 1$, $17 = 2^4 + 1$, $257 = 2^8 + 1$, and $65537 = 2^{16} + 1$. However, no other Fermat numbers are known to be prime. (See Chapter 7.) Thus, we know of five regular polygons with a prime number of sides that can be constructed using straightedge and compass, but (at present) no one knows whether there are more.

Exercises

3.78 Show how to carry out the construction of the regular pentagon.

3.79 Prove that if $2^l + 1$ is prime, then l is a power of 2.

$\diamond$**3.80** Show that 3 is a primitive root modulo 257 and 65537.

3.81 Given a regular m-gon and a regular n-gon, with $\gcd(m, n) = 1$, show that a regular mn-gon is constructible using straightedge and compass.

$\star$**3.82** Show that a regular heptagon (seven sides) may be constructed using straightedge, compass, and an angle trisecting device.

$\star$**3.83** Below is Pascal's triangle with each entry reduced modulo 2. Next to each row is the decimal representation of that row regarded as a binary number. The first 32 rows yield the only known odd values of n for which a circle can be divided into n equal parts using only straightedge and compass. Explain why this is so.

1									1
1	1								3
1	0	1							5
1	1	1	1						15
1	0	0	0	1					17
1	1	0	0	1	1				51
1	0	1	0	1	0	1			85
1	1	1	1	1	1	1	1		255
1	0	0	0	0	0	0	0	1	257
⋮									⋮

3.10 Notes

Leonhard Euler

Leonhard Euler (1707–1783) is generally regarded as the greatest mathematician of the 18th century and the most prolific mathematician of all time. His work touched all areas of mathematics and spilled over into the applied sciences as well. In addition to his original work, Euler wrote a series of foundational texts, including treatments of algebra and calculus, which are prototypes for the textbooks written today. With regard to Euler's skills as an expositor, Laplace said "Read Euler, read Euler. He is the master of us all." The collection of his published work fills 70 large volumes. Of these, four volumes are concerned with number theory.

At the time of Euler, the theory of numbers had been virtually stagnant for nearly a hundred years. Not since Fermat had significant progress in the area been made. Furthermore, given the fashion in which the work of Fermat had been recorded (many statements, not too many proofs), it was not clear exactly what the status of number theory was when Euler first came to it. He worked his way through the notes of Fermat, supplying proofs for many statements, and disproving others. This was a particularly arduous task as some of Fermat's deepest observations were alongside elementary ones, and so Euler had no one way of knowing which were which.

Euler's contributions to number theory include a proof of Fermat's Last Theorem in the case $n = 3$, introduction of the ϕ function, the generalization of Fermat's theorem known as Euler's theorem, the statement of quadratic reciprocity, and explicit computation of several values of the ζ function. He is credited with some of the first uses of analytic methods in the study of number theory. In his work on Diophantine equations, Euler made use of an extended system of numbers containing the rationals, and so his work also contains the seeds of algebraic number theory.

We will see more of Euler's work in Chapter 5.

Groups

Given a pair of residue classes $[a]$ and $[b]$ in $\mathbf{Z}_n$, we can define addition and multiplication of these classes as follows:

$$[a] + [b] = [a + b],$$

and

$$[a][b] = [ab].$$

Proposition 3.3 guarantees that these operations are well defined. The operation of addition endows $\mathbf{Z}_n$ with what we call a group structure.

DEFINITION

A group *is a nonempty set G together with an operation that assigns to any ordered pair of elements g_1 and g_2 of G an element $g_1 * g_2$ of G satisfying the following:*

(a) *(Associativity) For all g_1, g_2, and g_3 in G, we have $g_1 * (g_2 * g_3) = (g_1 * g_2) * g_3$.*

(b) *(Identity) There exists an element e in G (called the identity element of G) such that $e * g = g * e = g$ for all g in G.*

(c) *(Inverses) For every g_1 in G there exists an element g_2 in G (called the inverse of g_1) such that $g_1 * g_2 = g_2 * g_1 = e$.*

The inverse of an element g in a group is typically denoted g^{-1}, and the product $g * g * \cdots * g$ of n copies of g is denoted by g^n. We also write g^{-n} to indicate the product of n copies of g^{-1}, and g^0 is understood to be the identity. The usual rules of exponents hold under this notation (i.e., $(g^m)^n = g^{mn}$ and $g^m * g^n = g^{m+n}$ for all integers m and n). If the group operation is naturally thought of as addition, we may write $-g$ and ng in place of g^{-1} and g^n, respectively.

Examples of groups are the sets $\mathbf{Z}$ and $\mathbf{Q}$ of rational numbers. In both cases, the group operation is addition, the identity element is 0, and the inverse of an element x is $-x$. Another example is the set $\mathbf{Q} \setminus \{0\}$ of nonzero rational numbers under the operation multiplication. In this group, the identity element is 1, and the inverse of a number p/q is its reciprocal q/p. The set of nonzero integers does not form a group under multiplication. (Why not?)

Returning now to $\mathbf{Z}_n$, we see that $[0]$ serves as the identity element (for the operation of addition). The inverse of the element $[a]$ is $[n-a]$. (Notice that $[n-a] = [-a]$.) The associativity of addition follows from the associativity of the usual addition of integers in $\mathbf{Z}$. The set $\mathbf{Z}_n$ is not a group under multiplication as the residue class of 0 does not have an inverse. If n is a composite number, any class represented by a number not relatively prime to n will also fail to have an inverse. However, if we only include those classes in $\mathbf{Z}_n^*$, we do obtain a group under multiplication with identity element $[1]$. The inverse of an element $[a]$ is exactly the residue class of the inverse of a modulo n.

It follows from a result known as Lagrange's theorem that if G is a finite group having m elements, then

$$g^m = e$$

for all g in G. Notice that Fermat's little theorem and Euler's theorem are special cases of this result.

A group G is said to be *cyclic* if there exists an element g in G such that any element of G is of the form g^k for some integer k. The element g is called a "generator" of G. The group $\mathbf{Z}_n^*$ is cyclic exactly when there exists a primitive root modulo n. Notice that Theorem 3.56 gives a complete classification of the values of n for which $\mathbf{Z}_n^*$ is cyclic.

Straightedge and compass constructions

In Section 3.9, we showed that the regular pentagon and the regular 17-gon are both constructible using straightedge and compass. In each case, we proved that a complex number z representing one of the vertices of the polygon in the complex plane is given by a formula involving only rational numbers and the operations of addition, subtraction, multiplication, division, and square root extraction. We claimed that these operations can all be performed using straightedge and compass. We now give some justification for this assertion.

Suppose that a unit length is given in the plane. All segments that we might subsequently construct will be compared to this unit length. Now, if we are given segments of lengths a and b, we can clearly construct a segment of length $a+b$. We just construct segments congruent to the given segments along a line, touching at a point. We say that we "lay the segments" along a line. (In fact, a little further justification is needed, for straightedge and compass construction does not allow us to gauge a segment length with the compass and then "lift" that length and mark it elsewhere. However, we do this often in geometrical constructions. The justification is an early theorem in Euclid's *Elements* ([13, 14, 15]) that shows that this "lifting" can be accomplished in terms of allowable operations with straightedge and compass.) Subtraction of segments, i.e., construction of $a-b$, is carried out similarly. The constructions for segments $a \cdot b$ (multiplication), a/b (division), and $\sqrt{a}$ (square root extraction), all compared to the unit length, are also elementary theorems of Euclidean geometry, and perhaps we can do no better than refer readers to the seminal work, *The Elements*.

The result of all of this is that the numbers that can be constructed as lengths of segments form a field (the field of constructible numbers). A field is an algebraic structure that allows addition, subtraction, multiplication, and division. Other examples are $\mathbf{Q}$, $\mathbf{R}$, and $\mathbf{C}$. The field of constructible numbers contains the rational numbers and is closed under square root extraction. In fact, the field of constructible numbers is the *smallest* field containing the rational numbers and closed under square root extraction. This is proved by showing that each basic construction (finding the intersection of two lines, two circles, or a line and a circle), involves only the four basic field operations ($+$, $-$, $\times$, and $\div$) and square roots.

An aside: the Pythagoreans used a regular pentagram inscribed inside a regular pentagon as a sign of recognition in their society. Over two thousand years later, Gauss requested that a regular 17-gon, the symbol of one of his greatest mathematical achievements, be inscribed on his gravestone.

Chapter 4

Cryptography

The magic words are squeamish ossifrage.

PLAINTEXT OF RSA-129 (factored in 1994)

There is perhaps no greater application of number theory than its uses in many of the methods of modern cryptography. The fundamental problem in cryptography is for one party, the "sender," to transmit a message to another party, the "receiver," in such a way that no other party can gain knowledge of the message. Typically, the sender performs an encryption algorithm that converts the original "plaintext" message into an apparently unintelligible message called the "ciphertext." If all goes according to plan, only someone with knowledge of the encryption algorithm and the particular key used to produce the ciphertext can make sense of it. To help keep matters straight, we enlist the help of three fictional characters, Samantha, Robert, and Theresa. Samantha will be the individual wishing to send an encrypted message while Robert will be her intended recipient. Theresa represents a third party who is attempting to decipher the contents of the message that Samantha is sending to Robert (against their wishes).

4.1 Monoalphabetic substitution ciphers

We begin with a simple example.

Example 4.1 Suppose that Samantha wishes to send Robert the message "integer." (So "integer" is the plaintext version of the message.) She could replace each letter of the message according to Table 4.1. She would replace the "i" with "F," the "n" with "D," and so on to obtain the ciphertext "FDKATAB." Samantha sends Robert this ciphertext version of the message. Robert is also in possession of Table 4.1, and so he can reverse the process. Suppose that in a separate transmission, Robert receives the ciphertext "OPRLVP" from Samantha. He would replace the "O" with an "m," the

a	b	c	d	e	f	g	h	i	j	k	l	m
G	N	U	R	A	J	T	C	F	E	M	V	O
n	o	p	q	r	s	t	u	v	w	x	y	z
D	P	W	H	B	Y	K	L	Q	Z	X	S	I

TABLE 4.1: Encryption key

"P" with an "o," and so on to reveal the plaintext "modulo." This type of encryption algorithm is called a "monoalphabetic substitution cipher," and the table above is the key. Notice that if Samantha and Robert wish to generate a key for a monoalphabetic substitution cipher, they have 26 choices for how an "a" will be encrypted. After they make that decision, they have 25 choices remaining for how to encrypt a "b." Continuing in this way we see that there are 26! (about 4×10^{26}) possible keys for a monoalphabetic cipher. ∎

Monoalphabetic substitution ciphers date back to the days of Julius Caesar. His key was generated by simply shifting plaintext letters three places to the right to obtain the ciphertext letter. So an "a" would become a "D," a "b" would become an "E," etc. When we come to the end of the alphabet, we simply wrap around to the front. So an "x" becomes "A," "y" becomes "B," and "z" becomes "C." Notice that the operation can be thought of in terms of modulo 26 arithmetic. Each letter represents a residue class, as indicated in Table 4.2, and the encryption is performed by adding 3 to the residue class.

a	b	c	d	e	f	g	h	i	j	k	l	m
1	2	3	4	5	6	7	8	9	10	11	12	13
n	o	p	q	r	s	t	u	v	w	x	y	z
14	15	16	17	18	19	20	21	22	23	24	25	26

TABLE 4.2: Residue classes for letters

Any permutation of the residue classes mod 26 will likewise generate a key for the monoalphabetic substitution cipher. A linear function from $\mathbf{Z}_{26}$ to itself will provide such a permutation under certain assumptions.

PROPOSITION 4.2

If $\gcd(m, n) = 1$ *and* c *is any integer, then the map* f *from* $\mathbf{Z}_n$ *to itself defined by*

$$f(a) = ma + c$$

is a bijection.

PROOF We begin by noting that for any m and c, the map is well defined as a result of Proposition 3.3. That is, if a and a' are in the same residue class, then $ma + c \equiv ma' + c \pmod{n}$. The bijectivity of f follows directly from Theorem 3.10 under the assumption that $\gcd(m, n) = 1$. ∎

Thus, if we choose an integer m relatively prime to 26 and any integer c, then the map that sends the residue class of a modulo 26 to the class of $ma+c$ will produce a permutation of the residue classes and hence generate a key for the monoalphabetic substitution cipher. When a key is produced in this way, we call the cipher an *affine cipher*. We now outline the steps in the affine cipher encryption computation.

ALGORITHM 4.3 (Affine cipher: encryption)
Samantha encrypts her plaintext message.

1. *Samantha and Robert choose key values m such that* $\gcd(m, 26) = 1$ *and arbitrary c.*
2. *The letters of the plaintext are converted to numbers according to Table 4.2.*
3. *Each number a is replaced by the number $ma + c$. The resulting number is reduced to the least positive residue mod 26 and converted back to a letter, again using Table 4.2.*

Example 4.4 Suppose that Samantha wishes to encrypt the message "affine cipher" using $m = 9$ and $c = 18$. First, she converts the plaintext to numbers. This is easy enough to do using the table, but we will enlist help from a computer to speed things up.

Mathematica

```
toNumbers[message_] :=
Flatten[ToCharacterCode[
ToLowerCase[Select[Characters[message], LetterQ]]]] - 96

toNumbers["affine cipher"]

{1, 6, 6, 9, 14, 5, 3, 9, 16, 8, 5, 18}
```

The *Mathematica* command `ToCharacterCode` converts letters to numbers using the ASCII (American Standard Code for Information Interchange) in which an "a" becomes 97, a "b" becomes 98, etc. The `Select` command strips out any spacing or punctuation in the message. As a result, Robert may have to put in a little extra effort to make sense of the message after it has been decrypted.

Next, Samantha needs to multiply each number by 9, add 18, and reduce the result to the least positive residue mod 26. We would like to use the *Mathematica* command `Mod` to perform the reducing. However, this command returns the least *nonnegative* residue rather than the least *positive* residue. We define a new command to suit our needs in this case.

Mathematica

```
modpositive[m_, n_] := Mod[m - 1, n] + 1

modpositive[9*{1, 6, 6, 9, 14, 5, 3, 9, 16, 8, 5, 18} + 18,
26]

{1, 20, 20, 21, 14, 11, 19, 21, 6, 12, 11, 24}
```

Finally, Samantha converts the numbers back to letters.

Mathematica

```
toLetters[numbers_] := ToUpperCase[FromCharacterCode[numbers
+ 96]]

toLetters[{1, 20, 20, 21, 14, 11, 19, 21, 6, 12, 11, 24}]

ATTUNKSUFLKX
```

So the encrypted message is "ATTUNKSUFLKX."

Here are the corresponding *Maple* commands.

Maple

```
with(StringTools):
toNumbers:=message->map(x->x-96,
map(Ord,(Explode(Select(IsLower,LowerCase(message)))))):
toNumbers("affine cipher");

[1, 6, 6, 9, 14, 5, 3, 9, 16, 8, 5, 18]

modpositive:=(m,n)->modp(m-1,n)+1:
map(x->modpositive(9*x+18,26),[1, 6, 6, 9, 14, 5, 3, 9, 16,
8, 5, 18]);

[1, 20, 20, 21, 14, 11, 19, 21, 6, 12, 11, 24]

toLetters:=numbers->
UpperCase(Join(map(Char,map(x->x+96,numbers)),"")):

toLetters([1, 20, 20, 21, 14, 11, 19, 21, 6, 12, 11, 24]);

"ATTUNKSUFLKX"
```

□

To perform the decryption, Robert must find the inverse of the mapping $f(a) = ma + c$, and then proceed in a similar way. Obviously, we can invert the addition of c by subtracting c. To invert the multiplication by m, we must find a number m' such that $mm' \equiv 1 \pmod{26}$.

ALGORITHM 4.5 (Affine cipher: decryption)
Given a ciphertext message encrypted with affine cipher key (m, c), Robert recovers the plaintext message.

1. *Robert finds a number m' satisfying $mm' \equiv 1 \pmod{26}$.*
2. *He converts the letters of the ciphertext to numbers using Table 4.2.*
3. *Robert replaces each (ciphertext) number b with the number $m'(b - c)$. He reduces the resulting numbers to the least positive residues mod 26 and converts back to letters, again using Table 4.2.*

Suppose that Robert has received the ciphertext, `ATTUNKSUFLKX` from our example. First he must solve the congruence

$$9m' \equiv 1 \pmod{26}.$$

By trial and error (or inspection), we see that $m' \equiv 3 \pmod{26}$.

Mathematica

```
ToLowerCase[toLetters[
modpositive[3*(toNumbers["ATTUNKSUFLKX"] - 18), 26]]]

affinecipher
```

Maple

```
LowerCase(toLetters(
map(x->modpositive(3*(x-18),26),toNumbers("ATTUNKSUFLKX"))));

"affinecipher"
```

There are relatively few key choices available for the affine cipher (see Exercise 4.4). Even if Theresa is not in possession of the key, she could decrypt a message simply by testing every possible key. Such a "brute force" attempt to decrypt a message would not work for the general monoalphabetic substitution cipher (given the large number of keys). However, there are other ways for Theresa to make sense of a ciphertext encrypted with this cipher. With a sufficient amount of ciphertext, say 100–200 letters, she can apply frequency analysis to make reasonable guesses for the key. The most frequently occurring letter in the ciphertext likely represents the plaintext letter "e." The next few most frequently occurring letters are likely other vowels or perhaps the letter "t." With enough correct guesses she can begin to spot parts of words that lead to a determination of the rest of the key. Thus, even the general monoalphabetic substitution cipher cannot be regarded as a secure means of secret communication.

Exercises

4.1 The table below provides a key for the monoalphabetic substitution cipher.

a	b	c	d	e	f	g	h	i	j	k	l	m
F	E	L	C	O	W	P	H	T	A	Z	Y	I
n	o	p	q	r	s	t	u	v	w	x	y	z
S	G	M	D	N	X	R	Q	U	B	K	V	J

(a) Use the table to encrypt the plaintext "cryptography."

(b) Use the table to decrypt the ciphertext "XHGNREQRXBOOR."

4.2 (a) Use the affine cipher to encrypt the message "All is well" using the key $m = 11$ and $c = 8$.

(b) Use the affine cipher to encrypt the message "All is not well" using the key $m = 21$ and $c = 23$.

4.3 The first row of the table below contains the list of possible values for m in the key of the affine cipher.

m	1	3	5	7	9	11	15	17	19	21	23	25
m'	1	9			3							

The second row is for values m' satisfying $mm' \equiv 1 \pmod{26}$, which are used in the decryption of a ciphertext encrypted with the affine cipher. Fill in the missing values in the table.

4.4 Show that there are 311 keys (m, c) for the affine cipher (not counting $(1, 0)$ as a valid key).

4.5 (a) Decrypt the following message, which was encrypted with the affine cipher using the key $m = 17$ and $c = 14$.

AUUPAUKRPTUJKVHEHW

(b) Decrypt the following message, which was encrypted with the affine cipher using the key $m = 7$ and $c = 6$.

KOLOMDOMPSQHZQCJP

4.6 Many newspapers include a "Cryptoquip" in their puzzle section. A cryptoquip is a ciphertext for a famous saying or quote that has been encrypted using a monoalphabetic substitution cipher. To make the puzzle a little bit easier to solve, the punctuation is usually preserved and one letter from the key is given. In the following cryptoquip, the plaintext letter "a" was encrypted as "R." Determine the original plaintext.

```
VRCOYVRCQTQRZM ORFY CAQYU QZ FRQZ CB COQM URH CB
UQMTBFYA MBVY BAUYA QZ COY MYLEYZTY BW XAQVY
ZEVSYAM, RZU KY ORFY AYRMBZ CB SYPQYFY CORC QC
QM R VHMCYAH QZCB KOQTO COY OEVRZ VQZU KQPP ZYFYA
XYZYCARCY.
```

4.2 The Pohlig–Hellman cipher

We noted in the previous section that the monoalphabetic substitution cipher is susceptible to frequency analysis, and so is not a particularly useful encryption scheme. Samantha and Robert can make a more secure cipher if instead of substituting letters one at a time, they substitute groups of letters. For instance, they could divide the plaintext into blocks of four-letter strings. The key would explain what to substitute for every possible four-letter string. They could develop a key in which the plaintext "abcd" is replaced with the ciphertext "MNOP," but the plaintext "abce" is replaced with ciphertext "WXYZ." Such an encryption scheme is called a "polygraphic substitution cipher." Notice that there are $(26^4)!$ (about $10^{2000000}$) possible keys for a polygraphic substitution cipher in which we are substituting four letters at a time.

To simplify matters, we will once again convert letters to numbers using Table 4.2. If Samantha's message is "polygraphic," then the list of numbers is as follows.

Mathematica

```
toNumbers["polygraphic"]

{16, 15, 12, 25, 7, 18, 1, 16, 8, 9, 3}
```

Maple

```
toNumbers("polygraphic");

[16, 15, 12, 25, 7, 18, 1, 16, 8, 9, 3]
```

Samantha takes the first four numbers and concatenates them into one eight-digit number. Then she concatenates the next four numbers, adding a 0 in front of each one-digit number as a place holder. The remaining three numbers are concatenated into a single number (again prepending the single-digit numbers with a 0). The resulting number form of the plaintext is

```
16151225, 07180116, 080903.
```

Now Samantha and Robert need to agree on a way to scramble the set of eight-digit numbers.

If they have a bijective map

$$f\colon \mathbf{Z}_n \to \mathbf{Z}_n,$$

where $n > 10^8$, then they could encrypt every plaintext number a as $f(a)$. (We note that $f(a)$ may not be a concatenation of four numbers from 1 to 26, and so Samantha and Robert will have to be content to use the number $f(a)$ as the ciphertext.) Ideally, they (especially Robert) would like a map f that has an easy to compute inverse map. A linear function

$$f(a) = ma + c$$

would satisfy this requirement (assuming $\gcd(m, n) = 1$). However, there is a security weakness in this approach. Suppose that Theresa has intercepted a message that Samantha sent to Robert, and somehow Theresa has managed to decode part of it. (Perhaps she correctly guessed that the message began "Hello Robert.") Suppose that Theresa knows values a_1, a_2, b_1, and b_2, where

$$\begin{aligned} b_1 &= f(a_1) = ma_1 + c \\ b_2 &= f(a_2) = ma_2 + c. \end{aligned}$$

Then she could compute

$$b_1 - b_2 \equiv (a_1 - a_2)m \pmod{n}.$$

If $\gcd(a_1 - a_2, n) = 1$, then she can solve this congruence for m. Given the value for m, she can then solve for c. At this point, Theresa would be in possession of the key, and she could decrypt the remainder of the message.

As an alternative to a linear map, Samantha and Robert can get improved security by using an exponential map.

PROPOSITION 4.6
If p is a prime number and $\gcd(r, p-1) = 1$, then the map f from $\mathbf{Z}_p$ to itself defined by

$$f(a) = a^r$$

is a bijection. Furthermore, the inverse of this map is given by the function

$$g(b) = b^{r'}$$

where r' satisfies $rr' \equiv 1 \pmod{p-1}$.

PROOF We first note that since $\gcd(r, p-1) = 1$, there does exist an r' satisfying $rr' \equiv 1 \pmod{p-1}$. To prove the theorem, we just need to check that $g(f(a)) = a = f(g(a))$. We may write $rr' = 1 + k(p-1)$, for some number k, and so we have

$$g(f(a)) = a^{rr'} = a \cdot \left(a^{p-1}\right)^k.$$

Certainly, $g(f(0)) = 0$. If $a \not\equiv 0 \pmod{p}$, then by Fermat's theorem we conclude that $g(f(a)) \equiv a \pmod{p}$. Likewise, $f(g(a)) \equiv a \pmod{p}$. ∎

The proposition paves the way for us to use exponentiation to perform the encryption. The resulting algorithm is known as the Pohlig–Hellman cipher.

ALGORITHM 4.7 (Pohlig–Hellman cipher: encryption)
Samantha encrypts a plaintext message, which we assume to be a sequence of numbers.

1. *Samantha and Robert agree on a prime number p and exponent r satisfying* $\gcd(r, p-1) = 1$.

2. *Samantha encrypts each plaintext number a by computing* a^r *and reducing it modulo p.*

From Proposition 4.6, we see how to perform the decryption.

ALGORITHM 4.8 (Pohlig–Hellman cipher: decryption)
Robert decrypts a ciphertext message, which was encrypted with exponent r and modulus p.

1. *Robert solves the congruence* $rr' \equiv 1 \pmod{p}$ *for* r'.

2. *Robert decrypts each ciphertext number b by computing* $b^{r'}$ *and reducing it modulo p.*

Example 4.9 Let's continue with the example "polygraphic." Samantha has already converted this to the sequence of numbers

`16151225, 07180116, 080903.`

This list of numbers serves as the plaintext. Now she and Robert need to find a prime number with at least nine digits. They start with a random nine-digit number and increase it by 1 until a prime results.

Mathematica

```
n = 123456789;
While[Not[PrimeQ[n]], n = n + 1];
Print[n]

123456791
```

Maple

```
n := 123456789:
for i from n while not(isprime(n)) do
n:=n+1:
end do:
print(n);

123456791
```

So Samantha and Robert will use $p = 123456791$. Next, they must agree on the exponent r. They choose a random value for r, say, $r = 123$ and check that $\gcd(r, p-1) = 1$. Indeed, $\gcd(123, 123456790) = 1$.

Now Samantha is ready to encrypt her message. Before we perform this calculation, we note a computational issue. If r is a large number, it is not feasible to compute (even with a computer) a^r. However, since we are only interested in the result mod p, we can build up to a^r by computing smaller powers of a, reducing them mod p, and then putting the results together. We give details on how to perform this computation in the notes at the end of this chapter. For now, we make use of *Mathematica*'s built-in command `PowerMod[a,r,p]` and *Maple*'s built-in command `modp(a&^r,p)`, which compute the quantity a^r reduced mod p.

Mathematica

```
PowerMod[{16151225, 07180116, 080903}, 123, 123456791]

{85925373, 3137134, 6901637}
```

These three numbers serve as the ciphertext, which Samantha now transmits to Robert.

To perform the decryption, Robert must first solve the congruence

$$123r' \equiv 1 \pmod{123456790}.$$

This can be done using the method developed in Section 3.2. However, we will permit Robert to use the built-in `Solve` command.

Mathematica

```
Solve[{123rprime == 1, Modulus == 123456790}, rprime]

{{Modulus-> 123456790, rprime->50185687}}
```

Now Robert can decrypt the message by exponentiating the ciphertext with exponent $r' = 50185687$.

Mathematica

```
PowerMod[{85925373, 3137134, 6901637}, 50185687, 123456791]

{16151225, 7180116, 80903}
```

This is exactly the list of numbers that Samantha encrypted. Robert can now separate the numbers into a list of two-digit numbers and recover the original message.

Mathematica

```
toLetters[{16, 15, 12, 25, 7, 18, 1, 16, 8, 9, 3}]

POLYGRAPHIC
```

Here are the corresponding *Maple* commands.

Maple

```
gcd(123,123456790);

1

map(x->modp(x&^123,123456791),[16151225, 07180116, 080903]);

[85925373, 3137134, 6901637]

msolve({123*rprime=1},123456790);

{rprime = 50185687}

map(x->modp(x&^50185687,123456791),[85925373, 3137134,
6901637]);

[16151225, 7180116, 80903]

toLetters([16,15,12,25,7,18,1,16,8,9,3]);

"POLYGRAPHIC"
```

□

Let's examine the security of the Pohlig–Hellman cipher. Once again, suppose that Theresa has intercepted a message and knows a portion of both the plaintext and ciphertext of the message. For instance, suppose that she knows the values of the first ciphertext number and plaintext number from the last example. Then she would know that

$$16151225^r \equiv 85925373 \pmod{123456791}.$$

How can she determine the value of r from this congruence? One approach would be for her to simply test values of r. Does the congruence hold for $r = 1$? Does it hold for $r = 2$? Continuing in this way, she will eventually find the correct value of r for which the congruence holds (if we ignore the possibility of multiple solutions for r). Fortunately (or unfortunately for Theresa), this is not a practical approach. While it may lead to a solution in this case, Samantha and Robert could switch to a much larger prime modulus for which Theresa could not perform an exhaustive search for r in a reasonable amount of time.

In general, given values a, b, and p, the problem of finding a solution r to the congruence

$$a^r \equiv b \pmod{p}$$

is known as the *discrete logarithm problem*. There are a number of algorithms that can produce a solution to the problem, but in practice, none of them are fast enough to provide the solution in an acceptable amount of time when p is sufficiently large (for instance, when p is a several-hundred-digit number).

Exercises

4.7 A bijective linear map from $\mathbf{Z}_{100}$ to itself is given by

$$f(a) = ma + c.$$

If $f(12) \equiv 65 \pmod{100}$ and $f(29) \equiv 94 \pmod{100}$, determine m and c.

4.8 (a) Solve the congruence $11r' \equiv 1 \pmod{16}$.

(b) Use part (a) to solve the congruence

$$a^{11} \equiv 4 \pmod{17}.$$

4.9 (a) Solve the congruence $9r' \equiv 1 \pmod{22}$.

(b) Use part (a) to solve the congruence

$$a^9 \equiv 10 \pmod{23}.$$

◇**4.10** Use Proposition 4.6 (and a computer) to solve

$$a^{123} \equiv 7453 \pmod{10601}.$$

Note that 10601 is prime.

◇**4.11** Samantha wishes to send a PIN number to Robert using the Pohlig–Hellman cipher. The PIN number is 1357. They agree on the prime $p = 12157$ and exponent $r = 456$. What will the encrypted version of the PIN number be?

◇**4.12** Samantha sends her bank account number to Robert using the Pohlig–Hellman cipher. She and Robert have agreed upon the prime $p = 3097177$ and exponent $r = 77$. The encrypted version of the bank account number that Robert receives is 1264114. What is the unencrypted bank account number?

◇**4.13** Samantha wishes to send the message "send help" to Robert. They have agreed to use the Pohlig–Hellman cipher with $p = 123456791$ and exponent $r = 123$. Samantha will group blocks of four letters together and covert to numbers as in Example 4.9. What will the ciphertext be for this message?

◇**4.14** Robert replies to Samantha's message using the same key ($p = 123456791$ and $r = 123$). The ciphertext that he sends back is

`29451779, 99071351, 109079786.`

Decrypt this message.

◇**4.15** Using a calculator, find a number s such that

$$5^s \equiv 17 \pmod{23}.$$

◇**4.16** Using a computer, find a number s such that

$$126^s \equiv 716 \pmod{823}.$$

◇**4.17** Can you find a number s such that

$$23^s \equiv 4770459482380 \pmod{6451265877961}?$$

4.3 The Massey–Omura exchange

One shortcoming of the Pohlig–Hellman cipher is that it requires both Samantha and Robert to be in possession of the key (the prime and the exponent). If Samantha and Robert have not previously met to agree upon the key, then they have a problem. For instance, suppose that Robert sells math textbooks over the Internet, and Samantha wishes to purchase one with her credit card. Given that Theresa is able to monitor the Internet exchanges between Samantha and Robert, how can Samantha pass her credit card number to Robert without Theresa gaining access to the number?

In 1976 Whitfield Diffie and Martin Hellman introduced a procedure that would allow Samantha and Robert to exchange information publicly in such a way that they both would be able to make a computation to determine a key that Theresa could not discover. In this section, we will present a variation of this procedure introduced by James Massey and Jim Omura. The algorithm is based on the following simple idea. Suppose that Samantha wishes to send a secret message to Robert. She puts the message in a box and places a lock on the box for which only she has the key. She then sends the box to Robert.

Now Robert places his own lock on the box, for which only he has the key and sends the box back to Samantha. At this point, Samantha removes her lock and returns the box to Robert. Since his lock is the only one on the box, Robert may now remove it and access the message in the box.

The trick now is to find a "mathematical lock" that can be unlocked only by the one who applied it, and for which the locking and unlocking operations commute. To begin, we first assume that our message is a number. If our message consisted of letters, we could convert the letters to numbers using Table 4.2 as in the previous sections. In practice, the Massey–Omura exchange is most often used to exchange a key (which is itself a number) to be used with some other encryption procedure.

Given a message a, we apply a lock by raising a to some power and reducing mod p for some prime p. Using Fermat's theorem, we can remove the locks by raising the previous result to an appropriate power. We describe the process below.

ALGORITHM 4.10 (Massey–Omura exchange)
Samantha communicates a plaintext message, which we assume to be a number a (or sequence of numbers), to Robert through a sequence of exchanges.

1. *Samantha and Robert agree upon a large prime number p. (For our immediate purposes, p should be larger than a. In practice p should be a prime with at least 100 digits. We will describe a method for finding such large primes in Section 7.5.) The choice of this prime need not be a secret.*

2. *Samantha chooses a number s satisfying $\gcd(s, p-1) = 1$, and computes the number a^s. This number is reduced mod p (call the residue a_1) and transmitted to Robert.*

3. *Robert chooses a number r satisfying $\gcd(r, p-1) = 1$, and computes a_1^r. This result is reduced mod p (call the residue a_2) and returned to Samantha.*

4. *Since $\gcd(s, p-1) = 1$, Samantha can solve the congruence*

$$ss' \equiv 1 \pmod{p-1}$$

for s'. Now Samantha computes $a_2^{s'}$, reduces mod p, and transmits the result (call it a_3) back to Robert. At this point, Samantha has removed her "lock."

5. *Similarly, Robert can solve the congruence*

$$rr' \equiv 1 \pmod{p-1}.$$

By computing $a_3^{r'}$ and reducing the result mod p, he obtains a.

Let's examine these calculations.

$$\begin{aligned} a_3^{r'} &\equiv a_2^{s'r'} \pmod p \\ &\equiv a_1^{rs'r'} \pmod p \\ &\equiv a^{srs'r'} \pmod p \\ &\equiv (a^{ss'})^{rr'} \pmod p \end{aligned}$$

Since $ss' \equiv 1 \pmod{p-1}$, we can write $ss' = 1+x(p-1)$. Hence, by Fermat's theorem,

$$a^{ss'} = a(a^{p-1})^x \equiv a \pmod p.$$

Similarly, since $rr' \equiv 1 \pmod{p-1}$, we have

$$(a^{ss'})^{rr'} \equiv a^{rr'} \equiv a \pmod p.$$

Note: Both Samantha and Robert must have access to the prime p, and so this cannot be kept secret. However, only Samantha needs to know s and s' and only Robert needs to know r and r', and so these numbers can be kept secret.

We now use *Mathematica* to illustrate the process. We will transmit the message $a = 123456789$. Samantha and Robert agree to use the prime $p = 41889443039$. Samantha chooses a value for s, checks that it is relatively prime to $p-1$, and computes $a^s \bmod p$.

Mathematica

```
a=123456789;
p=41889443039;
s=85479892289;
GCD[s,p-1]

1

a1=PowerMod[a,s,p]

19155614901
```

Now Robert chooses r, checks that it is relatively prime to $p-1$, and computes $a_1^r \pmod p$.

Mathematica

```
r=46828919;
GCD[r,p-1]

1

a2=PowerMod[a1,r,p]

6133869575
```

Now Samantha determines s' using the `Solve` command.

Mathematica

```
Solve[{sprime s == 1, Modulus == p - 1}, sprime]

{{Modulus -> 41889443038, sprime -> 32914471425}}

a3 = PowerMod[a2, 32914471425, p]

2121551126
```

Now Robert performs similar calculations and arrives at the value of a.

Mathematica

```
Solve[{rprime r == 1, Modulus == p - 1}, rprime]

{{Modulus -> 41889443038, rprime -> 32579883087}}

PowerMod[a3, 32579883087, p]

123456789
```

Here are the corresponding *Maple* commands.

Maple

```
a := 123456789:
p := 41889443039:
s := 85479892289:
gcd(s,p-1);

1

a1 := modp(a&^s,p);
a1 := 19155614901

r := 46828919:
gcd(r,p-1);

1

a2 := modp(a1&^r,p);
a2 := 6133869575

msolve({s*sprime = 1}, p-1);

{sprime = 32914471425}

a3:=modp(a2&^32914471425,p);
a3 := 2121551126

msolve({r*rprime = 1}, p-1);

{rprime = 32579883087}

modp(a3&^32579883087,p);
123456789
```

Let's consider the security of this key exchange. Suppose that Theresa is able to eavesdrop and read all three transmissions that are made. Thus, she is in possession of the numbers a_1, a_2 and a_3. Since

$$a_1^{s'} \equiv a^{ss'} \equiv a \pmod{p},$$

Theresa can decipher the message if she can solve the congruence

$$a_2^{s'} \equiv a_3 \pmod{p}$$

for s'. Thus, Theresa must solve a discrete logarithm problem. While Theresa might be able to manage a solution for the value of p in the example, she would not be able to do so for sufficiently large (100-digit) prime numbers.

Exercises

◊**4.18** Let $a = 5$, $p = 23$, $s = 7$, and $r = 5$. Make the following calculations using only a calculator.

(a) Compute $a^s \bmod p$.

(b) Compute $a_1^r \bmod p$, where a_1 is the result of the calculation in part (a).

(c) Solve the congruence $ss' \equiv 1 \pmod{p-1}$.

(d) Compute $a_2^{s'} \bmod p$, where a_2 is the result of the calculation in part (b).

(e) Solve the congruence $rr' \equiv 1 \pmod{p-1}$.

(f) Verify that $a_3^{r'} \equiv 5 \pmod{p}$, where a_3 is the result of the calculation in part (d).

◊**4.19** Let $a = 126$, $p = 823$, $s = 125$, and $r = 521$. Make the following calculations using a computer.

(a) Compute $a^s \bmod p$.

(b) Compute $a_1^r \bmod p$, where a_1 is the result of the calculation in part (a).

(c) Solve the congruence $ss' \equiv 1 \pmod{p-1}$.

(d) Compute $a_2^{s'} \bmod p$, where a_2 is the result of the calculation in part (b).

(e) Solve the congruence $rr' \equiv 1 \pmod{p-1}$.

(f) Verify that $a_3^{r'} \equiv 126 \pmod{p}$, where a_3 is the result of the calculation in part (d).

◊**4.20** Samantha wishes to send Robert the secret combination to a safe. The combination is the number 135468. They have agreed to use the Massey–Omura exchange with prime $p = 1632899$. Samantha has chosen the exponent $s = 4321$.

(a) What number does Samantha send to Robert in her first transmission?

(b) Samantha receives the number 611092 back from Robert. What number does she send back to him?

4.4 The RSA algorithm

One of the drawbacks of the Massey–Omura exchange is that it requires several transmissions between the sender and receiver before the message can be read by the receiver. In 1977 at MIT, a new type of encryption algorithm was introduced by Ron Rivest, Adi Shamir, and Leonard Adelman (RSA) that requires only one transmission. (Actually, the algorithm was discovered in 1973 by Clifford Cocks, but not publicized.) The algorithm is similar to the Pohlig–Hellman cipher, except that in place of a prime modulus, we use a number that is a product of two distinct primes.

PROPOSITION 4.11

Let p and q be distinct prime numbers, and let $n = pq$. If r is a number satisfying $\gcd(r, \phi(n)) = 1$, then the map f from $\mathbf{Z}_n$ to itself defined by

$$f(a) = a^r$$

is a bijection. Furthermore, the inverse of this map is given by the function

$$g(b) = b^{r'},$$

where r' is a number satisfying $rr' \equiv 1 \pmod{\phi(n)}$.

PROOF First, we observe that there is a unique value of r' modulo $\phi(n)$ satisfying $rr' \equiv 1 \pmod{\phi(n)}$, since $\gcd(r, \phi(n)) = 1$. For such an r', write

$$rr' = 1 + k\phi(n),$$

for some number k. We have

$$g(f(a)) = f(g(a)) = a^{rr'} = a \cdot a^{k\phi(n)}. \tag{4.1}$$

Since p and q are distinct primes,

$$\phi(n) = \phi(p)\phi(q) = (p-1)(q-1).$$

Thus, we may rewrite (4.1) as

$$g(f(a)) = a \cdot (a^{p-1})^{k(q-1)}.$$

If $a \equiv 0 \pmod{p}$, then $g(f(a)) \equiv 0 \pmod{p}$. Otherwise, we may apply Fermat's theorem to conclude that

$$(a^{p-1})^{k(q-1)} \equiv 1 \pmod{p}.$$

In either case, we have $g(f(a)) \equiv a \pmod{p}$. Likewise, $g(f(a)) \equiv a \pmod{q}$. It follows by the Chinese remainder theorem that $g(f(a)) \equiv a \pmod{pq}$. ☐

Just like the Pohlig–Hellman cipher, encryption is performed by raising the plaintext a to the power r and reducing mod n, and decryption is performed by raising the ciphertext b to the power r' and reducing mod n. A subtle but crucial difference is that to determine r', one must first factor n to compute $\phi(n)$. However, it is not necessary to know the factorization of n to perform the encryption. Thus, Robert can determine the primes p and q on his own, but pass along only the values of n and r to Samantha.

ALGORITHM 4.12 (RSA algorithm: encryption)
Samantha encrypts a plaintext message, which we assume to be a sequence of numbers, to send to Robert.

1. *Robert chooses distinct large prime numbers p and q and sets $n = pq$.*
2. *Robert chooses an exponent r satisfying* $\gcd(r, (p-1)(q-1)) = 1$.
3. *Robert publicly announces the values of n and r, but does not reveal the primes p and q.*
4. *Samantha encrypts each plaintext number a by computing a^r and reducing it mod n.*

ALGORITHM 4.13 (RSA algorithm: decryption)
Robert decrypts a ciphertext message that was encrypted with his key consisting of exponent r and modulus n.

1. *Robert solves the congruence $rr' \equiv 1 \pmod{(p-1)(q-1)}$ for r'.*
2. *Robert decrypts each ciphertext number b by computing $b^{r'}$ and reducing it mod n.*

As with the Massey–Omura system, RSA is most often used for exchanging keys for some other encryption algorithm or for other applications of public key cryptography such as digital signatures.

Example 4.14 Robert begins by selecting his primes p and q. Let's take $p = 115518123229$ and $q = 8445806041$. (We could produce these primes as in Example 4.9.)

Mathematica

```
p = 115518123229;
q = 8445806041;
n = p q

975643663012470626389

m =(p-1)(q-1)

975643662888506697120

r = 875189387279;

GCD[m,r]

1
```

Maple

```
p := 115518123229:
q := 8445806041:
n := p*q;

n := 975643663012470626389:

m := (p-1)*(q-1);

m := 975643662888506697120:

r := 875189387279:

gcd(m,r);

1
```

Again, the values for p, q, and m are known only by Robert, who will receive the message. Suppose that Samantha wishes to transmit $a = 123456789$. She computes and transmits the following.

Mathematica

```
PowerMod[123456789, r, n]

164186964938094304597
```

To decrypt the message, Robert first solves the congruence $rr' \equiv 1 \pmod{m}$.

Mathematica

```
Solve[{r rprime == 1, Modulus == m}, rprime]

{{Modulus -> 975643662888506697120, rprime ->
305536865450871050159}}
```

Now the message can be decrypted with $r' = 305536865450871050159$.

Mathematica

```
PowerMod[164186964938094304597, 305536865450871050159, n]

123456789
```

Here are the corresponding *Maple* commands.

Maple

```
modp(123456789&^r,n);

164186964938094304597

msolve({r*rprime = 1}, m);

{rprime = 305536865450871050159}

modp(164186964938094304597&^305536865450871050159,n);

123456789
```

□

Exercises

◇**4.21** Using a computer, find primes p and q large enough so that your computer cannot factor the number pq in less than a couple of minutes.

4.22 Let $n = 3 \cdot 5$.

(a) Compute $7^{11} \pmod{n}$.

(b) Find an integer solution to the equation $11y + 8z = 1$ with $y > 0$.

(c) Check that $7 \equiv b^y \pmod{n}$ where b is your answer to part (a).

4.23 Let $n = 7 \cdot 11$. Note that $\gcd(13, \phi(n)) = 1$.

(a) Find an integer solution to the equation $13y + 60z = 1$ with $y > 0$.

(b) Solve the congruence $a^{13} \equiv 14 \pmod{n}$ by computing $14^y \pmod{n}$.

◇**4.24** Let $n = 809 \cdot 1069$. Note: 809 and 1069 are prime numbers.

(a) Find an integer solution to the equation $12121y + 862944z = 1$ with $y > 0$.

(b) Solve the congruence $a^{12121} \equiv 132246 \pmod{n}$.

◇**4.25** Use the RSA algorithm to encrypt the message "RSA test message" by first grouping the letters into blocks as in Example 4.9. Use the RSA key values

```
p=6184582007
q=49472019391
r=1234567.
```

◇**4.26** Use the RSA algorithm to decrypt the message

`236621645496453797068, 41864260073650084250.`

The RSA key values are

```
p=6184582007
q=49472019391
r=1234567.
```

◇**4.27** Use the RSA algorithm to decrypt the message

```
68845131345663643088, 14115385642073824003,
239820413777247302078, 22891779279906520513,
7159538012677256490.
```

The RSA *public* key values are

```
n=32879555389034738049
r=1234321.
```

4.5 Notes

Computing powers mod p

As we have seen throughout this chapter, it is sometimes necessary to compute $a^k \bmod n$ for large numbers a, k, and n. For instance, to encrypt the number 123456789, we computed

$$123456789^{85479892289} \equiv 19155614901 \pmod{41889443039}.$$

It is not feasible to compute the number $123456789^{85479892289}$ as it has over one billion digits. However, we can break the problem down into pieces and avoid the use of large numbers.

We illustrate with an example involving smaller numbers. Consider the problem of computing

$$57^{107} \bmod 541.$$

We begin by squaring 57 and reducing it mod 541:

$$57^2 = 3249 \equiv 3 \pmod{541}.$$

We can now easily compute the value of $57^4 \bmod 541$:

$$57^4 = (57^2)^2 \equiv 3^2 \equiv 9 \pmod{541}.$$

Continuing in this way, we obtain a list of values of 57^t for exponents t that are powers of 2.

t	$57^t \pmod{541}$
1	57
2	3
4	9
8	81
16	69
32	433
64	303

We observe that $107 = 64 + 32 + 8 + 2 + 1$, and so

$$57^{107} \equiv 303 \cdot 433 \cdot 81 \cdot 3 \cdot 57 \equiv 496 \pmod{541}.$$

In general, to compute $a^k \bmod n$, we write k as a sum of powers of 2, compute $a^t \bmod n$ for such powers of 2, and then multiply together the appropriate powers of a. Notice that we will need to compute at most $\log_2 k$ powers of a and perform at most $\log_2 k$ multiplications in combining the appropriate powers of a. Furthermore, we will never need to work with numbers larger than n^2. As a further consideration, we can compute our powers of a and assemble the appropriate product of these powers simultaneously, and thus we need to keep only the current power of a we are working with in memory. The following *Mathematica* procedure will carry out such a computation.

Mathematica

```
ourpowermod[a_, k_, n_] := (
  b = 1; c = a; ktemp = k;
  While[
    ktemp > 0,
    r = Mod[ktemp, 2];
    If[r == 1, b = Mod[b c, n]];
    c = Mod[c^2, n];
    ktemp = (ktemp - r)/2
  ];
  b
)
```

At any point in the `While` loop, c is the current power of a, and b is the product of the appropriate powers of a up to c. At each pass though the loop, we multiply b by c if the coefficient of the current power of 2 in the corresponding binary digit of k is 1, and we replace c with $c^2 \bmod n$. As an example, we use `ourpowermod` to compute $123456789^{85479892289} \bmod 41889443039$.

Mathematica

```
ourpowermod[123456789, 85479892289, 41889443039]

19155614901
```

The corresponding *Maple* computation is shown on the next page.

Maple

```
ourpowermod:=proc(a,k,n)
  local b,c,ktemp,r;
  b:=1;
  c:=a;
  ktemp:=k;
  while ktemp>0 do
    r:=modp(ktemp,2);
    if r=1 then b:=modp(b*c,n) end if;
    c:=modp(c^2,n);
    ktemp:=(ktemp-r)/2;
  end do;
  return(b);
end proc:

ourpowermod(123456789, 85479892289, 41889443039);

19155614901
```

RSA cryptography

The RSA encryption scheme is now built into the operating systems of Microsoft, Apple, Novell, and Sun. At the time of publication, an estimated 300 million installations of RSA encryption engines had been made around the world. Thus, with ever changing technology and new research in the theory of numbers, it is critical to have an up-to-date understanding of what is secure and what is not. How large must the modulus used in an RSA encryption scheme be so that it cannot be factored in a reasonable amount of time?

In an effort to monitor such issues, in 1991 RSA Labs issued a list of numbers as part of a factoring challenge with cash prizes awarded for each number factored. (See <http://www.rsa.com/rsalabs/>.) The numbers are products of two primes of roughly the same number of digits and, hence, suitable for use in an RSA encryption key. The smallest number on the list, called RSA-100 (a 100-digit number), was factored in 1991. RSA-120 was factored in 1993 and RSA-140 in 1999. The factorization of RSA-140 took nine weeks to complete, under the guidance of a team of mathematicians and computer scientists using 185 computers. Later that year, RSA-155 was factored by a group that included those involved in the factorization of RSA-140, making use of almost 300 computers over the course of seven months. In 2005 a 193-digit RSA challenge number was factored. The current recommended key size for implementation of the RSA scheme is 309 digits.

RSA-129 (the code referred to in the quotation at the beginning of the chapter) is mentioned in an episode of the popular television program *Numb3rs*.

Chapter 5

Quadratic Residues

Pauca sed Matura.

[Few but ripe.]

CARL FRIEDRICH GAUSS (1777–1855)

In this chapter we develop the basic theory of quadratic congruences. When are quadratic congruences solvable? How do we find their solutions? We discover that the key to these questions is the set of quadratic residues of a prime number. The theory of quadratic residues is quite elegant. Also, quadratic residues have certain interesting properties that make them useful for studying "random-looking" configurations in combinatorics and geometry.

5.1 Quadratic congruences

We have already discovered (in Chapter 2 and Section 3.4) how to solve the *general linear congruence*,

$$ax + b \equiv 0 \pmod{n}.$$

Now let's consider the *general quadratic congruence*,

$$ax^2 + bx + c \equiv 0 \pmod{n},$$

where $a \not\equiv 0 \pmod{n}$. We will use the method of "completing the square." We multiply the congruence by $4a$:

$$4a^2x^2 + 4abx + 4ac \equiv 0 \pmod{n}.$$

Hence,

$$(2ax + b)^2 \equiv b^2 - 4ac \pmod{n}.$$

Making the substitution $y \equiv 2ax + b \pmod{n}$, we see that if n is odd and $\gcd(a, n) = 1$, then the above congruence has a solution if and only if the congruence

$$y^2 \equiv b^2 - 4ac \pmod{n}$$

has a solution.

PROPOSITION 5.1

The congruence

$$ax^2 + bx + c \equiv 0 \pmod{n},$$

where n *is odd and* $\gcd(a, n) = 1$, *has a solution if and only if the congruence*

$$y^2 \equiv b^2 - 4ac \pmod{n}$$

has a solution.

Let $d = b^2 - 4ac$. You are probably familiar with this expression from solving quadratic equations. It is called a "discriminant." When is the congruence $y^2 \equiv d \pmod{n}$ solvable? If $d \equiv 0 \pmod{n}$, then there is the obvious solution $y \equiv 0 \pmod{n}$. Another special case has been covered in Chapter 3, for $n = p$, a prime. If $d \equiv -1 \pmod{p}$, then the congruence is solvable when $p = 4k+1$ and not solvable when $p = 4k + 3$. Moreover, in the solvable case there are exactly two solutions. For if $x^2 \equiv y^2 \pmod{p}$, then $p \mid x^2 - y^2 = (x-y)(x+y)$ and hence $p \mid x - y$ or $p \mid x + y$, from which it follows that $x \equiv y \pmod{p}$ or $x \equiv -y \pmod{p}$.

From these considerations, we see that to solve quadratic congruences we need to distinguish between squares and nonsquares. We will explore this issue in the next section.

Exercises

5.1 Find all solutions (if any) to the following congruences:

(a) $2x^2 + 2x + 1 \equiv 0 \pmod{7}$
(b) $2x^2 + 2x + 1 \equiv 0 \pmod{17}$
(c) $2x^2 + 3x + 1 \equiv 0 \pmod{7}$
(d) $2x^2 + 3x + 1 \equiv 0 \pmod{17}$.

5.2 Find all solutions (if any) to the congruence

$$x^2 + x + 1 \equiv 0 \pmod{2}.$$

5.3 For what values of b and c in $\mathbf{Z}_2$ does the congruence

$$x^2 + bx + c \equiv 0 \pmod{2}$$

have solutions?

◇**5.4** Use a computer to find all solutions (if any) to the congruence

$$5x^2 + 21x + 7 \equiv 0 \pmod{1999}.$$

5.2 Quadratic residues and nonresidues

For p an odd prime, the $p-1$ elements of $\mathbf{Z}_p^*$ fall into two classes: "quadratic residues" and "quadratic nonresidues."

DEFINITION 5.2 *Let $a \in \mathbf{Z}_p^*$. If $a \equiv x^2$ modulo p for some nonzero x, then a is a* quadratic residue *(modulo p); if not, then a is a* quadratic nonresidue *(modulo p). Thus, the set of quadratic residues modulo p is*

$$R = \{x^2 : x \in \mathbf{Z}_p^*\},$$

and the set of quadratic nonresidues is

$$N = \mathbf{Z}_p^* \setminus R.$$

Example 5.3 Let $p = 7$. Then $R = \{1, 2, 4\}$ and $N = \{3, 5, 6\}$. □

We now show that $|R| = |N|$. Consider the map $f \colon \mathbf{Z}_p^* \to \mathbf{Z}_p^*$ defined by $f(x) = x^2$. By the above considerations, the range of this function is R, and each element of the range is mapped onto by two elements of the domain. Therefore, $|R| = (p-1)/2$. It follows that $|N| = (p-1)/2$.

PROPOSITION 5.4
If p is an odd prime, then $|R| = |N| = (p-1)/2$.

When is the congruence

$$x^2 \equiv a \pmod{p}$$

solvable? Suppose that g is a primitive root modulo p. Clearly, $1, g^2, g^4, \ldots, g^{p-1}$ are elements of R, and since R has order $(p-1)/2$, this must be the complete list of elements of R. Hence, quadratic residues are even powers of a primitive root.

THEOREM 5.5 (Euler's criterion)
Suppose that p is an odd prime and $p \nmid a$. Then the congruence $x^2 \equiv a \pmod{p}$ has two solutions if $a^{(p-1)/2} \equiv 1 \pmod{p}$ and no solution if $a^{(p-1)/2} \equiv -1 \pmod{p}$.

PROOF Assume that g is a primitive root modulo p. Suppose that $a \in R$. Then $a \equiv g^{2k} \pmod{p}$, for some k, and

$$a^{(p-1)/2} \equiv g^{k(p-1)} \equiv 1 \pmod{p}.$$

On the other hand, suppose that $a \in N$. Then $a \equiv g^{2k+1} \pmod{p}$, for some k, so that

$$a^{(p-1)/2} \equiv g^{(2k+1)(p-1)/2} \equiv g^{k(p-1)} \cdot g^{(p-1)/2} \equiv g^{(p-1)/2} \equiv -1 \pmod{p},$$

since the square of $g^{(p-1)/2}$ is 1 but $g^{(p-1)/2}$ is not itself 1. □

Example 5.6 Let's use Euler's criterion to determine the number of solutions to the congruence $x^2 \equiv 10 \pmod{101}$. We calculate: $10^{(101-1)/2} \equiv 10^{50} \equiv (10^2)^{25} \equiv (-1)^{25} \equiv -1 \pmod{101}$. By Euler's criterion, the given congruence has no solution. □

DEFINITION 5.7 *The* Legendre symbol $\left(\frac{a}{p}\right)$ *is defined by*

$$\left(\frac{a}{p}\right) = \begin{cases} 0 \textit{ if } a \equiv 0 \pmod{p} \\ 1 \textit{ if } a \in R \\ -1 \textit{ if } a \in N. \end{cases}$$

The Legendre symbol was introduced by Adrien-Marie Legendre (1752–1833).

PROPOSITION 5.8 (Properties of the Legendre symbol)
For p an odd prime,

(i) $\left(\frac{0}{p}\right) = 0$;

(ii) $\left(\frac{1}{p}\right) = 1$;

(iii) $\left(\frac{a^2}{p}\right) = 1$;

(iv) $\left(\frac{-1}{p}\right) = \begin{cases} 1 \textit{ if } p = 4k+1 \\ -1 \textit{ if } p = 4k+3 \end{cases}$;

(v) $\left(\frac{a}{p}\right)\left(\frac{b}{p}\right) = \left(\frac{ab}{p}\right)$;

(vi) $\sum_{a \in \mathbf{Z}_p} \left(\frac{a}{p}\right) = 0$;

(vii) $\left(\frac{a}{p}\right) \equiv a^{(p-1)/2} \pmod{p}$.

Assertion (v) follows from Euler's criterion (Theorem 5.5). Assertion (vi) follows from the fact that $|R| = |N|$. Assertion (vii) follows from Euler's criterion.

Example 5.9 Is the congruence $x^2 \equiv 301 \pmod{1999}$ solvable?

We use *Mathematica* to list the elements of R.

Mathematica

```
Table[Mod[x^2,1999],{x,1,1998}]//Union

{1,2,4,5,8,9,10,11,13,16,18,20,21,22,23,25,...,301,...,1996}
```

We see that 301 is on this list. Hence, 301 is a quadratic residue modulo 1999, and the congruence is solvable.

We will check this fact in two other ways. First, we use Euler's criterion.

Mathematica

```
PowerMod[301,(1999-1)/2,1999]

1
```

According to Euler's criterion, the congruence is solvable.

Finally, we compute the Legendre symbol $\left(\frac{301}{1999}\right)$ using *Mathematica*. In fact, *Mathematica* does not have a function called `LegendreSymbol`. Instead, we use a more general function called `JacobiSymbol`. We will discuss the Jacobi symbol is in Section 5.4.

Mathematica

```
JacobiSymbol[301,1999]
1
```

Here are the corresponding *Maple* calculations.

Maple

```
{seq(modp(x^2,1999),x=1..1998)};

{1,2,4,5,8,9,10,11,13,16,18,20,21,22,23,25,...,301,...,1996}

modp(301&^((1999-1)/2),1999);

1

numtheory[jacobi](301,1999);

1
```

In summary, we have found that 301 is a quadratic residue modulo 1999. ∎

What kind of set is R? It turns out that R has a "pseudorandom" structure. We investigate some properties of R in the sections that follow. For now, we provide an example of the random appearance of R.

Example 5.10

Mathematica

```
Table[JacobiSymbol[x,19],{x,1,18}]

{1,-1,-1,1,1,1,1,-1,1,-1,1,-1,-1,-1,-1,1,1,-1}
```

Maple

```
[seq(numtheory[jacobi](x,19), x = 1..18)];

{1,-1,-1,1,1,1,1,-1,1,-1,1,-1,-1,-1,-1,1,1,-1}
```

□

The sequence elements follow patterns. One pattern is that of "negative symmetry," that is, $\left(\frac{x}{19}\right) = -\left(\frac{19-x}{19}\right)$. In general, $\left(\frac{x}{p}\right) = \left(\frac{p-x}{p}\right)$, if $p = 4k+1$, and $\left(\frac{x}{p}\right) = -\left(\frac{p-x}{p}\right)$, if $p = 4k+3$.

Exercises

5.5 (a) Determine the set of quadratic residues mod 17.

(b) Solve the congruence $x^2 \equiv 8 \pmod{17}$.

5.6 Show that 2 is a quadratic residue mod 7 but a quadratic nonresidue mod 11.

5.7 Compute

(a) $\left(\dfrac{5}{7}\right)$

(b) $\left(\dfrac{144}{863}\right)$

(c) $\left(\dfrac{-1}{41}\right)$

(d) $\left(\dfrac{96}{97}\right)$

(e) $\left(\frac{51}{67}\right)$.

5.8 Let $p > 5$ be a prime number. Show that at least one of the numbers 2, 3, or 6 must be a quadratic residue mod p.

5.9 If p is a prime number and $p > 3$, show that the sum of all quadratic residues mod p is divisible by p.

5.10 Let p be an odd prime.

(a) Show that if $p \nmid a$, then the congruence

$$x^2 \equiv a \pmod{p^k}$$

has a solution if and only if $\left(\frac{a}{p}\right) = 1$.

(b) Determine the number of congruence classes a of $\mathbf{Z}_{p^k}^*$ for which

$$x^2 \equiv a \pmod{p^k}$$

has a solution.

(c) Let n be an odd integer greater than 1 and a an integer relatively prime to n. Prove that the congruence

$$x^2 \equiv a \pmod{n}$$

has a solution if and only if $\left(\frac{a}{p}\right) = 1$ for all primes p dividing n.

5.11 Suppose that $p \equiv 3 \pmod 4$. Show that if the congruence

$$x^2 \equiv a \pmod{p}$$

has a solution, then it must be of the form $x \equiv \pm a^{(p+1)/4} \pmod{p}$.

5.12 Let p be an odd prime number and suppose that a is an element of $\mathbf{Z}_p^*$ of odd order. Show that

$$\left(\frac{a}{p}\right) = 1.$$

5.13 Determine the number of irreducible degree 2 polynomials in $\mathbf{Z}_p[x]$, that is, the number of polynomials $ax^2 + bx + c$ (with $a, b, c \in \mathbf{Z}_p$), where $a \not\equiv 0 \pmod{p}$, such that the congruence

$$ax^2 + bx + c \equiv 0 \pmod{p}$$

does not have a solution.

5.14 Prove that no Fibonacci number f_n, with n odd, is divisible by a prime of the form $4k + 3$.

Hint: Use Cassini's identity from Chapter 1.

5.3 Quadratic reciprocity

In light of the fact that $\left(\frac{ab}{p}\right) = \left(\frac{a}{p}\right)\left(\frac{b}{p}\right)$, to compute the value of the Legendre symbol $\left(\frac{x}{p}\right)$, it is enough to compute $\left(\frac{q}{p}\right)$ for all primes q that divide x. We will first consider the case $q = 2$. We will show that the value of Legendre symbol $\left(\frac{2}{p}\right)$ depends only on the congruence class of p modulo 8. To deduce this we will use a result that provides us with another method of computing Legendre symbols.

PROPOSITION 5.11 (Gauss's lemma)
Let p be an odd prime number, and a an integer not divisible by p. Let S_a be the set of least positive residues of the set of elements

$$\left\{a,\ 2a,\ 3a,\ \ldots,\ \left(\frac{p-1}{2}\right) a\right\}.$$

Let s be the number of elements of S_a that are greater than $p/2$. Then

$$\left(\frac{a}{p}\right) = (-1)^s.$$

PROOF We will show that

$$a^{(p-1)/2} \equiv (-1)^s \pmod{p},$$

and the claim will follow from Euler's criterion. Let $b_1, \ldots, b_s$ be the elements of S_a that are greater than $p/2$ and $l_1, \ldots, l_t$ be the elements of S_a that are less than $p/2$ (the b_i's are the "big" residues and l_i's are the "little" residues). We observe that

$$b_1 b_2 \ldots b_s l_1 l_2 \ldots l_t \equiv \left(\frac{p-1}{2}\right)!\, a^{(p-1)/2} \pmod{p}. \tag{5.1}$$

The product on the left side of (5.1) is the same modulo p as the product

$$(-1)^s (p - b_1)(p - b_2) \ldots (p - b_s) l_1 l_2 \ldots l_t. \tag{5.2}$$

Notice that the numbers $p - b_1, \ldots, p - b_s, l_1, \ldots, l_t$ are all positive and less than $p/2$. We claim that the numbers in this list are distinct. Suppose that $p - b_i = l_j$, for some i and j. Then

$$p - m_i a = m_j a,$$

for some m_i and m_j satisfying $0 < m_i, m_j \leq (p-1)/2$. Rearranging the equality and reducing mod p, we obtain $a(m_i + m_j) \equiv 0 \pmod{p}$, which implies that p divides $m_i + m_j$. Since $0 < m_i, m_j < p/2$, this cannot be. It is clear that no $l_i = l_j$ or $b_i = b_j$ for i, j distinct. We conclude that the list of numbers $p - b_1, \ldots, p - b_s, l_1, \ldots, l_t$ is just the (reordered) list $1, 2, \ldots, (p-1)/2$. Hence, the product (5.2) is exactly $(-1)^s(\frac{p-1}{2})!$. Now equation (5.1) becomes

$$(-1)^s \left(\frac{p-1}{2}\right)! \equiv \left(\frac{p-1}{2}\right)! \, a^{(p-1)/2} \pmod{p}.$$

Canceling $(\frac{p-1}{2})!$ from both sides, we obtain $(-1)^s \equiv a^{(p-1)/2} \pmod{p}$, as desired. ∎

We apply Gauss's lemma to determine when 2 is a square mod p for any odd prime p.

THEOREM 5.12

Let p be an odd prime. Then $\left(\frac{2}{p}\right) = 1$ if and only if $p \equiv \pm 1 \pmod{8}$. Equivalently, $\left(\frac{2}{p}\right) = (-1)^{(p^2-1)/8}$.

PROOF We must count the number, s, of elements in the set

$$2,\ (2)2,\ \ldots,\ ((p-1)/2)2$$

whose least positive residues are greater than $p/2$. However, all of the numbers in this set are less than p, and so we just need to count the number of elements that are greater than $p/2$. We see that $(m)2 > p/2$ if and only if $m > p/4$. It follows that

$$s = (p-1)/2 - \lfloor p/4 \rfloor. \tag{5.3}$$

We now determine when the quantities $(p-1)/2$ and $\lfloor p/4 \rfloor$ are even. Observe that $(p-1)/2$ is even if and only if $p-1$ is divisible by 4. Thus, $(p-1)/2$ is even exactly when $p \equiv 1 \pmod{4}$. Any odd prime can be written in the form $8k + r$, where $r = 1$, 3, 5, or 7. We compute

$$\left\lfloor \frac{p}{4} \right\rfloor = \left\lfloor \frac{8k+r}{4} \right\rfloor = 2k + \left\lfloor \frac{r}{4} \right\rfloor.$$

Since $\lfloor r/4 \rfloor = 0$ when $r = 1$ or 3, but $\lfloor r/4 \rfloor = 1$ when $r = 5$ or 7, we see that $\lfloor p/4 \rfloor$ is even exactly when $p \equiv 1$ or $3 \pmod{8}$. We summarize our observations in the table below.

$p \pmod{8}$	$(p-1)/2$	$\lfloor p/4 \rfloor$
1	even	even
3	odd	even
5	even	odd
7	odd	odd

From (5.3), we see that s is even if and only if $p \equiv 1$ or $7 \pmod 8$. □

Computing $\left(\frac{3}{p}\right)$ for a prime number greater than 3 using Gauss's lemma is more difficult. However, there is a way to determine the value of this Legendre symbol from the congruence class of p mod 12. We will show how to relate $\left(\frac{3}{p}\right)$ to the Legendre symbol $\left(\frac{p}{3}\right)$. Since $\left(\frac{p}{3}\right) = 1$ if and only if $p \equiv 1 \pmod 3$, we will have solved the problem. The relation between these two Legendre symbols is given by the celebrated result known as quadratic reciprocity. The theorem tells us that for distinct odd primes p and q, the Legendre symbols $\left(\frac{p}{q}\right)$ and $\left(\frac{q}{p}\right)$ are equal if either p or q is congruent to 1 mod 4; otherwise, the symbols are unequal. The theorem was first conjectured by Euler and subsequently reformulated by Legendre. The first correct proof was given by Gauss, who managed to provide several different proofs over the course of his career.

THEOREM 5.13 (Quadratic reciprocity)

Let p and q be distinct odd primes. Then

$$\left(\frac{p}{q}\right) = \begin{cases} \left(\frac{q}{p}\right) & \text{if } p \equiv 1 \pmod 4 \text{ or } q \equiv 1 \pmod 4 \\ -\left(\frac{q}{p}\right) & \text{if } p \equiv 3 \pmod 4 \text{ and } q \equiv 3 \pmod 4. \end{cases}$$

Equivalently, $\left(\frac{p}{q}\right)\left(\frac{q}{p}\right) = (-1)^{(p-1)(q-1)/4}$.

To prove this theorem, we use Gauss's lemma to deduce yet another formula for computing the Legendre symbol $\left(\frac{a}{p}\right)$.

LEMMA 5.14

Let p be an odd prime number and a an odd integer not divisible by p. Set

$$v = \sum_{m=1}^{(p-1)/2} \lfloor ma/p \rfloor .$$

Then $\left(\frac{a}{p}\right) = (-1)^v$.

PROOF We will use the same notation as in the proof of Gauss's lemma. Our goal is to show that $v \equiv s \pmod 2$. First observe that the integer ma is the sum of the quantity $\lfloor ma/p \rfloor p$ and the least positive residue of ma mod p. It follows that

$$\sum_{m=1}^{(p-1)/2} ma = \sum_{m=1}^{(p-1)/2} \lfloor ma/p \rfloor p + \sum_{i=1}^{s} b_i + \sum_{j=1}^{t} l_j.$$

The first term on the right hand side is just the quantity vp. As we noted in the proof of Gauss's lemma, the list of integers $p - b_1, \ldots, p - b_s, l_1, \ldots, l_t$ is just the (reordered) list of numbers 1, 2, ..., $(p-1)/2$. Thus, we may rewrite the equation above as

$$\sum_{m=1}^{(p-1)/2} ma = vp + \left(\sum_{m=1}^{(p-1)/2} m\right) + 2\left(\sum_{i=1}^{s} b_i\right) - sp.$$

Rearranging, we obtain

$$(a-1)\sum_{m=1}^{(p-1)/2} m = p(v-s) + 2\sum_{i=1}^{s} b_i.$$

Since a is odd, the left hand side is an even integer. Reducing the above equality mod 2, and using the fact that p is odd we get

$$0 \equiv v - s \pmod{2}.$$

Hence, $v \equiv s \pmod{2}$. □

Note: The proof above can be modified to give a second proof of Theorem 5.12. If we take $a = 2$, everything in the proof but the last three statements still holds. In particular, we have

$$(2-1)\sum_{m=1}^{(p-1)/2} m = p(v-s) + 2\sum_{i=1}^{s} b_i.$$

Now the left hand side is the quantity $(p^2-1)/8$. For $1 \le m \le (p-1)/2$, we have $\lfloor 2m/p \rfloor = 0$, and so $v = 0$. We conclude that

$$\frac{p^2-1}{8} \equiv s \pmod{2}.$$

PROOF (Proof of quadratic reciprocity) We will show that

$$\left(\frac{q}{p}\right)\left(\frac{p}{q}\right) = (-1)^{(p-1)(q-1)/4}.$$

Using Lemma 5.14, it is enough to show that

$$\sum_{m=1}^{(p-1)/2} \left\lfloor \frac{mq}{p} \right\rfloor + \sum_{n=1}^{(q-1)/2} \left\lfloor \frac{np}{q} \right\rfloor = \frac{(p-1)(q-1)}{4}.$$

To achieve this, we will count the number of ordered pairs in the set

$$\{(x,y) : 1 \le x \le (p-1)/2, 1 \le y \le (q-1)/2\}.$$

On the one hand, there are $(p-1)/2$ possibilities for x, and for each x, there are $(q-1)/2$ possibilities for y. Hence, our set has $((p-1)/2)((q-1)/2)$ elements. Now consider our set as the disjoint union of the subset A that contains ordered pairs (x, y), where $yp < xq$, and the subset B that contains ordered pairs (x, y), where $yp > xq$. Note that $xq \neq yp$ for ordered pairs in our set because p and q are distinct primes. Given an ordered pair (x, y) in A, there are $(p-1)/2$ possibilities for x. For each x, the corresponding possibilities for y must satisfy

$$y < \frac{xq}{p}.$$

Since

$$\frac{xq}{p} < \frac{(p/2)q}{p} = \frac{q}{2},$$

all y that satisfy the previous inequality do give rise to elements of A. Thus, for each x, there are exactly $\lfloor xq/p \rfloor$ choices for y. As a result, the total number of elements in A is

$$\sum_{x=1}^{(p-1)/2} \left\lfloor \frac{xq}{p} \right\rfloor.$$

A similar argument shows that the number of elements in B is

$$\sum_{y=1}^{(q-1)/2} \left\lfloor \frac{yp}{q} \right\rfloor.$$

Thus, the total number of ordered pairs is the sum of these two sums, and so we have shown that

$$\sum_{x=1}^{(p-1)/2} \left\lfloor \frac{xq}{p} \right\rfloor + \sum_{y=1}^{(q-1)/2} \left\lfloor \frac{yp}{q} \right\rfloor = \frac{(p-1)(q-1)}{4}.$$

This completes the proof. □

Provided that the numbers are of a size that can be readily factored, quadratic reciprocity makes computing Legendre symbols easy.

Example 5.15 Let's compute $\left(\frac{561}{659}\right)$. As 561 factors as $3 \cdot 11 \cdot 17$, we have

$$\left(\frac{561}{659}\right) = \left(\frac{3}{659}\right)\left(\frac{11}{659}\right)\left(\frac{17}{659}\right).$$

Applying quadratic reciprocity, the right side becomes

$$\left(-\left(\frac{659}{3}\right)\right)\left(-\left(\frac{659}{11}\right)\right)\left(\frac{659}{17}\right).$$

Now reducing 659 modulo 3, 11, and 17, respectively, gives us

$$\left(-\left(\frac{2}{3}\right)\right)\left(-\left(\frac{10}{11}\right)\right)\left(\frac{13}{17}\right).$$

We know that $\left(\frac{2}{3}\right) = -1$. Since $11 \equiv 3 \pmod 4$, we see that $\left(\frac{10}{11}\right) = \left(\frac{-1}{11}\right) = -1$. We can apply quadratic reciprocity once again to the last term, and we find that

$$\begin{aligned}\left(\frac{13}{17}\right) &= \left(\frac{17}{13}\right)\\ &= \left(\frac{4}{13}\right)\\ &= \left(\left(\frac{2}{13}\right)\right)^2\\ &= 1.\end{aligned}$$

Putting our calculations together,

$$\left(\frac{561}{659}\right) = (-(-1))(-(-1))(1) = 1.$$

∎

We now return to the issue of computing $\left(\frac{3}{p}\right)$ for an odd prime p. By quadratic reciprocity,

$$\left(\frac{3}{p}\right) = (-1)^{(p-1)/2}\left(\frac{p}{3}\right).$$

We know that $\left(\frac{p}{3}\right) = 1$ if and only if $p \equiv 1 \pmod 3$. Thus $\left(\frac{3}{p}\right) = 1$ if and only if p is congruent to both $1 \pmod 3$ and $1 \pmod 4$ or p is congruent to both $2 \pmod 3$ and $3 \pmod 4$. We can summarize in terms of congruence classes mod 12.

PROPOSITION 5.16

For $p \neq 2, 3$,

$$\left(\frac{3}{p}\right) = \begin{cases} 1 \text{ if } p \equiv 1, 11 \pmod{12} \\ -1 \text{ if } p \equiv 5, 7 \pmod{12}. \end{cases}$$

Exercises

5.15 Compute $\left(\frac{5}{17}\right)$ and $\left(\frac{5}{19}\right)$ three ways using

(a) Gauss's lemma

(b) Lemma 5.14

(c) Quadratic reciprocity.

5.16 Use quadratic reciprocity to compute the following Legendre symbols by hand.

(a) $\left(\frac{5}{89}\right)$

(b) $\left(\frac{5}{103}\right)$

(c) $\left(\frac{7}{89}\right)$

(d) $\left(\frac{7}{103}\right)$

5.17 Compute by hand

$$\left(\frac{8569}{8831}\right).$$

Hint: $8569 = 11 \cdot 19 \cdot 41$ and 8831 is prime.

5.18 Let p be an odd prime number. Prove that

$$\left(\frac{-2}{p}\right) = \begin{cases} 1 & \text{if } p \equiv 1 \text{ or } 3 \pmod 8 \\ -1 & \text{if } p \equiv 5 \text{ or } 7 \pmod 8 \end{cases}.$$

5.19 (a) Let p be an odd prime number not equal to 5. Show that $\left(\frac{5}{p}\right) = 1$ if and only if $p \equiv \pm 1 \pmod 5$.

(b) Determine which odd primes p not equal to 7 satisfy $\left(\frac{7}{p}\right) = 1$.

5.20 (a) Determine which odd primes p not equal to 3 satisfy $\left(\frac{-3}{p}\right) = 1$.

(b) Prove that there are infinitely many primes of the form $3n + 1$. Hint: Suppose that there are only finitely many primes of this form and that their product is N. Consider the Legendre symbol $\left(\frac{-3}{p}\right)$ where p is a prime dividing $4N^2 + 3$.

5.21 For which odd primes p does the congruence

$$x^2 + 3x + 1 \equiv 0 \pmod p$$

have a solution?

†⋆**5.22** (Pepin's test) Let $F_m = 2^{2^m} + 1$ be the mth Fermat number, where $m \geq 1$.

(a) Show that if F_m is prime then

$$3^{(F_m-1)/2} \equiv -1 \pmod{F_m}.$$

(b) Show that if F_m is prime then 3 is a primitive root modulo F_m.

(c) Prove the converse of part (a). That is, show that if

$$3^{(F_m-1)/2} \equiv -1 \pmod{F_m},$$

then F_m is prime.

5.4 The Jacobi symbol

The Jacobi symbol is a generalization of the Legendre symbol. It is due to Carl Jacobi (1804–1851).

DEFINITION 5.17 *Let a be any integer and n a positive odd integer with prime factorization $n = p_1 p_2 \dots p_t$ (where the p_i are not necessarily distinct). The Jacobi symbol $\left(\frac{a}{n}\right)$ is defined as*

$$\left(\frac{a}{n}\right) = \prod_{i=1}^{t} \left(\frac{a}{p_i}\right)$$

where the factors on the right side are Legendre symbols.

Example 5.18 We compute the Jacobi symbol $\left(\frac{2}{585}\right)$. First, we observe that $585 = 3 \cdot 3 \cdot 5 \cdot 13$. Hence,

$$\left(\frac{2}{585}\right) = \left(\frac{2}{3}\right)\left(\frac{2}{3}\right)\left(\frac{2}{5}\right)\left(\frac{2}{13}\right).$$

Now using Theorem 5.12, we see that

$$\left(\frac{2}{585}\right) = (-1)(-1)(-1)(-1) = 1.$$

□

Note: If the congruence $x^2 \equiv a \pmod{n}$ has a solution, then $x^2 \equiv a \pmod{p}$ must have a solution for all primes p dividing n. In this case, we see that $\left(\frac{a}{n}\right) = 1$. However, the converse is not true. For instance, $\left(\frac{2}{585}\right) = 1$, but the congruence $x^2 \equiv 2 \pmod{585}$ cannot have a solution as the congruence $x^2 \equiv 2 \pmod{3}$ has no solution.

PROPOSITION 5.19 (Properties of the Jacobi symbol)
For odd positive integers m and n and any integers a and b,

(i) $\left(\frac{a}{n}\right) = 0$ *if and only if* $\gcd(a,n) \neq 1$;

(ii) $\left(\frac{1}{n}\right) = 1$;

(iii) $\left(\frac{a^2}{n}\right) = 1$;

(iv) $\left(\frac{ab}{n}\right) = \left(\frac{a}{n}\right)\left(\frac{b}{n}\right)$;

(v) $\left(\frac{a}{mn}\right) = \left(\frac{a}{m}\right)\left(\frac{a}{n}\right)$;

(vi) $\left(\frac{-1}{n}\right) = 1$ *if and only if* $n \equiv 1 \pmod 4$; *equivalently,* $\left(\frac{-1}{n}\right) = (-1)^{(n-1)/2}$;

(vii) $\left(\frac{2}{n}\right) = 1$ *if and only if* $n \equiv \pm 1 \pmod 8$; *equivalently,* $\left(\frac{2}{n}\right) = (-1)^{(n^2-1)/8}$.

Properties (i)–(v) follow directly from properties of the Legendre symbol and the definition of the Jacobi symbol. Properties (vi) and (vii) are left as exercises. The proof of property (vi) uses the following lemma.

LEMMA 5.20
Let n be a positive odd integer with prime factorization $n = p_1 p_2 \dots p_t$ (where the p_i are not necessarily distinct). Then

$$\sum_{i=1}^{t} \frac{p_i - 1}{2} \equiv \frac{n-1}{2} \pmod 2,$$

and as a result,

$$\prod_{i=1}^{t} (-1)^{(p_i-1)/2} = (-1)^{(n-1)/2}.$$

PROOF Reordering the p_i, if necessary, we can assume that the first j primes are congruent to 3 (mod 4) and the remaining primes are congruent to 1 (mod 4). We see that $n \equiv 1 \pmod 4$ if and only if j is even. Hence, $(n-1)/2 \equiv 0 \pmod 2$ if and only if j is even.

Now $(p_i - 1)/2 \equiv 1 \pmod 2$, for $1 \leq i \leq j$, and $(p_i - 1)/2 \equiv 0 \pmod 2$, for $j+1 \leq i \leq k$. Thus,

$$\sum_{i=1}^{k} \frac{p_i - 1}{2} \equiv j \pmod 2,$$

and so the sum is congruent to 0 (mod 2) if and only if j is even. ∎

Given the lemma, we can now quickly see that quadratic reciprocity generalizes to Jacobi symbols.

THEOREM 5.21 (Quadratic reciprocity for Jacobi symbols)
Let a and b be positive odd integers. Then

$$\left(\frac{a}{b}\right)\left(\frac{b}{a}\right) = (-1)^{(a-1)(b-1)/4}.$$

PROOF We begin by writing prime factorizations $a = p_1p_2\dots p_s$ and $b = q_1q_2\dots q_t$. By the definition of the Jacobi symbol and the multiplicativity of the Legendre symbol, we see that

$$\left(\frac{a}{b}\right) = \prod_{i=1}^{s}\prod_{j=1}^{t}\left(\frac{p_i}{q_j}\right).$$

We may apply quadratic reciprocity for Legendre symbols to each of the symbols in the product on the right to obtain

$$\begin{aligned}\left(\frac{a}{b}\right) &= \prod_{i=1}^{s}\prod_{j=1}^{t}(-1)^{((p_i-1)/2)((q_j-1)/2)}\left(\frac{q_j}{p_i}\right)\\ &= (-1)^{\sum_{i=1}^{s}\sum_{j=1}^{t}((p_i-1)/2)((q_j-1)/2)}\left(\prod_{i=1}^{s}\prod_{j=1}^{t}\left(\frac{q_j}{p_i}\right)\right)\\ &= (-1)^{\sum_{i=1}^{s}((p_i-1)/2)\sum_{j=1}^{t}((q_j-1)/2)}\left(\frac{b}{a}\right).\end{aligned}$$

Now we may apply Lemma 5.20 to the two sums to obtain the formula. ∎

An important consequence of this generalization is that we can now use reciprocity to compute values of Legendre symbols without having to perform any factoring (other than pulling off factors of 2). This is particularly helpful if we wish to compute a Legendre symbol in which the numbers that arise are large and not easily factored.

Example 5.22 We now compute the Legendre symbol $\left(\frac{403}{829}\right)$ by regarding it as a Jacobi symbol and making use of quadratic reciprocity for Jacobi symbols.

$$\begin{aligned}
\left(\frac{403}{829}\right) &= \left(\frac{829}{403}\right) && \text{(since } 829 \equiv 1 \pmod 4\text{)}\\
&= \left(\frac{23}{403}\right)\\
&= -\left(\frac{403}{23}\right) && \text{(since } 403 \equiv 23 \equiv 3 \pmod 4\text{)}\\
&= -\left(\frac{12}{23}\right)\\
&= -\left(\frac{4}{23}\right)\left(\frac{3}{23}\right)\\
&= -(1)\left(-\left(\frac{23}{3}\right)\right) && \text{(since } 23 \equiv 3 \pmod 4\text{)}\\
&= \left(\frac{2}{3}\right)\\
&= -1
\end{aligned}$$

□

Exercises

5.23 Compute the values of the following Jacobi symbols.

(a) $\left(\frac{5}{12345}\right)$
(b) $\left(\frac{49}{12345}\right)$
(c) $\left(\frac{12345}{49}\right)$
(d) $\left(\frac{12344}{12345}\right)$
(e) $\left(\frac{2}{12345}\right)$
(f) $\left(\frac{12343}{12345}\right)$

5.24 Compute the values of the following Jacobi symbols using quadratic reciprocity for Jacobi symbols.

(a) $\left(\frac{55}{77}\right)$
(b) $\left(\frac{55}{79}\right)$
(c) $\left(\frac{481}{3977}\right)$

†5.25 Prove Proposition 5.19 (vi) and (vii).

Hint: For (vi), use Lemma 5.20 and the fact that $\left(\frac{-1}{p}\right) = (-1)^{(p-1)/2}$ when p is an odd prime. For (vii), show the congruence result by grouping the prime factors of n according to whether they are $\pm 3 \pmod 8$ or $\pm 1 \pmod 8$.

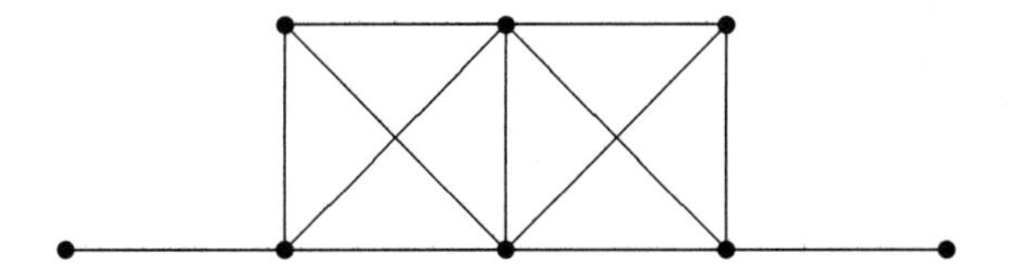

FIGURE 5.1: A graph.

5.5 Application: Construction of tournaments

In this section we exploit the pseudorandom nature of the set R of quadratic residues modulo p to construct a tournament that has a special property.

DEFINITION 5.23 *A* graph *is a set of* vertices *and a set of* edges *joining some pairs of vertices.*

The vertices of a graph are points that may be located anywhere in space, and the edges are lines joining the points that may be curved or straight and may cross. The only relevant thing is the incidence between vertices and edges, that is, whether or not a particular edge contains a particular vertex. In our definition of graph, we do not allow an edge to connect a vertex to itself and we do not allow multiple edges between vertices. Figure 5.1 shows an example of a graph with eight vertices and thirteen edges.

DEFINITION 5.24 *A* complete *graph is one in which every pair of vertices is joined by an edge.*

Note: The complete graph on n vertices has $\binom{n}{2}$ edges.

DEFINITION 5.25 *An* oriented *graph is one in which each edge has been replaced by an arrow.*

Figure 5.2 shows an oriented graph based on the graph of Figure 5.1.

DEFINITION 5.26 *A* tournament *is a complete oriented graph.*

Figure 5.3 shows a cyclic tournament on three vertices that we call the

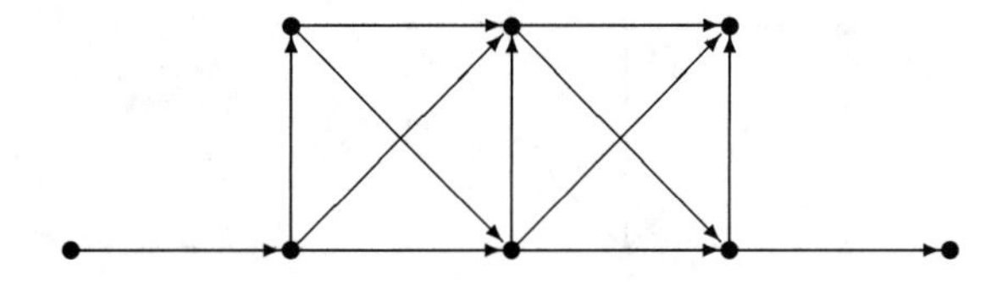

FIGURE 5.2: An oriented graph.

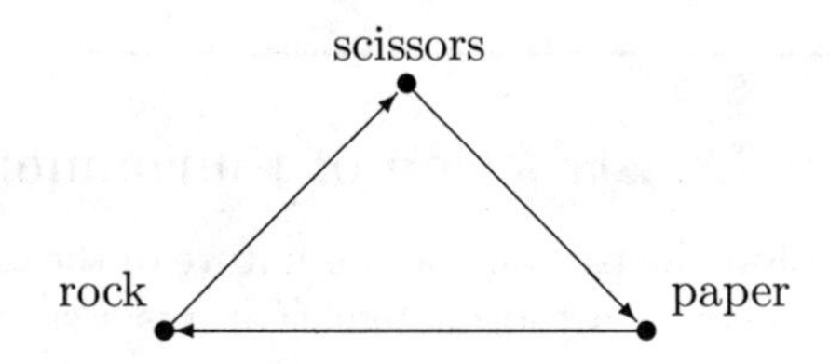

FIGURE 5.3: Rock–paper–scissors tournament.

rock–paper–scissors tournament (because it is based on the popular game of the same name). In this tournament, each vertex (rock, paper, scissors) is "beaten" by some other vertex. We wish to generalize this concept.

Let T_n be a tournament on n vertices. We say that T_n has *property* S_k if, for every set S of k vertices in T_n, there is a vertex $v \in T_n - S$ such that all edges between v and S are directed from v to S. For example, the rock–paper–scissors tournament has property S_1 but not property S_2. The following theorem, called Schütte's theorem but actually proved by Erdős in 1963, shows that, for each k, there is a tournament T_n with property S_k.

THEOREM 5.27

Let k be a positive integer. Then, for some integer n, there exists a tournament T_n with property S_k.

PROOF Let k be specified and n be a number that we will determine later. As the complete graph on n vertices has $\binom{n}{2}$ edges, there are $2^{\binom{n}{2}}$ different tournaments on n vertices. We will show that, for n sufficiently large, *most* of these tournaments have property S_k.

Let t be the number of tournaments on n vertices that fail to have property S_k. For each subset S of k vertices, let A_S be the collection of tournaments for which there exists *no* vertex directed to S (in other words, tournaments that fail to have property S_k because S is a "bad" subset). Then $t = |\bigcup_S A_S|$, and this quantity satisfies the inequality

$$t \leq \sum_S |A_S|. \tag{5.4}$$

This is an inequality rather than an equality since the sum on the right involves some potential double-counting of the terms in the union.

Now we must evaluate $\sum_S |A_S|$. There are $\binom{n}{k}$ terms in the sum. Once a particular subset S is chosen, there is complete freedom in choosing the directions of arrows between the k vertices of S and between the $n-k$ vertices of the complement of S; there are $2^{\binom{n}{2}-k(n-k)}$ choices in total. If any vertex in the complement of S is directed to all vertices in S, then S is not a bad subset. There are $2^k - 1$ choices for the directions of such arrows from each vertex, and hence, $(2^k-1)^{n-k}$ choices altogether. Therefore,

$$\sum_S |A_S| = \binom{n}{k} 2^{\binom{n}{2}-k(n-k)}(2^k-1)^{n-k}.$$

We rewrite the expression on the right and combine it with the inequality (5.4):

$$t \leq \binom{n}{k}\left(1-2^{-k}\right)^{n-k} 2^{\binom{n}{2}}. \tag{5.5}$$

As we have mentioned, there are $2^{\binom{n}{2}}$ different tournaments on n vertices. We see the expression $2^{\binom{n}{2}}$ on the right side of (5.5). This term is multiplied by two other terms: $\binom{n}{k}$ and $\left(1-2^{-k}\right)^{n-k}$. What is the effect of this multiplication? As k is fixed, the term $\binom{n}{k}$ is a polynomial in n (of degree k), while the term $\left(1-2^{-k}\right)^{n-k}$ is an exponential function in n (with base less than 1). As n tends to infinity, the exponential function dominates, and the product of the two terms tends to 0. Hence, the upper bound in (5.5) tends to an arbitrarily small fraction of $2^{\binom{n}{2}}$. In other words, as n tends to infinity, *almost all* tournaments on n vertices have property S_k. Therefore, for n sufficiently large, there exists a tournament with property S_k. □

Although we have shown that almost all tournaments of order n (for n sufficiently large) have property S_k, we have not shown how to actually construct such a tournament. The above proof suggests that a "random" tournament is very likely to have property S_k (when n is large), but how do we construct a random tournament? We will see presently that we can use quadratic residues to accomplish this.

Let's consider the case of finding a tournament with property S_2. A little experimenting shows that there is no such tournament on six vertices. We define a function $f(n,k)$ equal to the factor multiplied by $2^{\binom{n}{2}}$ in (5.5):

$$f(n,k) = \binom{n}{k}(1-2^{-k})^{n-k}.$$

We want to find a value of n for which $f(n, 2) < 1$, for then, according to the method of proof above, there must be a tournament on n vertices with property S_2. We calculate that $f(20, 2) \doteq 1.07116$, while $f(21, 2) \doteq 0.88794$. Hence, $n = 21$ vertices forces the matter. However, even with this value, randomization is not guaranteed to produce such a tournament with high probability. We would have to choose a larger value of n for this. Then we could randomize a tournament (by flipping a coin for each edge) that is likely to have property S_2. Even so, it might require quite a lot of checking to establish that the tournament really has the desired property. In fact, there is a tournament on only seven vertices that has property S_2. The tournament is easily constructed using quadratic residues. Furthermore, the structure of the tournament makes it easy to carry out the required checks.

A tournament on seven vertices with property S_2 can be constructed from the set of quadratic residues modulo 7 as follows. Let the vertices of the tournament be 0, 1, 2, 3, 4, 5, and 6. Let R and N be the set of quadratic residues and nonresidues modulo 7, respectively; that is, $R = \{1, 2, 4\}$ and $N = \{3, 5, 6\}$ (see Example 5.3). Put a directed edge from vertex i to vertex j if $j - i \in R$ and a directed edge from j to i if $j - i \in N$. (Such a tournament is called a *quadratic residue tournament*.) Note that, since $-1 = 6 \in N$, the choice of direction of edges is well-defined. The tournament is shown in Figure 5.4. We check that the tournament has the desired property. Choose any two vertices. Since the tournament has cyclic symmetry, we may take one vertex to be 0 and the other v. Now we need a vertex w such that $w \in N$ and $w - v \in N$. We simply exhibit such a w for each choice of v:

v	1	2	3	4	5	6
w	6	5	6	3	3	5

Note: This construction of quadratic residue tournaments with property S_2 generalizes to all primes greater than or equal to 7 and congruent to 3 (mod 4). A proof of this fact is called for in the exercises. In 1971 Joel Spencer and Ronald Graham proved that a quadratic residue tournament T_p, for a prime $p \equiv 3 \pmod 4$, has property S_k if $p > k^2 2^{2k-2}$.

Exercises

5.26 Show that no tournament on six vertices has property S_2.

Hint: Let the "outdegree" ("indegree") of a vertex in a tournament be the number of arrows that leave (enter) the vertex. Examine the possibilities for the outdegree and indegree of a vertex in a tournament on six vertices.

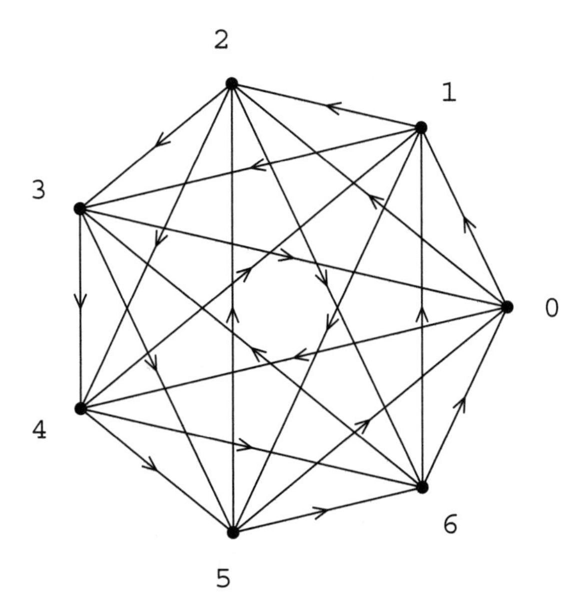

FIGURE 5.4: A tournament with property S_2.

5.27 Prove that the construction of quadratic residue tournaments with property S_2 given in Theorem 5.5 generalizes to all primes greater than or equal to 7 and congruent to 3 (mod 4).

5.28 Find a lower bound for the probability that a random tournament on 100 vertices has property S_3.

5.29 Find a quadratic residue tournament with property S_3.

$\star$**5.30** (An example of Ramsey theory [10]) Use quadratic residues and nonresidues to solve the following problem: Let $V = \{0, 1, 2, 3, \ldots, 16\}$. Color the edges of the complete graph on V using two colors in such a way that no complete subgraph on four vertices of V has all its edges the same color.

5.6 Consecutive quadratic residues and nonresidues

Some of the properties of R and N extend to the quadratic character of consecutive numbers modulo p.

Define

$$\begin{aligned} RR &= \{a \in \mathbf{Z}_p^* \colon a \in R,\, a+1 \in R\} \\ RN &= \{a \in \mathbf{Z}_p^* \colon a \in R,\, a+1 \in N\} \\ NR &= \{a \in \mathbf{Z}_p^* \colon a \in N,\, a+1 \in R\} \\ NN &= \{a \in \mathbf{Z}_p^* \colon a \in N,\, a+1 \in N\}. \end{aligned}$$

Example 5.28 Let $p = 13$. Then

$$\begin{aligned} R &= \{1, 3, 4, 9, 10, 12\} \\ N &= \{2, 5, 6, 7, 8, 11\}, \end{aligned}$$

and

$$\begin{aligned} RR &= \{3, 9\} \\ RN &= \{1, 4, 10\} \\ NR &= \{2, 8, 11\} \\ NN &= \{5, 6, 7\}. \end{aligned}$$

□

We will show that, for a given value of p, each of the sets RR, RN, NR, and NN contains approximately 1/4 of the members of $\mathbf{Z}_p^*$.

We evaluate $|RR|$ by observing that, for $a \neq p - 1$,

$$\left[\left(\frac{a}{p}\right) + 1\right]\left[\left(\frac{a+1}{p}\right) + 1\right] = \begin{cases} 4 & \text{if } a \in RR \\ 0 & \text{if } a \notin RR. \end{cases}$$

Therefore,

$$|RR| = \frac{1}{4}\sum_{a=1}^{p-2}\left[\left(\frac{a}{p}\right) + 1\right]\left[\left(\frac{a+1}{p}\right) + 1\right]. \tag{5.6}$$

Example 5.29 We evaluate $|RR|$ when $p = 1999$, using (5.6).

Mathematica

```
f[k_]:= (JacobiSymbol[k,1999]+1)(JacobiSymbol[k+1,1999]+1)

(1/4)Sum[f[k],{k,1,1997}]

499
```

Maple

```
f := k-> (numtheory[jacobi](k,1999)+1)
*(numtheory[jacobi](k+1,1999)+1):

1/4*sum(f(k), k = 1..1997);

499
```

∎

Now we state and prove a simple formula for $|RR|$.

PROPOSITION 5.30

For p an odd prime,

$$|RR| = \frac{p - 4 - \left(\frac{-1}{p}\right)}{4}.$$

PROOF From equation (5.6) and basic properties of the Legendre symbol (Proposition 5.8), we obtain

$$\begin{aligned}
|RR| &= \frac{1}{4}\sum_{a=1}^{p-2}\left[\left(\frac{a}{p}\right)+1\right]\left[\left(\frac{a+1}{p}\right)+1\right] \\
&= \frac{1}{4}\sum_{a=1}^{p-2}\left[\left(\frac{a}{p}\right)\left(\frac{a+1}{p}\right)+\left(\frac{a}{p}\right)+\left(\frac{a+1}{p}\right)+1\right] \\
&= \frac{1}{4}\left[\sum_{a=1}^{p-2}\left(\frac{a}{p}\right)\left(\frac{a+1}{p}\right)+\sum_{a=1}^{p-2}\left(\frac{a}{p}\right)+\sum_{a=1}^{p-2}\left(\frac{a+1}{p}\right)+\sum_{a=1}^{p-2}1\right] \\
&= \frac{1}{4}\left[\sum_{a=1}^{p-2}\left(\frac{a^2}{p}\right)\left(\frac{1+a^{-1}}{p}\right)-\left(\frac{-1}{p}\right)-\left(\frac{1}{p}\right)+(p-2)\right] \\
&= \frac{1}{4}\left[\sum_{a=1}^{p-2}\left(\frac{1+a^{-1}}{p}\right)-\left(\frac{-1}{p}\right)+p-3\right].
\end{aligned}$$

As a ranges over all elements of $\mathbf{Z}_p^* \setminus \{-1\}$, so does a^{-1}. Hence, $1+a^{-1}$ ranges over $\mathbf{Z}_p^* \setminus \{1\}$, and

$$\sum_{a=1}^{p-2}\left(\frac{1+a^{-1}}{p}\right) = -\left(\frac{1}{p}\right) = -1.$$

Therefore,

$$|RR| = \frac{1}{4}\left[-1 - \left(\frac{-1}{p}\right) + p - 3\right] = \frac{p - 4 - \left(\frac{-1}{p}\right)}{4}.$$

□

Note: In Example 5.29, we found that $|RR| = 499$, for $p = 1999$. This agrees with the formula in Proposition 5.30: $|RR| = (p-4-(\frac{-1}{p}))/4 = 1996/4 = 499$.

Combining Proposition 5.30 with the formulas $|R| = (p-1)/2$ and $|RR| + |RN| = |R| - ((\frac{-1}{p}) + 1)/2$, we obtain a formula for $|RN|$.

COROLLARY 5.31

For p an odd prime,

$$|RN| = \frac{p - \left(\frac{-1}{p}\right)}{4}.$$

We leave the evaluation of $|NR|$ and $|NN|$ as an exercise.

Exercises

5.31 (a) List the elements of the sets RR, RN, NR, and NN, for $p = 17$.

(b) List the elements of the sets RR, RN, NR, and NN, for $p = 19$.

5.32 Determine $|NR|$ and $|NN|$.

5.33 Let p be an odd prime. Compute $\displaystyle\sum_{a=1}^{p-1}\left(\frac{a^2+1}{p}\right)$.

†⋆5.34 Suppose that $p = 2^l + 1$ is a prime and g is a primitive root mod p. Let $z = e^{2\pi i/p}$ and

$$\alpha = \sum_k z^{g^{2k}}, \quad \beta = \sum_k z^{g^{2k+1}}.$$

Show that $\alpha + \beta = 1$ and $\alpha \cdot \beta = -(p-1)/4$. Hence, show that α and β are roots of a quadratic equation with integer coefficients.

This result is part of the theory developed in Section 3.9.

5.7 Application: Hadamard matrices

How large can the determinant of a matrix be? In 1893 Hadamard posed and answered a refinement of the question in which the entries of the matrix are real numbers between -1 and 1.

QUESTION 5.32 *Let A be an $n \times n$ matrix whose entries a_{ij} are real numbers satisfying $-1 \leq a_{ij} \leq 1$. What is the largest possible value of* $\det A$ *?*

You have probably learned in a class on calculus and analytic geometry that the determinant of a 3×3 matrix is equal (up to a sign) to the volume of the parallelepiped whose edges are the three row vectors of the matrix. This rule holds for square matrices of any size: the determinant of an $n \times n$ matrix is equal (up to a sign) to the volume of the parallelepiped whose edges are the row vectors of the matrix.

The lengths of the row vectors of our matrix A are all at most $\sqrt{n}$. The volume of a parallelepiped determined by these vectors is at most equal to the volume of a parallelepiped whose edges all have length $\sqrt{n}$ and are orthogonal (dot product equals 0). In that case, the volume is $\sqrt{n}^n$.

THEOREM 5.33 (Hadamard)
If A is an $n \times n$ matrix with entries a_{ij} satisfying $-1 \leq a_{ij} \leq 1$, then $|\det A| \leq \sqrt{n}^n$. Equality is only possible when each $a_{ij} = \pm 1$ and the rows of A are pairwise orthogonal vectors.

DEFINITION 5.34 *A* Hadamard matrix *A of* order *n is an $n \times n$ matrix satisfying the conditions of equality in Hadamard's theorem. That is, A has entries $a_{ij} = \pm 1$ and the rows of A are pairwise orthogonal.*

We will show that for certain orders, Hadamard matrices can be constructed using quadratic residues.

Figure 5.5 shows examples of Hadamard matrices of orders 2 and 4.

$$\begin{bmatrix} 1 & 1 \\ -1 & 1 \end{bmatrix} \qquad \begin{bmatrix} 1 & 1 & 1 & 1 \\ -1 & 1 & -1 & 1 \\ -1 & -1 & 1 & 1 \\ 1 & -1 & -1 & 1 \end{bmatrix}$$

FIGURE 5.5: Hadamard matrices of orders 2 and 4.

The following result is credited to Raymond Paley (1907–1933).

THEOREM 5.35
Let p be a prime of the form $4k+3$. Let R and N be the set of quadratic residues and quadratic nonresidues modulo p, respectively. Define a $p \times p$ matrix $B = [b_{ij}]$ by the rules $b_{ij} = 1$ if $j - i \in R$, $b_{ij} = -1$ if $j - i \in N$, and $b_{ii} = -1$. Let A be the $(p+1) \times (p+1)$ matrix whose first column and first row consist of all 1's, and which contains B as a submatrix in its lower-right corner. Then A is a Hadamard matrix of order $p+1$.

The proof is similar to that of Proposition 5.30. We leave it as an exercise.

Example 5.36 We will show how to construct a Hadamard matrix of order 12.

Note that 11 is a prime of the form $4k+3$. We have $R = \{1, 3, 4, 5, 9\}$ and $N = \{2, 6, 7, 8, 10\}$, and we define B by the recipe given in the theorem:

$$B = \begin{bmatrix} -1 & 1 & -1 & 1 & 1 & 1 & -1 & -1 & -1 & 1 & -1 \\ -1 & -1 & 1 & -1 & 1 & 1 & 1 & -1 & -1 & -1 & 1 \\ 1 & -1 & -1 & 1 & -1 & 1 & 1 & 1 & -1 & -1 & -1 \\ -1 & 1 & -1 & -1 & 1 & -1 & 1 & 1 & 1 & -1 & -1 \\ -1 & -1 & 1 & -1 & -1 & 1 & -1 & 1 & 1 & 1 & -1 \\ -1 & -1 & -1 & 1 & -1 & -1 & 1 & -1 & 1 & 1 & 1 \\ 1 & -1 & -1 & -1 & 1 & -1 & -1 & 1 & -1 & 1 & 1 \\ 1 & 1 & -1 & -1 & -1 & 1 & -1 & -1 & 1 & -1 & 1 \\ 1 & 1 & 1 & -1 & -1 & -1 & 1 & -1 & -1 & 1 & -1 \\ -1 & 1 & 1 & 1 & -1 & -1 & -1 & 1 & -1 & -1 & 1 \\ 1 & -1 & 1 & 1 & 1 & -1 & -1 & -1 & 1 & -1 & -1 \end{bmatrix}.$$

Now, we create the Hadamard matrix:

$$A = \begin{bmatrix} 1 & 1 & 1 & 1 & 1 & 1 & 1 & 1 & 1 & 1 & 1 & 1 \\ 1 & -1 & 1 & -1 & 1 & 1 & 1 & -1 & -1 & -1 & 1 & -1 \\ 1 & -1 & -1 & 1 & -1 & 1 & 1 & 1 & -1 & -1 & -1 & 1 \\ 1 & 1 & -1 & -1 & 1 & -1 & 1 & 1 & 1 & -1 & -1 & -1 \\ 1 & -1 & 1 & -1 & -1 & 1 & -1 & 1 & 1 & 1 & -1 & -1 \\ 1 & -1 & -1 & 1 & -1 & -1 & 1 & -1 & 1 & 1 & 1 & -1 \\ 1 & -1 & -1 & -1 & 1 & -1 & -1 & 1 & -1 & 1 & 1 & 1 \\ 1 & 1 & -1 & -1 & -1 & 1 & -1 & -1 & 1 & -1 & 1 & 1 \\ 1 & 1 & 1 & -1 & -1 & -1 & 1 & -1 & -1 & 1 & -1 & 1 \\ 1 & 1 & 1 & 1 & -1 & -1 & -1 & 1 & -1 & -1 & 1 & -1 \\ 1 & -1 & 1 & 1 & 1 & -1 & -1 & -1 & 1 & -1 & -1 & 1 \\ 1 & 1 & -1 & 1 & 1 & 1 & -1 & -1 & -1 & 1 & -1 & -1 \end{bmatrix}.$$

□

It is easy to show that the order of a Hadamard matrix is 1, 2, or a multiple of 4. It is conjectured that there is a Hadamard matrix for every multiple of 4.

CONJECTURE
There is a Hadamard matrix of every order $4n$.

The smallest number for which it is not known whether there exists a Hadamard matrix of that order is 668.

OPEN PROBLEM *Is there a Hadamard matrix of order* 668*?*

Exercises

5.35 Show that the only possible orders for a Hadamard matrix are 1, 2, and $4n$.

5.36 (a) Given a Hadamard matrix of order m and a Hadamard matrix of order n, show how to construct a Hadamard matrix of order mn.

(b) Given the technique in (a), Theorem 5.35, and Exercise 5.35, what is the least feasible order for which we have not established the existence of a Hadamard matrix?

(c) Prove that there is a Hadamard matrix of order 984000.

5.37 Prove Theorem 5.35.

5.38 Use Theorem 5.35 to construct a Hadamard matrix of order 20.

5.39 Show that the columns of a Hadamard matrix are pairwise orthogonal.

5.8 Notes

Carl Friedrich Gauss

Carl Friedrich Gauss (1777–1855) introduced the notion of congruence in his monumental treatise on number theory, *Disquisitiones Arithmeticae* (1801). While much of the mathematics in this work had already been known, historical records indicate that it may not have been known to Gauss at the time he wrote *Disquisitiones*, or at least that he had worked out many of the results himself before encountering them in the literature.

Gauss's mathematical ability was apparent early in his life. According to one story, at age three Gauss found an error in a wage calculation performed by his father. In another famous episode, an eight-year-old Gauss foils the attempt of a teacher to keep him busy for a while by instantly computing the sum of 100 consecutive numbers. By age 19, Gauss had performed the construction of the regular 17-gon. Until this point in time, Gauss had been torn between two career paths: mathematics and philology. The breakthrough led him to devote himself completely to mathematics.

Gauss also had interests in other areas of mathematics as well as other areas of science. In fact, he was one of the leading astronomers of his time. He was also among the first mathematicians to seriously consider the possible existence of non-Euclidean geometries. The quote at the beginning of the chapter, "Pauca sed Matura" (few but ripe), served as Gauss's motto. By this he meant that he published few results but those that he published were significant and complete. Many other discoveries made by Gauss are referred to in his own diary. Unfortunately, many of the entries are not comprehensible, and even Gauss acknowledged he could not always reconstruct their original meanings.

Part II

Further Topics

Chapter 6

Arithmetic Functions

> If as many numbers as we please beginning from a unit be set out continuously in double proportion, until the sum of all becomes a prime, and if the sum multiplied into the last make some number, the product will be perfect.
>
> EUCLID, *The Elements* (c. 300 B.C.E.)

In this chapter we develop several functions and formulas that will serve us in our number theory investigations. We also discover that these functions exist in a harmonious relationship that is interesting in its own right. As a motivating problem, we describe what are called perfect numbers and we establish a few of their properties. Along the way, we use *Mathematica* and *Maple* to perform some computations that would be very difficult or impossible to carry out by hand.

6.1 Perfect numbers

One of the oldest concepts in number theory, dating back to Pythagoras, is that of a "perfect number." In this section we define perfect numbers, introduce some functions that help us to analyze them, and prove a few results about them.

DEFINITION 6.1 *A* perfect number *is a positive integer equal to the sum of its proper divisors.*

Note: Recall that a "proper divisor" of n is a positive divisor of n that is less than n.

We see that 6 is a perfect number, for $6 = 1 + 2 + 3$. In fact, 6 is the smallest perfect number. (Just check that 1, 2, 3, 4, 5 are not perfect.) With a little further searching, we find that 28 is the next perfect number: $28 = 1 + 2 + 4 + 7 + 14$.

We now introduce a function to help us analyze perfect numbers. Let $\sigma(n)$ be the sum of all the positive divisors of n. Thus,

$$\sigma(n) = \sum_{d|n} d. \tag{6.1}$$

Observe that n is perfect if and only if $\sigma(n) = 2n$.

Example 6.2 Is 49 a perfect number? Since the positive divisors of 49 are 1, 7, and 49, we find that

$$\sigma(49) = 1 + 7 + 49 = 57.$$

As $\sigma(49) \neq 2 \cdot 49$, the number 49 is not perfect. ▯

The *Mathematica* command `DivisorSigma` calculates $\sigma(n)$.

Mathematica

```
DivisorSigma[1, 49]

57
```

Note: The command `DivisorSigma[k,n]` gives the sum of the kth powers of the positive divisors of n. Since we are interested in the sum of the divisors, we set the value of the first argument of this command to 1.

We now use the `DivisorSigma` command to create a procedure that searches for perfect numbers over a range of integers. In the procedure, each value of n is tested in turn to see if it satisfies the condition $\sigma(n) = 2n$. If it does, then the value of n is printed.

Mathematica

```
Do[If[DivisorSigma[1, n] == 2 n, Print[n]], {n, 1, 10^6}]

6
28
496
8128
```

Here are the corresponding *Maple* commands.

Maple

```
numtheory[sigma](49);
57

for n from 1 to 1000000 do
  if numtheory[sigma](n) = 2*n then
    print(n);
  end if;
end do:

6
28
496
8128
```

We have found four perfect numbers: 6, 28, 496, and 8128 (we knew about the first two already). It is perhaps surprising that there are no perfect numbers between 8128 and 10^6. In fact, the next perfect number is greater than $3 \cdot 10^7$ (see Exercises). If we want to find more perfect numbers, we need to develop some theory.

But let's pause for a moment to contemplate the prime factorizations of the four perfect numbers we have found:

$$\begin{aligned} 6 &= 2 \cdot 3 \\ 28 &= 2^2 \cdot 7 \\ 496 &= 2^4 \cdot 31 \\ 8128 &= 2^6 \cdot 127. \end{aligned}$$

You may wish to try to find a pattern in these factorizations and compare your conjecture with the assertion of Euclid cited at the beginning of the chapter and recorded as Theorem 6.6.

Let's find a formula for $\sigma(n)$ in terms of the prime factorization of n. We examine some cases. First, assume that n is prime, $n = p$. Then the positive divisors of n are 1 and p, and therefore,

$$\sigma(n) = \sigma(p) = 1 + p. \tag{6.2}$$

Second, assume that n is a prime power, $n = p^k$. In this case, the divisors of n are the powers of p between 1 and p^k, and

$$\sigma(n) = \sigma(p^k) = 1 + p + p^2 + \cdots + p^k.$$

This sum is a geometric series, so we can give a concise formula for it:

$$\sigma(p^k) = \frac{p^{k+1} - 1}{p - 1}. \tag{6.3}$$

Third, assume that n is the product of two distinct primes, $n = pq$. Then the divisors of n are 1, p, q, and pq, so

$$\sigma(n) = \sigma(pq) = 1 + p + q + pq.$$

Notice that this last expression can be written as a product:

$$\sigma(pq) = (1 + p)(1 + q). \tag{6.4}$$

Hence,

$$\sigma(pq) = \sigma(p)\sigma(q). \tag{6.5}$$

PROPOSITION 6.3
If $\gcd(m, n) = 1$, *then*

$$\sigma(mn) = \sigma(m)\sigma(n).$$

PROOF Suppose that $\gcd(m, n) = 1$. If $d \mid mn$, then there exist unique positive integers d_m and d_n with $d_m \mid m$, $d_n \mid n$, and $d = d_m d_n$. Indeed, $d_m = \gcd(d, m)$ and $d_n = \gcd(d, n)$. Conversely, if $d_m \mid m$ and $d_n \mid n$, then $d_m d_n \mid mn$. It follows that

$$\begin{aligned}\sigma(mn) &= \sum_{d \mid mn} d \\ &= \sum_{d_m \mid m,\ d_n \mid n} d_m d_n \\ &= \sum_{d_m \mid m} d_m \sum_{d_n \mid n} d_n \\ &= \sigma(m)\sigma(n).\end{aligned}$$

□

PROPOSITION 6.4
If n has the canonical factorization $\prod_{i=1}^{k} p_i^{e_i}$, *then*

$$\sigma(n) = \prod_{i=1}^{k} \frac{p_i^{e_i+1} - 1}{p_i - 1}.$$

PROOF By Proposition 6.3,

$$\sigma(n) = \prod_{i=1}^{k} \sigma(p_i^{e_i}).$$

The result now follows immediately from equation (6.3). □

Example 6.5 Let's calculate $\sigma(1000)$. The prime factorization of 1000 is $2^3 \cdot 5^3$, so Proposition 6.4 tells us that

$$\sigma(1000) = \frac{2^4 - 1}{2 - 1} \cdot \frac{5^4 - 1}{5 - 1} = 15 \cdot 156 = 2340.$$

□

Using our formula for $\sigma(n)$, we can prove Euclid's assertion about perfect numbers cited at the beginning of this chapter. In modern notation, the claim is that if $1 + 2 + 2^2 + \cdots + 2^k$ is a prime number, then

$$2^k(1 + 2 + 2^2 + \cdots + 2^k)$$

is a perfect number. Since $1 + 2^2 + \cdots + 2^k = 2^{k+1} - 1$, we can restate the claim as follows (letting $n = k + 1$).

THEOREM 6.6 (Euclid)
If $2^n - 1$ is a prime number, then $2^{n-1}(2^n - 1)$ is a perfect number.

PROOF Assume that $2^n - 1$ is prime. Since $2^n - 1$ and 2^{n-1} are relatively prime (any divisor of both must divide 1),

$$\sigma(2^{n-1}(2^n - 1)) = \sigma(2^{n-1})\sigma(2^n - 1).$$

From (6.2) and (6.3), respectively, we find that $\sigma(2^n - 1) = 2^n$ and $\sigma(2^{n-1}) = 2^n - 1$. Hence,

$$\sigma(2^{n-1}(2^n - 1)) = (2^n - 1)2^n = 2 \cdot 2^{n-1}(2^n - 1),$$

and $2^{n-1}(2^n - 1)$ is perfect. □

The four perfect numbers we have found fit the pattern of Euclid's theorem.

n	$2^{n-1}(2^n - 1)$
2	6
3	28
5	496
7	8128

Note that n is prime in each case.

PROPOSITION 6.7
If $2^n - 1$ is a prime, then n is a prime.

A proof of Proposition 6.7 is called for in the exercises. Primes of the form $2^n - 1$ are called "Mersenne primes." We will have more to say about them in the next chapter. The converse of Proposition 6.7 is false, the smallest counterexample being

$$2^{11} - 1 = 2047 = 23 \cdot 89.$$

Theorem 6.6 and Proposition 6.7 give us a method for searching for perfect numbers: We check whether $2^n - 1$ is a prime, where n is a prime number; if it is a prime, then $2^{n-1}(2^n - 1)$ is a perfect number.

Mathematica

```
Do[
  n=Prime[k];
  If[PrimeQ[2^n-1], Print[{n,2^(n-1)(2^n-1)}]],
  {k,1,10}
]

{2,6}
{3,28}
{5,496}
{7,8128}
{13,33550336}
{17,8589869056}
{19,137438691328}
```

Maple

```
for k from 1 to 10 do
  n := ithprime(k);
  if isprime(2^n-1) then
    print([n, 2^(n-1)*(2^n-1)]);
  end if;
end do:

[2, 6]
[3, 28]
[5, 496]
[7, 8128]
[13, 33550336]
[17, 8589869056]
[19, 137438691328]
```

In the pairs $(n, 2^{n-1}(2^n - 1))$ listed above, n is a prime for which $2^n - 1$ is prime and $2^{n-1}(2^n - 1)$ is the corresponding perfect number given by Euclid's theorem. We have found three new perfect numbers: 33550336, 8589869056, and 137438691328.

This method of searching will not yield any odd perfect numbers (because $2^{n-1}(2^n - 1)$ is an even number for $n > 1$). It will, however, produce *all* even perfect numbers. This is because the converse of Euclid's observation is true for even perfect numbers. It is surprising that this converse was not proved until about two thousand years after Euclid.

THEOREM 6.8 (Euler)
Every even perfect number is of the form $2^{n-1}(2^n - 1)$, where $2^n - 1$ is a prime number.

PROOF Let m be an even perfect number. Since m is even, we can write $m = 2^{n-1}m'$, for some $n \geq 2$ and an odd integer m'. From Proposition 6.3, $\sigma(m) = \sigma(2^{n-1})\sigma(m')$. Because m is perfect, and from (6.3), we have

$$2m = (2^n - 1)\sigma(m'),$$

or

$$2^n m' = (2^n - 1)\sigma(m').$$

As $2^n - 1$ and 2^n are relatively prime, $(2^n - 1) \mid m'$. We now have two divisors of m': m' itself and $m'/(2^n - 1)$. Note that these numbers are different, since $n \geq 2$. The sum of these divisors is

$$m' + \frac{m'}{2^n - 1} = \frac{2^n m'}{2^n - 1} = \sigma(m').$$

By definition of σ, these are the only positive divisors of m'. Hence m' is a prime and $m' = 2^n - 1$. Therefore, $m = 2^{n-1}(2^n - 1)$, where $2^n - 1$ is a prime number. $\square$

Two fundamental questions about perfect numbers have remained unanswered for over two thousand years.

QUESTION
Does there exist an odd perfect number?

Again, we have an echo of the theme introduced at the beginning of the book; namely, that in number theory relatively straightforward problems (the classification of even perfect numbers) stand next to very difficult ones (the existence of odd perfect numbers).

It is known that there are no odd perfect numbers less than 10^{300}.

QUESTION
Do there exist infinitely many perfect numbers?

Currently, 44 perfect numbers are known. The largest of these is

$$2^{32582656} \times (2^{32582657} - 1).$$

See Notes, Chapter 7, and [6].

Exercises

◇**6.1** Use a computer to show that there are no perfect numbers between 8128 and 33550336.

6.2 Calculate the following.

(a) $\sigma(10)$

(b) $\sigma(100)$

(c) $\sigma(10^6)$

6.3 Show that the first two digits of $\sigma(10^n)$, for $n \geq 4$, are 24.

6.4 Prove that $\sigma(mn) \leq \sigma(m)\sigma(n)$, for all positive integers m, n, with equality if and only if m and n are relatively prime.

6.5 Given $m \geq 1$, prove that the equation $\sigma(n) = m$ has only finitely many solutions.

6.6 Prove that every even perfect number ends in a 6 or 8.

6.7 Prove that a perfect square cannot be a perfect number.

◇**6.8** An *abundant number* is a positive integer that is less than the sum of its proper divisors. A *deficient number* is a positive integer that is greater than the sum of its proper divisors. These terms were introduced by Nicomachus, in his *Introduction to Arithmetic* (c. 100).

(a) Prove that every multiple of an abundant number is abundant and every proper multiple of a perfect number is abundant.

(b) Prove that every divisor of a deficient number is deficient and every proper divisor of a perfect number is deficient.

(c) How prevalent are abundant and deficient numbers? Write a computer procedure to count the number of abundant and deficient numbers between 1 and 10^6. Which is more prevalent?

It has been proved that the perfect numbers have density 0 (see [26]). That is,

$$\lim_{n\to\infty} \frac{|\{m \leq n \text{ and } m \text{ is perfect}\}|}{n} = 0.$$

6.9 Prove Proposition 6.7.

Hint: Prove the contrapositive form of the theorem; i.e., if n is composite then $2^n - 1$ is composite.

6.10 (a) Show that for any positive integer n,

$$\sum_{d|n} \frac{1}{d} = \frac{\sigma(n)}{n}.$$

(b) Show that for n a perfect number,

$$\sum_{d|n} \frac{1}{d} = 2.$$

This yields a "partition of unity" (a set of reciprocals of distinct positive integers whose sum is 1):

$$\sum_{\substack{d|n \\ 1<d}} \frac{1}{d} = 1.$$

For example, the perfect number 6 yields the partition

$$\frac{1}{2} + \frac{1}{3} + \frac{1}{6} = 1.$$

(c) Find an odd abundant number.

Hint: The smallest is less than 1000.

(d) Use your answer in (c) to find distinct odd integers $d_1, d_2, \ldots, d_n$, $n > 1$, with

$$\sum_{i=1}^{n} \frac{1}{d_i} = 1.$$

Note that if no such collection existed, then there could be no odd perfect number.

6.11 Prove that an odd perfect number must have at least three distinct prime factors.

6.2 The group of arithmetic functions

The function σ of the preceding section is an example of an "arithmetic function." We now define arithmetic functions in general, give more examples, and prove some of their properties.

DEFINITION 6.9 *An* arithmetic function *is a function defined for all positive integers. (The range of the function can be any set of numbers, although it is typically* **Z** *or* **C**.*)*

Example 6.10 The function σ of the preceding section is an arithmetic function. ▯

Example 6.11 Euler's ϕ-function (see equation (3.7)) is an arithmetic function. ▯

We define another arithmetic function. Let $\tau(n)$ be the number of positive divisors of n. Thus,

$$\tau(n) = \sum_{d|n} 1. \tag{6.6}$$

Example 6.12 Let's evaluate $\tau(49)$. Since 49 has three divisors, namely, 1, 7, 49, we have $\tau(49) = 3$. ▯

We define two families of arithmetic functions. For $k \in \mathbf{R}$, let

$$\theta_k(n) = n^k, \quad n \geq 1. \tag{6.7}$$

Note that θ_1 is the identity function. For convenience, set $\theta = \theta_0$. Also, for $k \in \mathbf{R}$, let

$$\sigma_k(n) = \sum_{d|n} d^k, \quad n \geq 1. \tag{6.8}$$

Note that $\sigma_1 = \sigma$ and $\sigma_0 = \tau$.

The functions σ, ϕ, τ, θ_k, and σ_k have an important property indicated in the next definition.

DEFINITION 6.13 *A* multiplicative function f *is an arithmetic function, not identically* 0*, for which* $f(mn) = f(m)f(n)$ *for all relatively prime positive integers* m, n.

The multiplicativity of σ and ϕ is shown in Propositions 6.3 and 3.29, respectively. It is obvious that θ_k is multiplicative for each k. The fact that τ, and in general σ_k, is multiplicative will be proved shortly.

If f is a multiplicative function, then $f(1) = 1$ (see Exercises).

It is easy to show (by induction) that if $n = \prod_{i=1}^{k} p_i^{e_i}$ (canonical factorization), then, for f a multiplicative function,

$$f(n) = \prod_{i=1}^{k} f(p_i^{e_i}). \tag{6.9}$$

Thus, the values of f are determined by its values for prime powers. (We have seen examples of this fact in the formulas for σ and ϕ.)

PROPOSITION 6.14
Let f be a multiplicative function and $g(n) = \sum_{d|n} f(d)$. Then g is also multiplicative.

The proof is similar to the proof of Proposition 6.3 and is left as an exercise.

PROPOSITION 6.15
The function σ_k is multiplicative for each $k \in \mathbf{R}$.

PROOF The result follows directly from equation (6.8) and Proposition 6.14, as θ_k is a multiplicative function. ∎

COROLLARY 6.16
The function τ is multiplicative.

PROPOSITION 6.17
Let n have the canonical factorization $\prod_{i=1}^{k} p_i^{e_i}$. Then

$$\tau(n) = \prod_{i=1}^{k} (e_i + 1).$$

PROOF Since τ is multiplicative,

$$\tau(n) = \prod_{i=1}^{k} \tau(p_i^{e_i}).$$

The result now follows from the fact that $p_i^{e_i}$ has $e_i + 1$ divisors, namely, the powers of p_i from 1 to $p_i^{e_i}$. ∎

Example 6.18 Let's calculate $\tau(1000)$. The prime factorization of 1000 is $2^3 \cdot 5^3$. Hence, $\tau(1000) = (3 + 1)(3 + 1) = 16$. ∎

We now describe how to multiply arithmetic functions. Although we could simply multiply functions f and g together by the rule $(f \cdot g)(n) = f(n)g(n)$, for $n \geq 1$, this definition is not fruitful. A better way is provided by "Dirichlet multiplication."

DEFINITION 6.19 (Dirichlet multiplication) *Let f and g be arithmetic functions. The product of f and g, denoted $f * g$, is given by the formula*

$$(f * g)(n) = \sum_{d|n} f(d)g(n/d), \quad n \geq 1.$$

Example 6.20 Let's compute $\theta * \theta$. By Definition 6.19,

$$\begin{aligned}(\theta * \theta)(n) &= \sum_{d|n} \theta(d)\theta(n/d) \\ &= \sum_{d|n} 1.\end{aligned}$$

This last sum is equal to $\tau(n)$. Therefore, $\theta * \theta = \tau$. □

Example 6.21 Let's compute $\phi * \theta$:

$$\begin{aligned}(\phi * \theta)(n) &= \sum_{d|n} \phi(d)\theta(n/d) \\ &= \sum_{d|n} \phi(d) \\ &= n \quad \text{(recall Exercise 3.51).}\end{aligned}$$

Therefore, $\phi * \theta = \theta_1$ (the identity function). □

The operation of Dirichlet multiplication satisfies some algebraic properties on a certain set of arithmetic functions.

THEOREM 6.22

The operation of Dirichlet multiplication on the set of arithmetic functions f for which $f(1) \neq 0$ satisfies the following.

- *(Closure) Given functions f and g in the set, $f * g$ is also in the set.*
- *(Commutativity) If f and g are in the set, then $f * g = g * f$.*
- *(Associativity) If f, g, and h are in the set, then $f * (g * h) = (f * g) * h$.*

- *(Identity element) There exists an* identity *element e in our set such that $e * f = f * e = f$, for all f in the set.*

- *(Inverses) For every f in the set, there exists an element f^{-1} in the set such that $f * f^{-1} = f^{-1} * f = e$, where e is the identity.*

PROOF Closure: Let f and g be arithmetic functions such that $f(1) \neq 0$ and $g(1) \neq 0$. Then $f * g$ is clearly an arithmetic function also, and $(f * g)(1) = f(1)g(1) \neq 0$.

Commutativity: The Dirichlet product of f and g may be written in the symmetric form

$$(f * g)(n) = \sum_{(d_1, d_2):\, d_1 d_2 = n} f(d_1)g(d_2).$$

The sum is taken over all ordered pairs of positive integers whose product is n. With this formulation, the commutativity relation, $f * g = g * f$, is obvious.

Associativity: We need to show that $(f * g) * h = f * (g * h)$, for all arithmetic functions f, g, and h. Let $\alpha = f * g$ and $\beta = g * h$. Then

$$\begin{aligned} ((f * g) * h)(n) &= (\alpha * h)(n) \\ &= \sum_{cd=n} \alpha(d)h(c) \\ &= \sum_{cd=n} \sum_{ab=d} f(a)g(b)h(c) \\ &= \sum_{(a,b,c):\, abc=n} f(a)g(b)h(c), \end{aligned}$$

and

$$\begin{aligned} (f * (g * h))(n) &= (f * \beta)(n) \\ &= \sum_{ad=n} f(a)\beta(d) \\ &= \sum_{ad=n} \sum_{bc=d} f(a)g(b)h(c) \\ &= \sum_{(a,b,c):\, abc=n} f(a)g(b)h(c). \end{aligned}$$

Hence, $(f * g) * h = f * (g * h)$.

Identity element: We define e by the rule

$$e(n) = \begin{cases} 1 \text{ if } & n = 1 \\ 0 \text{ if } & n > 1. \end{cases} \tag{6.10}$$

We check that e has the desired property:

$$(f * e)(n) = \sum_{d|n} f(d)e(n/d) = f(n)e(1) = f(n).$$

Inverses: Assume that f is an arithmetic function for which $f(1) \neq 0$. (Here is where the condition $f(1) \neq 0$ is needed.) We will show that the inverse g of f exists. The proof is by induction on n. We want $g(1)f(1) = 1$, that is, $g(1) = 1/f(1)$. This expression is defined since $f(1) \neq 0$. Now suppose that $g(m)$ has been defined for all $m < n$. Then we want

$$\sum_{d|n} g(d)f(n/d) = e(n) = 0,$$

which holds if

$$g(n)f(1) + \sum_{\substack{d|n \\ d<n}} g(d)f(n/d) = 0$$

or

$$g(n) = -g(1) \sum_{\substack{d|n \\ d<n}} g(d)f(n/d).$$

Thus, $g(n)$ is defined in terms of previously defined values of g. The induction is complete. Hence, the inverse of f exists. ∎

Recall from the Chapter 3 Notes that a group G is a set endowed with an associative binary operation for which an identity element and inverses exist. If the operation is commutative, then G is called a commutative (or abelian) group. The set of arithmetic functions f for which $f(1) \neq 0$ is a commutative group that we refer to as "the group of arithmetic functions." The subset of multiplicative functions f also forms a commutative group. Recall that if f is a multiplicative function, then $f(1) = 1$.

PROPOSITION 6.23

If f and g are multiplicative functions, then so is $f * g$.

The proof of Proposition 6.23 is similar to that of Proposition 6.3 and is called for in the exercises.

Note: Proposition 6.14 follows as a corollary since the relation there may be written $g = f * \theta$.

THEOREM 6.24

If f is a multiplicative function, then so is f^{-1}.

PROOF The proof is by induction. Since $f^{-1}(1) = 1/f(1) = 1$, we have $f^{-1}(1 \cdot 1) = f^{-1}(1)f^{-1}(1)$. Now suppose that $\gcd(m, n) = 1$, with $m > 1$ or $n > 1$. Assume that $f^{-1}(ab) = f^{-1}(a)f^{-1}(b)$ for all a, b with $ab < mn$ and $\gcd(a, b) = 1$. Then

$$\begin{aligned}
0 &= (f * f^{-1})(mn) \\
&= \sum_{d|mn} f(d)f^{-1}(mn/d) \\
&= \sum_{d_1|m,\ d_2|n} f(d_1 d_2)f^{-1}(mn/d_1 d_2) \\
&= \sum_{\substack{d_1|m,\ d_2|n \\ 1<d_1 d_2}} f(d_1)f(d_2)f^{-1}(m/d_1)f^{-1}(n/d_2) + f(1)f^{-1}(mn) \\
&= \sum_{d_1|m} f(d_1)f^{-1}(m/d_1) \sum_{d_2|n} f(d_2)f^{-1}(n/d_2) \\
&\quad -f(1)f^{-1}(m)f(1)f^{-1}(n) + f(1)f^{-1}(mn) \\
&= (f * f^{-1})(m)(f * f^{-1})(n) - f^{-1}(m)f^{-1}(n) + f^{-1}(mn) \\
&= -f^{-1}(m)f^{-1}(n) + f^{-1}(mn).
\end{aligned}$$

It follows that

$$f^{-1}(mn) = f^{-1}(m)f^{-1}(n).$$

This completes the induction. ▯

THEOREM 6.25

The multiplicative functions f form a commutative group under Dirichlet multiplication.

PROOF We have already established that e is multiplicative, the product of multiplicative functions is multiplicative, and f^{-1} is multiplicative whenever f is. ▯

Recall that θ is the arithmetic function defined by the rule $\theta(n) = 1$, for $n \geq 1$. We define the inverse of θ to be μ (the "Möbius function"). We will investigate μ further in the next section.

Let's recapitulate some of our functions and formulas. We have defined the functions e, σ, θ_k, μ, σ_k, ϕ, and τ. Each of these functions is multiplicative. We have derived several identities that hold for these functions, and from these identities we can sometimes derive more identities. For example, since θ and μ are inverses,

$$\theta * \mu = e.$$

We also know (Example 6.21) that

$$\phi * \theta = \theta_1.$$

It follows from these facts that

$$\phi = \theta_1 * \mu,$$

and hence,

$$\phi(n) = \sum_{d|n} \mu(d)(n/d). \tag{6.11}$$

Exercises

6.12 A number has exactly 13 positive divisors. What can we say about the number?

◇**6.13** (a) What is the smallest positive integer that has exactly 32 positive divisors?

(b) What are the possible types of prime factorizations of such numbers?

(c) Use a computer to find all positive integers less than 10^4 that have exactly 32 positive divisors.

6.14 Prove that if f is a multiplicative function, then $f(1) = 1$.

†**6.15** Prove Proposition 6.14.

6.16 Prove that

$$\prod_{d|n} d = n^{\tau(n)/2}.$$

6.17 Prove that

$$\prod_{p|n}\left(1 - \frac{1}{p}\right) = \sum_{d|n} \frac{\mu(d)}{d}.$$

6.18 Prove that

$$\sum_{d|n} \tau(d)^3 = \left(\sum_{d|n} \tau(d)\right)^2.$$

6.19 Find a formula for $\sigma_k(n)$, where $k \neq 0$, in terms of the canonical factorization of n.

6.20 Prove that $\sigma_k = \theta_k * \theta$, for all $k \in \mathbf{R}$, and hence, in particular, $\sigma = \theta_1 * \theta$.

6.21 Prove that $(\theta_k * \theta_k)(n) = n^k \tau(n)$, for all $k \in \mathbf{R}$.

6.22 Prove that $\sigma = \phi * \tau$.

6.23 Prove that $\sigma(n) \geq \tau(n) n^{1/2}$.

†6.24 Prove Proposition 6.23.

6.25 What is wrong with the following argument to show that the inverse of a multiplicative function is multiplicative?

Suppose that f is a multiplicative function. Then

$$\begin{aligned} f^{-1}(ab) &= f^{-1} * (f(f^{-1}(a))f(f^{-1}(b))) \\ &= f^{-1} * (f(f^{-1}(a)f^{-1}(b))) \\ &= f^{-1}(a)f^{-1}(b). \end{aligned}$$

Hence, f^{-1} is a multiplicative function.

6.26 An arithmetic function f is called *additive* if

$$f(mn) = f(m) + f(n)$$

for all relatively prime positive integers m, n.

(a) Let $\omega(n)$ be the number of distinct prime divisors of n. Prove that ω is an additive function.

(b) Let $\Omega(n)$ be the number of prime factors of n (counting multiplicity). Prove that Ω is an additive function.

6.27 (a) Let f be an additive function and c a positive real number. Let $g(n) = c^{f(n)}$. Show that g is a multiplicative function.

(b) Let g be a multiplicative function and c a positive real number not equal to 1. Let $f(n) = \log_c g(n)$. Show that f is an additive function.

(c) Let $\lambda(n) = (-1)^{\Omega(n)}$. Prove that λ is a multiplicative function. (The function λ is called *Liouville's function.*)

†6.28 Prove that

$$\sum_{d|n} \mu(d) = \begin{cases} 1 \text{ if } n = 1 \\ 0 \text{ if } n > 1. \end{cases}$$

6.3 Möbius inversion

Since θ is a multiplicative function, we know (from Theorem 6.24) that the Möbius function μ is also multiplicative. Therefore, we need only determine the values of μ for prime powers.

Since μ is multiplicative, $\mu(1) = 1$. For p prime,

$$\sum_{d|p} \theta(d)\mu(p/d) = e(p),$$

and, therefore, $\theta(1)\mu(p) + \theta(p)\mu(1) = 0$, and $\mu(p) = -1$. Furthermore,

$$\sum_{d|p^2} \theta(d)\mu(p^2/d) = e(p^2),$$

and, therefore, $\theta(1)\mu(p^2)+\theta(p)\mu(p)+\theta(p^2)\mu(1) = 0$, so that $\mu(p^2)-1+1 = 0$, and $\mu(p^2) = 0$. Continuing in this manner, we see that $\mu(p^e) = 0$, for $e \geq 2$.

To summarize:

$$\mu(1) = 1$$
$$\mu(p^e) = \begin{cases} -1 \text{ if } e = 1 \\ \ \ 0 \text{ if } e \geq 2. \end{cases}$$

Now we have a formula for $\mu(n)$.

PROPOSITION 6.26

For $n \geq 1$,

$$\mu(n) = \begin{cases} 1 & \text{if } n = 1 \\ (-1)^k & \text{if } n \text{ is a product of } k \text{ distinct primes} \\ 0 & \text{otherwise.} \end{cases}$$

Example 6.27 Let's calculate $\mu(30)$. Since $30 = 2 \cdot 3 \cdot 5$, we obtain

$$\mu(30) = (-1)^3 = -1.$$

□

The fact that $\mu = \theta^{-1}$ leads to a useful summation formula.

THEOREM 6.28 (Möbius inversion formula)

Let g be an arithmetic function with $g(1) \neq 0$. If

$$f(n) = \sum_{d|n} g(d), \quad n \geq 1,$$

then

$$g(n) = \sum_{d|n} f(d)\mu(n/d), \quad n \geq 1.$$

PROOF Suppose that

$$f(n) = \sum_{d|n} g(d), \quad n \geq 1.$$

Interpreting this equation as a Dirichlet product,

$$f(n) = (g * \theta)(n),$$

or

$$f = g * \theta.$$

It follows that

$$g = f * \mu,$$

and writing this result as a sum yields

$$g(n) = \sum_{d|n} f(d)\mu(n/d), \quad n \geq 1.$$

□

The Möbius μ function is due to August Möbius (1790–1868). The Möbius function is a special case in a general "theory of inversion." See [30].

We now consider an application of Möbius inversion to "primitive strings." A "string" is a finite sequence of symbols from some fixed set. The length of a string is the number of symbols in it (counting repetitions).

Example 6.29 Let the set of symbols be $\{a, b, c\}$. Two examples of strings are *cabc* and *bbbaaabbb*. These strings have lengths 4 and 9, respectively. □

A "primitive string" is a string that is not the concatenation of identical smaller strings.

Example 6.30 Consider again the two strings of Example 6.29. The string *cbabc* is primitive but the string *bbbaaabbbaaa* is not (it is the concatenation of two copies of the string *bbbaaa*). □

How many primitive strings of a given length and from a given finite symbol set are there? Suppose that our symbol set has k elements. Then, clearly, there are k^n possible strings of length n. Let $f(k, n)$ be the number of primitive strings of length n. We can calculate $f(k, n)$ using Möbius inversion.

Each string of length n is the concatenation of copies of *some* primitive string of length d, where $d \mid n$. It follows that

$$k^n = \sum_{d|n} f(k, d).$$

Applying Möbius inversion, we obtain a formula for $f(k, n)$.

PROPOSITION 6.31
For $k, n \geq 1$,

$$f(k,n) = \sum_{d|n} \mu(n/d)k^d.$$

Example 6.32 Let's calculate $f(2,4)$, the number of primitive strings of length 4 from a two-element set (say, $\{a, b\}$). From Proposition 6.31,

$$\begin{aligned} f(2,4) &= \mu(4/1)2^1 + \mu(4/2)2^2 + \mu(4/4)2^4 \\ &= \mu(4)2 + \mu(2)4 + \mu(1)16 \\ &= -4 + 16 \\ &= 12. \end{aligned}$$

We verify by listing the twelve primitive strings.

aaab aaba aabb abaa abba abbb
baaa baab babb bbaa bbab bbba

□

Exercises

6.29 Determine the following values.

(a) $\mu(100)$
(b) $\mu(30)$
(c) $\mu(1999)$
(d) $\mu(221)$

6.30 How many primitive strings of length 6 are there from the set $\{a, b\}$?

†⋆**6.31** Prove that the number of monic irreducible polynomials of degree n over $\mathbf{Z}_p$ is

$$\frac{1}{n}\sum_{d|n} \mu(n/d)p^d.$$

6.32 (a) Prove that

$$\sum_{d|n} \mu^2(d) = 2^{\omega(n)}.$$

(b) Prove that

$$\phi(n) \geq \frac{n}{\omega(n)+1}.$$

Hint: Use induction on the number of prime divisors of n.

(c) Prove that

$$2^{\omega(n)} \leq \tau(n) \leq n.$$

(d) Prove that

$$\phi(n) \geq \frac{n}{\log_2 n + 1}, \quad n \geq 2.$$

★**6.33** Prove that for p prime the sum of all the primitive roots modulo p is $\mu(p-1)$.

†**6.34** The *von Mangoldt function* Λ is an arithmetic function defined by the formula

$$\Lambda(n) = \begin{cases} \log p & \text{if } n = p^k, \text{ where } p \text{ is prime and } k \geq 1 \\ 0 & \text{otherwise.} \end{cases}$$

(a) Prove that

$$\sum_{d|n} \Lambda(d) = \log n.$$

(b) Use Möbius inversion to prove that

$$\Lambda(n) = -\sum_{d|n} \mu(d) \log d.$$

†★**6.35** The Riemann zeta function $\zeta(s)$ is defined by

$$\zeta(s) = \sum_{n=1}^{\infty} \frac{1}{n^s},$$

for $s > 1$ (or, more generally, for complex s with $\mathrm{Re}(s) > 1$). Prove that

$$\zeta(s) = \prod_p \frac{1}{1 - p^{-s}}$$

and

$$\sum_{n=1}^{\infty} \frac{\mu(n)}{n^s} = \prod_p (1 - p^{-s}),$$

and hence,

$$\frac{1}{\zeta(s)} = \sum_{n=1}^{\infty} \frac{\mu(n)}{n^s}.$$

6.4 Application: Cyclotomic polynomials

Recall the concept of a "root of unity" from Section 3.9.

DEFINITION 6.33 *An* nth root of unity *is a complex number* z *such that* $z^n = 1$. *A* primitive nth root of unity *is an* n*th root of unity* z *for which* $z^k = 1$ *holds for no* k *with* $1 \le k \le n-1$.

Example 6.34 The complex number i is a 4th root of unity, since $i^4 = 1$. In fact, i is a primitive 4th root of unity, since $i^k = 1$ does not hold for $k = 1$, 2, or 3. It is easy to check that $-i$ is also a primitive 4th root of unity. □

For $n \ge 2$, define $\zeta = e^{2\pi i/n}$, a primitive nth root of unity. The powers of ζ, that is, ζ^k, for $1 \le k \le n$, are all nth roots of unity. How many are primitive roots? It turns out (see Exercises) that ζ^k is a primitive nth root of unity if and only if $\gcd(k, n) = 1$. Hence, there are $\phi(n)$ primitive nth roots of unity.

DEFINITION 6.35 *The* order n cyclotomic polynomial Φ_n *is the monic polynomial whose roots are the distinct primitive* n*th roots of unity. That is,*

$$\Phi_n(z) = \prod_{\substack{1 \le k \le n \\ \gcd(k,n)=1}} (z - \zeta^k).$$

Note: The degree of $\Phi_n(z)$ is $\phi(n)$.

Example 6.36 The cyclotomic polynomial of order 4 is the monic polynomial whose roots are $\pm i$. Thus,

$$\Phi_4(z) = (z-i)(z+i) = z^2 + 1.$$

□

THEOREM 6.37
For $n \ge 1$,

$$z^n - 1 = \prod_{d \mid n} \Phi_d(z).$$

PROOF Every root of $z^n - 1$ is a primitive dth roots of unity for *some unique* value of d, with $d \mid n$. □

COROLLARY 6.38
For p prime,

$$\Phi_p(z) = z^{p-1} + z^{p-2} + \cdots + z + 1.$$

PROOF From the theorem,

$$z^p - 1 = \Phi_1(z)\Phi_p(z) = (z-1)\Phi_p(z),$$

and hence

$$\Phi_p(z) = \frac{z^p - 1}{z - 1} = z^{p-1} + z^{p-2} + \cdots + z + 1.$$

∎

We can use Theorem 6.37 to recursively compute cyclotomic polynomials.

Example 6.39 Let's compute Φ_8, the cyclotomic polynomial of order 8. From Theorem 6.37 we obtain

$$z^8 - 1 = \Phi_1(z)\Phi_2(z)\Phi_4(z)\Phi_8(z).$$

We know from Example 6.36 that $\Phi_4(z) = z^2 + 1$. Since $\Phi_1(z) = z - 1$ and $\Phi_2(z) = z + 1$, we obtain

$$\Phi_8(z) = \frac{z^8 - 1}{(z-1)(z+1)(z^2+1)} = z^4 + 1.$$

∎

THEOREM 6.40
For $n \geq 1$,

$$\Phi_n(z) = \prod_{d|n} (z^d - 1)^{\mu(n/d)}.$$

PROOF We use Möbius inversion on the relation in Theorem 6.37, first putting it into summation form by taking logarithms:

$$\log(z^n - 1) = \sum_{d|n} \log(\Phi_d(z)).$$

Now by Möbius inversion,

$$\log(\Phi_n(z)) = \sum_{d|n} \mu(n/d) \log(z^d - 1).$$

The result follows immediately by taking "anti-logs" (i.e., the exponential function). □

We can deduce from Theorem 6.40 the fact that the coefficients of Φ_n are integers, because the possible values of μ are 1, 0, and -1. The -1 terms combine to produce a polynomial in the denominator. Since that polynomial is monic, the resulting quotient polynomial has all integer coefficients (think about the way that long division works).

COROLLARY 6.41

The coefficients of Φ_n are integers, for each $n \geq 1$.

Note: The cyclotomic polynomial $\Phi_n(z)$ is irreducible over $\mathbf{Q}$.

Example 6.42 Let's compute Φ_{15}, the cyclotomic polynomial of order 15. From Theorem 6.40 we find that

$$\begin{aligned}\Phi_{15}(z) &= (z-1)^{\mu(15)}(z^3-1)^{\mu(5)}(z^5-1)^{\mu(3)}(z^{15}-1)^{\mu(1)}\\ &= (z-1)(z^3-1)^{-1}(z^5-1)^{-1}(z^{15}-1)\\ &= z^8 - z^7 + z^5 - z^4 + z^3 - z + 1.\end{aligned}$$

□

Note: The roots of a cyclotomic polynomial can be obtained using radicals and rational operations. As a specific case, recall that in Chapter 3 we showed that the roots of the cyclotomic polynomial of order 17 are expressible in terms of *square* roots and rational operations, and this fact shows that the regular 17-gon is constructible with straightedge and compass.

What does the polynomial $\Phi_{pq}(z)$, for p and q distinct primes, look like? From Theorem 6.37, we find that

$$\Phi_{pq}(z) = \frac{(z^{pq}-1)(z-1)}{(z^p-1)(z^q-1)}.$$

We know that this rational function is equal to a polynomial of degree $\phi(n) = (p-1)(q-1)$ with integer coefficients. We will show that the coefficients of this polynomial are all -1, 0, or $+1$. (Compare Exercise 6.42.) Our treatment is based on [21].

Let r and s be nonnegative integers such that

$$(p-1)(q-1) = rp + sq.$$

The existence of r and s is obtained by writing the above equation as

$$pq + 1 = (r+1)p + (s+1)q.$$

Since $\gcd(p,q)=1$ and $pq+1>pq$, we are guaranteed, by the forthcoming Theorem 9.1, a positive solution $r+1$, $s+1$ and, hence, a nonnegative solution r, s.

Suppose that ζ is a primitive pqth root of unity. Then

$$\sum_{i=0}^{r}(\zeta^p)^i = -\sum_{i=r+1}^{q-1}(\zeta^p)^i, \quad \sum_{j=0}^{s}(\zeta^q)^j = -\sum_{j=s+1}^{p-1}(\zeta^q)^j,$$

and hence,

$$\left(\sum_{i=0}^{r}\zeta^{ip}\right)\left(\sum_{j=0}^{s}\zeta^{jq}\right) - \zeta^{-pq}\left(\sum_{i=r+1}^{q-1}\zeta^{ip}\right)\left(\sum_{j=s+1}^{p-1}\zeta^{jq}\right) = 0.$$

Now consider the polynomial

$$f(z) = \left(\sum_{i=0}^{r}z^{ip}\right)\left(\sum_{j=0}^{s}z^{jq}\right) - z^{-pq}\left(\sum_{i=r+1}^{q-1}z^{ip}\right)\left(\sum_{j=s+1}^{p-1}z^{jq}\right).$$

We know that (every primitive pqth root of unity) ζ is a root of this polynomial. Also, the degree of the polynomial is $(p-1)(q-1)=\phi(pq)$. Hence, $f(z)=\Phi_{pq}(z)$. The monomials in this expression are distinct, and we see that each coefficient of the polynomial is -1, 0, or 1.

In 2006 Gennady Bachman [3] proved the existence of an infinite family of cyclotomic polynomials of order pqr (where p, q, r are distinct odd primes) with coefficients only -1, 0, and 1.

Exercises

6.36 Let $\zeta = e^{2\pi i/n}$. Prove that ζ^k, where $1 \le k \le n$, is a primitive nth root of unity if and only if $\gcd(k,n)=1$.

6.37 Calculate the cyclotomic polynomial $\Phi_{16}(z)$ "by hand."

6.38 Prove that cyclotomic polynomials of order greater than 1 are palindromic (the coefficients are the same read left-to-right as right-to-left).

6.39 Prove that for n odd, $\Phi_{2n}(z)=\Phi_n(-z)$.

6.40 Prove that for p prime and $k\ge 1$,

$$\Phi_{p^k}(z) = \Phi_p(z^{p^{k-1}}).$$

6.41 Prove that if $n = p_1^{\alpha_1}\dots p_k^{\alpha_k}$ (canonical factorization), then

$$\Phi_n(z) = \Phi_{p_1\dots p_k}\left(z^{p_1^{\alpha_1-1}\dots p_k^{\alpha_k-1}}\right).$$

◇**6.42** Use a computer to find the cyclotomic polynomial of least order that has a coefficient different from 0, −1, and 1. How does your answer relate to the previous three exercises?

For a cyclotomic polynomial of order n in the variable z, the *Mathematica* command is `Cyclotomic[n,z]`, and the *Maple* command is `numtheory[cyclotomic](n,z)`.

Issai Schur (1875–1941) proved the existence of cyclotomic polynomials with arbitrarily large coefficients (see [22]).

◇**6.43** Find distinct odd primes p, q, and r such that $\phi_{pqr}(z)$ has coefficients only −1, 0, and 1.

Hint: There is a choice of p, q, and r with $pqr < 1000$.

6.5 Partitions of an integer

In this section we introduce partitions of an integer and investigate some of their number-theoretic properties.

DEFINITION 6.43 *A* partition *of a positive integer n is a summation of positive integers (order unimportant) that equals n. The summands in a partition are called* parts. *The number of partitions of n is denoted $p(n)$. The number of partitions of n into exactly k parts is denoted $p(n,k)$. The values of $p(n)$ and $p(n,k)$ are called* partition numbers.

Clearly,

$$p(n) = \sum_{k=1}^{n} p(n,k). \tag{6.12}$$

Example 6.44

$$\begin{aligned} p(4,1) &= 1 \qquad 4 \\ p(4,2) &= 2 \qquad 2+2,\ 3+1 \\ p(4,3) &= 1 \qquad 2+1+1 \\ p(4,4) &= 1 \qquad 1+1+1+1 \end{aligned}$$

$$\begin{aligned} p(4) &= p(4,1) + p(4,2) + p(4,3) + p(4,4) \\ &= 1+2+1+1 = 5 \end{aligned}$$

□

It is convenient to represent a partition of n with the parts in monotonically decreasing order:

$$n = \lambda_1 + \lambda_2 + \cdots + \lambda_k, \qquad \lambda_1 \geq \lambda_2 \geq \cdots \geq \lambda_k. \tag{6.13}$$

The partitions of 4 in Example 6.44 are so represented.

How can we determine $p(n,k)$ and $p(n)$ for larger values of n and k? We can use the recurrence relation formulas

$$\begin{aligned} &p(1,1) = 1, \\ &p(n,k) = 0, \quad k > n \text{ or } k = 0, \\ &p(n,k) = p(n-1,k-1) + p(n-k,k), \quad n \geq 2 \text{ and } 1 \leq k \leq n. \end{aligned} \tag{6.14}$$

The value $p(1,1) = 1$ is obvious. The formula $p(n,k) = 0$, for $k > n$ or $k = 0$, says that there are no partitions of n into more than n parts or into 0 parts. The formula $p(n,k) = p(n-1,k-1) + p(n-k,k)$, for $n \geq 2$ and $1 \leq k \leq n$, is proved by observing that there are two possibilities for the least part, λ_k, in a partition of n into k parts: either $\lambda_k = 1$ or $\lambda_k > 1$. In the former case, there are $p(n-1,k-1)$ partitions of the remaining number $n-1$ into $k-1$ parts. In the latter case, the partitions of n into k parts are equinumerous with the partitions of $n-k$ into k parts (just subtract 1 from each part in the partition of n).

The recurrence relations (6.14) may be implemented with the following *Mathematica* code. Tables 6.1 and 6.2 list the values of $p(n,k)$ and $p(n)$.

Mathematica

```
n = 100;
Do[
  p[i, 1] = 1; p[i] = 1;
  Do[
    If[i - j >= j,
    p[i, j] = p[i - 1, j - 1] + p[i - j, j],
    p[i, j] = p[i - 1, j - 1]];
    p[i] = p[i] + p[i, j],
    {j, 2, i - 1}
  ],
  {i, 1, n}
]

Table[p[n, k], {n, 1, 10}, {k, 1, n}] // TableForm
Table[p[n], {n, 1, 100}] // TableForm
```

Here are the corresponding *Maple* commands.

n \ k	1	2	3	4	5	6	7	8	9	10
1	1									
2	1	1								
3	1	1	1							
4	1	2	1	1						
5	1	2	2	1	1					
6	1	3	3	2	1	1				
7	1	3	4	3	2	1	1			
8	1	4	5	5	3	2	1	1		
9	1	4	7	6	5	3	2	1	1	
10	1	5	8	9	7	5	3	2	1	1

TABLE 6.1: Partition numbers $p(n, k)$ for $1 \leq k \leq n \leq 10$.

Maple

```
n:=100:
for i from 1 to n do:
  p(i,1):=1;
  par(i):=1;
  for j from 2 to i do:
    if i-j>=j then
      p(i,j):=p(i-1,j-1)+p(i-j,j);
      else
      p(i,j):=p(i-1,j-1);
    end if;
    par(i):=par(i)+p(i,j);
  end do:
end do:

Matrix(10,p);
seq(par(n),n=1..100);
```

It is easy to check (for instance, using the *Mathematica* command `PrimeQ` or the *Maple* command `isprime`) that $p(n)$ is a prime number for $n = 2$, 3, 4, 5, 6, 13, 36, and 77. It is not known whether $p(n)$ is prime for infinitely many n. It is not even known if infinitely many values of $p(n)$ are multiples of 3 (although it is known that infinitely many values of $p(n)$ are odd and infinitely many are even).

Srinivasa Ramanujan (1887–1920) noticed that 5 divides every 5th value of $p(n)$ starting with $n = 4$; that is to say, $5 \mid p(5n + 4)$, for all $n \geq 0$. Similarly, $7 \mid p(7n + 5)$, for all $n \geq 0$, and $11 \mid p(11n + 6)$, for all $n \geq 0$. These

n	$p(n)$	n	$p(n)$	n	$p(n)$	n	$p(n)$
1	1	26	2436	51	239943	76	9289091
2	2	27	3010	52	281589	77	10619863
3	3	28	3718	53	329931	78	12132164
4	5	29	4565	54	386155	79	13848650
5	7	30	5604	55	451276	80	15796476
6	11	31	6842	56	526823	81	18004327
7	15	32	8349	57	614154	82	20506255
8	22	33	10143	58	715220	83	23338469
9	30	34	12310	59	831820	84	26543660
10	42	35	14883	60	966467	85	30167357
11	56	36	17977	61	1121505	86	34262962
12	77	37	21637	62	1300156	87	38887673
13	101	38	26015	63	1505499	88	44108109
14	135	39	31185	64	1741630	89	49995925
15	176	40	37338	65	2012558	90	56634173
16	231	41	44583	66	2323520	91	64112359
17	297	42	53174	67	2679689	92	72533807
18	385	43	63261	68	3087735	93	82010177
19	490	44	75175	69	3554345	94	92669720
20	627	45	89134	70	4087968	95	104651419
21	792	46	105558	71	4697205	96	118114304
22	1002	47	124754	72	5392783	97	133230930
23	1255	48	147273	73	6185689	98	150198136
24	1575	49	173525	74	7089500	99	169229875
25	1958	50	204226	75	8118264	100	190569292

TABLE 6.2: Partition numbers $p(n)$ for $1 \leq n \leq 100$.

observations are called "Ramanujan's congruences":

$$\begin{aligned} p(5n+4) &\equiv 0 \pmod 5 \\ p(7n+5) &\equiv 0 \pmod 7 \\ p(11n+6) &\equiv 0 \pmod{11}. \end{aligned}$$

A partition $n = \lambda_1 + \lambda_2 + \cdots + \lambda_k$, with $\lambda_1 \geq \lambda_2 \geq \cdots \geq \lambda_k$, may be pictured by a "Ferrers diagram" (named after Norman Ferrers (1829–1903)) consisting of k rows of dots with λ_i dots in row i, for $1 \leq i \leq k$. The Ferrers diagram for the partition $12 = 7 + 3 + 1 + 1$ is shown in Figure 6.1.

The "transpose" of a Ferrers diagram is created by writing each row of dots as a column. The resulting partition is called the "conjugate" of the original partition. For example, the partition $12 = 7 + 3 + 1 + 1$ of Figure 6.1 is transposed to create the conjugate partition $12 = 4 + 2 + 2 + 1 + 1 + 1 + 1$ of Figure 6.2.

The reader may enjoy matching each partition of 4 of Example 6.44 with its conjugate. (One partition is self-conjugate.)

$$\begin{array}{l} \lambda_1 \bullet \bullet \bullet \bullet \bullet \bullet \bullet \\ \lambda_2 \bullet \bullet \bullet \\ \lambda_3 \bullet \\ \lambda_4 \bullet \end{array}$$

FIGURE 6.1: The Ferrers diagram of a partition of 12.

$$\begin{array}{l} \lambda_1 \bullet \bullet \bullet \bullet \\ \lambda_2 \bullet \bullet \\ \lambda_3 \bullet \bullet \\ \lambda_4 \bullet \\ \lambda_5 \bullet \\ \lambda_6 \bullet \\ \lambda_7 \bullet \end{array}$$

FIGURE 6.2: A transpose Ferrers diagram.

We now give the generating function for the partition numbers $p(n)$. The generating function is an "infinite polynomial" such that the coefficient of x^n is $p(n)$. For convenience, we set $p(0) = 1$. Thus, the generating function begins

$$1 + x + 2x^2 + 3x^3 + 5x^4 + 7x^5 + 11x^6 + 15x^7 + 22x^8 + \cdots .$$

We can give a neat algebraic representation of the generating function, as the next result shows.

THEOREM 6.45

$$\sum_{n=0}^{\infty} p(n)x^n = \prod_{k=1}^{\infty} (1 - x^k)^{-1}.$$

PROOF We need to show that the coefficients of x^n on the two sides of the equation are equal. The coefficient of x^n on the left side is $p(n)$. On the right side, the product may be written as

$$\prod_{k=1}^{\infty} (1 - x^k)^{-1} = \prod_{k=1}^{\infty} (1 + x^k + x^{2k} + x^{3k} + x^{4k} + \cdots).$$

To calculate the contribution to x^n from this product, suppose that the term $x^{j(k)k}$ is selected from the kth factor, for $1 \le k \le n$, and that these terms are multiplied together to yield $x^{j(1)+j(2)2+\cdots+j(n)n}$. If this expression equals x^n, then

$$n = j(1) \cdot 1 + j(2) \cdot 2 + \cdots + j(n) \cdot n. \tag{6.15}$$

Contributions to x^n correspond to solutions of (6.15). These solutions correspond to Ferrers diagrams for partitions of n. With m the greatest integer for which $j(m)$ is nonzero, we form the Ferrers diagram with $j(m)$ rows of m dots, followed by $j(m-1)$ rows of $m-1$ dots, etc. This correspondence between solutions to (6.15) and partitions of n completes the proof. □

Next we will determine the generating function for $p(n, k)$ with k fixed. We make an elementary observation from the transposes of Ferrers diagrams.

PROPOSITION 6.46

The number of partitions of n into exactly k parts is the same as the number of partitions of n where the size of the greatest summand is k.

Now we can give the generating function for $p(n, k)$.

THEOREM 6.47

$$\sum_{n=k}^{\infty} p(n, k)x^n = x^k \prod_{j=1}^{k} (1 - x^j)^{-1}.$$

PROOF Let $p(n, \leq k)$ be the number of partitions of n into at most k parts. From the previous proposition, $p(n, \leq k)$ is equal to the number of partitions of n into parts of size at most k. Clearly,

$$p(n, k) = p(n, \leq k) - p(n, \leq k - 1).$$

By the same reasoning as that given in the proof of Theorem 6.45,

$$\sum_{n=k}^{\infty} p(n, \leq k)x^n = \prod_{j=1}^{k} (1 + x^j + x^{2j} + x^{3j} + \cdots) = \prod_{j=1}^{k} (1 - x^j)^{-1}.$$

With some algebraic juggling, we have

$$\begin{aligned}
\sum_{n=k}^{\infty} p(n,k)x^n &= \sum_{n=0}^{\infty} p(n,k)x^n \\
&= \sum_{n=k}^{\infty} [p(n, \leq k) - p(n, \leq k-1)]x^n \\
&= \prod_{j=1}^{k} (1-x^j)^{-1} - \prod_{j=1}^{k-1} (1-x^j)^{-1} \\
&= [1-(1-x^k)] \prod_{j=1}^{k} (1-x^j)^{-1} \\
&= x^k \prod_{j=1}^{k} (1-x^j)^{-1}.
\end{aligned}$$

□

Let $p(n \mid \text{distinct parts})$ denote the number of partitions of n into distinct parts. The next result (the proof is called for in the Exercises) gives the generating function for these numbers.

PROPOSITION 6.48

$$\sum_{n=0}^{\infty} p(n \mid \text{distinct parts})x^n = \prod_{k=1}^{\infty} (1+x^k).$$

Let $p(n \mid \text{odd parts})$ denote the number of partitions of n into summands each of which is an odd number. The following delightful and unexpected relation was found by Euler.

PROPOSITION 6.49

$$p(n \mid \text{odd parts}) = p(n \mid \text{distinct parts}).$$

PROOF As in the proof of Theorem 6.45, we have

$$\begin{aligned}\sum_{n=0}^{\infty} p(n \mid \text{odd parts})x^n &= \frac{1}{(1-x)(1-x^3)(1-x^5)\dots} \\ &= \frac{(1-x^2)}{(1-x)(1-x^2)} \cdot \frac{(1-x^4)}{(1-x^3)(1-x^4)} \cdot \frac{(1-x^6)}{(1-x^5)(1-x^6)} \cdot \dots \\ &= \frac{(1-x^2)}{(1-x)} \cdot \frac{(1-x^4)}{(1-x^2)} \cdot \frac{(1-x^6)}{(1-x^3)} \cdot \dots \\ &= (1+x)(1+x^2)(1+x^3)\dots \\ &= \sum_{n=0}^{\infty} p(n \mid \text{distinct parts})x^n.\end{aligned}$$

We obtain the desired identity by comparing coefficients of the two generating functions. ∎

Example 6.50

$$p(5 \mid \text{odd parts}) = 3 \qquad (5,\ 3+1+1,\ 1+1+1+1+1)$$
$$p(5 \mid \text{distinct parts}) = 3 \qquad (5,\ 4+1,\ 3+2)$$

∎

Proposition 6.48 says that the coefficients of the polynomial

$$(1+x)(1+x^2)(1+x^3)\dots$$

count partitions of n into distinct parts. It is natural to ask what is counted by the coefficients of the polynomial

$$(1-x)(1-x^2)(1-x^3)\dots.$$

Just as in the case of the previous polynomial, there are contributions to the term x^n only from partitions with distinct parts. However, some of the contributions are $+1$ (if the number of parts is even) and some are -1 (if the number of parts is odd). This results in the following formula for the coefficients.

PROPOSITION 6.51

$$\prod_{n=1}^{\infty}(1-x^n) = \sum_{n=0}^{\infty}(p(n \mid \text{even no. distinct parts}) - p(n \mid \text{odd no. distinct parts}))x^n$$

Let's use a computer to generate a list of coefficients. We may use the *Mathematica* command

```
Table[{j, Coefficient[Product[1 - x^k, {k,1,50}], x^j]}, {j,1,50}]
// TableForm
```

or the *Maple* command

```
seq([j,coeff(product(1-x^k,k=1..50),x,j)],j=1..50);
```

to produce the following array.

1	−1	11	0	21	0	31	0	41	0
2	−1	12	−1	22	1	32	0	42	0
3	0	13	0	23	0	33	0	43	0
4	0	14	0	24	0	34	0	44	0
5	1	15	−1	25	0	35	−1	45	0
6	0	16	0	26	1	36	0	46	0
7	1	17	0	27	0	37	0	47	0
8	0	18	0	28	0	38	0	48	0
9	0	19	0	29	0	39	0	49	0
10	0	20	0	30	0	40	−1	50	0

The coefficients appear to be only 0, −1, and 1. The pattern is given in terms of "pentagonal numbers," which are figurate numbers (see Figure 6.3), just as are square numbers and triangular numbers (see Chapter 9 Notes). Pentagonal numbers have a geometric interpretation as the number of dots in a pentagonal array of dots with given side length. The pentagonal numbers are thus

$$1, 5, 12, 22, 35, 51, 70, 92, 117, \ldots.$$

Consulting the diagram, we obtain a recurrence relation for a_k, the kth pentagonal number:

$$a_k = a_{k-1} + 3k - 2, \quad k \geq 2.$$

Working backwards, we arrive at the formula

$$\begin{aligned} a_k &= a_1 + 3(2 + 3 + \cdots + k) - 2(k-1) \\ &= 1 + 3\left(\frac{k(k+1)}{2} - 1\right) - 2(k-1) \\ &= \frac{k(3k-1)}{2}. \end{aligned}$$

DEFINITION 6.52 *A* pentagonal number *is a number of the form*

$$k(3k-1)/2, \quad k \geq 1.$$

We may also take k to be negative, obtaining the numbers (sometimes also referred to as pentagonal numbers or as "pseudopentagonal numbers")

$$2, 7, 15, 26, 40, 57, 77, 100, 126, \ldots.$$

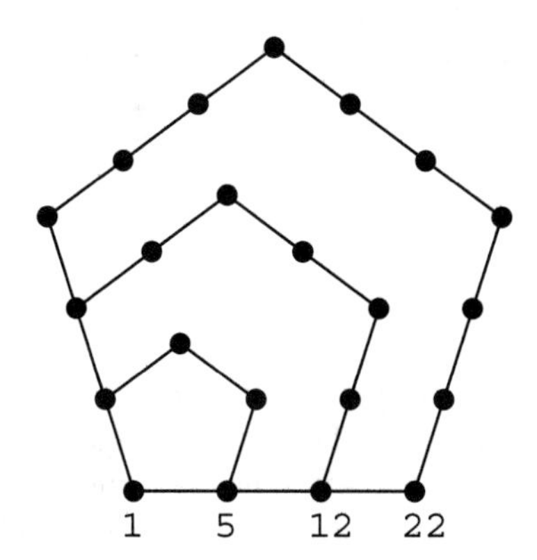

FIGURE 6.3: Pentagonal numbers.

These numbers are given by the expression $(-k)(-3k-1)/2 = k(3k+1)/2$, for $k \geq 1$. The two types of pentagonal numbers, together with 0, are called "generalized pentagonal numbers" and are given as

$$n = k(3k \pm 1)/2, \quad k \geq 0.$$

PROPOSITION 6.53

$$p(n \mid \text{even number of distinct parts}) - p(n \mid \text{odd number of distinct parts})$$
$$= \begin{cases} (-1)^k \text{ if } n = k(3k \pm 1)/2, & \text{for } k \geq 1 \\ 0 \quad\quad \text{otherwise} \end{cases}$$

PROOF We will prove this relation using Ferrers diagrams. Specifically, we will exhibit a correspondence between partitions of n with an even number of distinct parts and partitions of n with an odd number of distinct parts. Of course, this correspondence cannot be a bijection for all n, or else the difference we are computing would be identically 0. However, it is a bijection for all n that are not pentagonal numbers, and for pentagonal numbers there is one extra partition in one of the sets.

Let's illustrate the correspondence in the case $n = 17$ (not a generalized pentagonal number). Consider the Ferrers diagrams of the partitions $17 = 6+5+4+2$ and $17 = 7+6+4$.

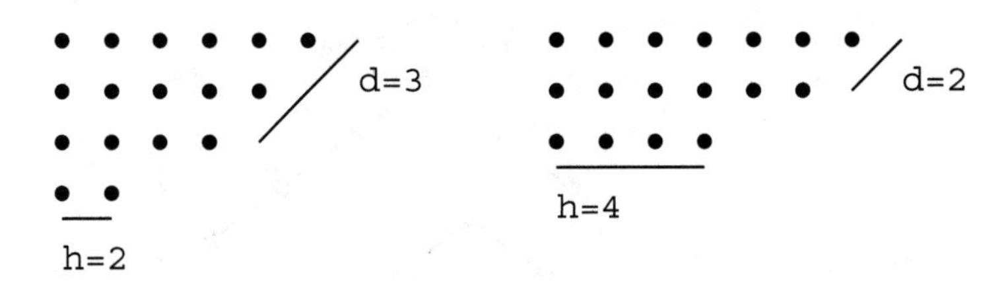

Let H (for horizontal) be the bottom row of dots in a partition (i.e., the smallest part of the partition). Let D (for diagonal) be the longest diagonal of dots in a partition, starting with the right-most dot in the top row. Let h be the number of dots in H and d the number of dots in D. If $h \leq d$, then move H so that it forms a diagonal to the right of D, with its top dot in the top row. If $h > d$, then move D to the bottom row (left justified, of course). The picture above shows the correspondence for our two specific partitions of 17. In general, the correspondence changes the parity of the number of parts in the partition. The correspondence breaks down in two cases: when H and D have a dot in common and $h = d$ or $h = d + 1$. In the first case,

$$n = d^2 + 1 + \cdots + (d - 1) = \frac{d(3d - 1)}{2},$$

and in the second case,

$$n = d^2 + 1 + \cdots + d = \frac{d(3d + 1)}{2}.$$

For these generalized pentagonal numbers, the difference between the number of partitions with an even number of distinct parts and the number of partitions with an odd number of distinct parts is $(-1)^d$, as d is the number of parts. ∎

Example 6.54

$p(10 \mid \text{even number of distinct parts}) = 5$
$(6 + 4,\ 7 + 3,\ 8 + 2,\ 9 + 1,\ 4 + 3 + 2 + 1)$
$p(10 \mid \text{odd number of distinct parts}) = 5$
$(10,\ 7 + 2 + 1,\ 6 + 3 + 1,\ 5 + 4 + 1,\ 5 + 3 + 2)$
$p(10 \mid \text{even number of distinct parts}) - p(10 \mid \text{odd number of distinct parts})$
$= 5 - 5 = 0$ (10 is not of the form $k(3k \pm 1)/2$)

$p(12 \mid \text{even number of distinct parts}) = 7$
$(11+1,\ 10+2,\ 9+3,\ 8+4,\ 7+5,\ 6+3+2+1,\ 5+4+2+1)$
$p(12 \mid \text{odd number of distinct parts}) = 8$
$(12,\ 9+2+1,\ 8+3+1,\ 7+4+1,\ 6+5+1,\ 7+3+2,\ 6+4+2,\ 5+4+3)$
$p(12 \mid \text{even number of distinct parts}) - p(12 \mid \text{odd number of distinct parts})$
$= 7 - 8 = -1 = (-1)^3 \quad (12 = 3(3 \cdot 3 - 1)/2)$

□

Putting together the last two propositions, we have proved the following famous theorem of Euler.

THEOREM 6.55 (Euler's pentagonal number theorem)

$$\prod_{n=1}^{\infty}(1-x^n) = \sum_{n=-\infty}^{\infty}(-1)^n x^{n(3n-1)/2}.$$

COROLLARY 6.56

$p(n) = p(n-1) + p(n-2) - p(n-5) - p(n-7) + p(n-12) + p(n-15) - \cdots$

The expression on the right includes all terms $p(n-m)$, where m is a generalized pentagonal number for which $n - m \geq 0$. The terms occur in pairs of like signs, one pentagonal and the other "pseudopentagonal," and the signs alternate.

PROOF Putting Theorems 6.45 and 6.55 together, we obtain

$$1 = \sum_{n=0}^{\infty} p(n)x^n \prod_{n=1}^{\infty}(1-x^n) = \sum_{n=0}^{\infty} p(n)x^n \sum_{n=-\infty}^{\infty}(-1)^n x^{n(3n-1)/2}.$$

The desired recurrence relation follows. Note that for $n \geq 1$, the coefficient of x^n must be 0. We get contributions to the coefficient of x^n from products of the form $p(r)x^r(-1)^s x^{s(3s-1)/2}$. When $s = 0$, we have $r = n$ and the contribution is $p(n)$. Otherwise, $s(3s-1)/2$ is a generalized pentagonal number m, and $r = n - m$. The contribution in this case is $\pm p(n-r)$, with the sign depending on whether s is even or odd. □

Example 6.57 We use the recurrence relation of Corollary 6.56 to compute $p(n)$, for $1 \leq n \leq 10$.

$$\begin{aligned}
p(1) &= 1 \\
p(2) &= p(1) + p(0) = 1 + 1 = 2 \\
p(3) &= p(2) + p(1) = 2 + 1 = 3 \\
p(4) &= p(3) + p(2) = 3 + 2 = 5 \\
p(5) &= p(4) + p(3) - p(0) = 5 + 3 - 1 = 7 \\
p(6) &= p(5) + p(4) - p(1) = 7 + 5 - 1 = 11 \\
p(7) &= p(6) + p(5) - p(2) - p(0) = 11 + 7 - 2 - 1 = 15 \\
p(8) &= p(7) + p(6) - p(3) - p(1) = 15 + 11 - 3 - 1 = 22 \\
p(9) &= p(8) + p(7) - p(4) - p(2) = 22 + 15 - 5 - 2 = 30 \\
p(10) &= p(9) + p(8) - p(5) - p(3) = 30 + 22 - 7 - 3 = 42
\end{aligned}$$

These values are in agreement with Table 6.2. ◻

We implement our recurrence formula using a computer. Notice that we can compute $p(n)$ for large n.

Mathematica

```
p[0]=1;
Do[
  p[n]=0;
  k=1;
  While[n-k(3k-1)/2>=0,
    p[n]=p[n]+p[n-k(3k-1)/2](-1)^(k+1);k++];
  k=1;
  While[n-k(3k+1)/2>=0,
    p[n]=p[n]+p[n-k(3k+1)/2](-1)^(k+1);k++],
  {n,1,1000}
];

Table[p[n],{n,1,10}]
{1,2,3,5,7,11,15,22,30,42}

p[1000]
24061467864032622473692149727991
```

Maple

```
p(0):=1:
for n from 1 to 1000 do
  p(n):=0;
  k:=1;
  while (n-k*(3*k-1)/2) >= 0 do
    p(n):=p(n)+p(n-k*(3*k-1)/2)*(-1)^(k+1);
    k:=k+1;
  end do;
  k:=1;
  while (n-k*(3*k+1)/2) >= 0 do
    p(n):=p(n)+p(n-k*(3*k+1)/2)*(-1)^(k+1);;
    k:=k+1;
  end do;
end do:

seq(p(n),n=1..10);
1, 2, 3, 5, 7, 11, 15, 22, 30, 42

p(1000);
24061467864032622473692149727991
```

For convenience, we list the generating functions we have found.

$$\sum_{n=0}^{\infty} p(n)x^n = \prod_{k=1}^{\infty} (1-x^k)^{-1}$$

$$\sum_{n=k}^{\infty} p(n,k)x^n = x^k \prod_{j=1}^{k} (1-x^j)^{-1}$$

$$\sum_{n=k}^{\infty} p(n, \leq k)x^n = \prod_{j=1}^{k} (1-x^j)^{-1}$$

$$\sum_{n=0}^{\infty} p(n \mid \text{distinct parts})x^n = \prod_{k=1}^{\infty} (1+x^k)$$

$$\sum_{n=0}^{\infty} p(n \mid \text{odd parts})x^n = \frac{1}{(1-x)(1-x^3)(1-x^5)\ldots}$$

$$\prod_{n=1}^{\infty} (1-x^n) = \sum_{n=-\infty}^{\infty} (-1)^n x^{n(3n-1)/2}$$

Exercises

6.44 List (by hand) the partitions of 5, 6, and 7.

6.45 Show that $p(n,k) = \sum_{j=1}^{k} p(n-k,j)$.

6.46 Prove Proposition 6.49 by establishing a bijection between the set of partitions of n into odd parts and the set of partitions of n into distinct parts.

⋆6.47 Find formulas for $p(n,1)$, $p(n,2)$, and $p(n,3)$. Conjecture an asymptotic estimate for $p(n,k)$ (with k fixed).

⋄6.48 Use a computer to validate Theorem 6.45 for $0 \le n \le 20$.

6.49 Let $p(n \mid \text{even no. of parts})$ and $p(n \mid \text{odd no. of parts})$ be the number of partitions of n into an even number of parts and into an odd number of parts, respectively. Let $p(n \mid \text{distinct odd parts})$ be the number of partitions of n with distinct odd parts. Let $c(n)$ be the number of self-conjugate partitions of n. Prove that

$$\begin{aligned} c(n) &= p(n \mid \text{distinct odd parts}) \\ &= (-1)^n (p(n \mid \text{even no. of parts}) - p(n \mid \text{odd no. of parts})). \end{aligned}$$

⋄6.50 (a) Use a computer and an appropriate generating function to determine the number of ways of making change for \$1. The change may include pennies, nickels, dimes, quarters, half-dollars, and silver dollars.

(b) How many ways are there to make change for a dollar using an even number of coins?

6.51 Prove Proposition 6.48.

6.52 Show that the number of partitions of n into summands none of which occurs exactly once is the same as the number of partitions of n into summands none of which is congruent to 1 or 5 modulo 6.

⋆6.53 Let $t(n)$ be the number of incongruent triangles with integer side lengths and perimeter n. Find a formula for $t(n)$. What is $t(100)$?

6.54 Prove that the two sequences of "pentagonal numbers" have no elements in common.

6.55 Prove that every pentagonal number is 1/3 of a triangular number (see Notes to Chapter 9).

6.56 Show the correspondence in Proposition 6.53 for the partitions of $n = 10$ with distinct parts.

†⋆**6.57** (For those who have studied group theory) Let S_n be the group of permutations of the set $\{1, 2, 3, \ldots, n\}$. We say that two permutations α and β of S_n are in the same *conjugacy class* if there exists a permutation $\sigma \in S_n$ such that $\beta = \sigma\alpha\sigma^{-1}$. Show that two permutations are in the same conjugacy class if and only if they have the same cycle structure (multiset of cycle lengths). How many conjugacy classes of S_n are there?

⋆**6.58** (For those who have studied group theory)

(a) How many nonisomorphic abelian (commutative) groups of order 2700 are there?

(b) How many ways may one make \$2.32 postage using 1-cent, 2-cent, 3-cent, 10-cent, 20-cent, \$1, and \$2 stamps, and not more than three of any one denomination?

Notice that the answers to parts (a) and (b) are the same. Why is this so?

6.6 Notes

The lore of perfect numbers

Many early mathematical thinkers asserted that numbers have a scientific significance and even a mystical one. (See, for example, [4].) The Pythagoreans defined a "perfect number" to be one equal to the sum of its proper divisors. Perhaps because of their rarity, great significance was attached to perfect numbers.

The number six has held a special fascination for people throughout history. For example, in the well-known Hindu parable immortalized as "The Blind Men and the Elephant" by John Godfrey Saxe (1816–1887), we have:

It was six men of Indostan
To learning much inclined,
Who went to see the Elephant
(Though all of them were blind),
That each by observation
Might satisfy his mind.

This theme of six persons confronting a mystery is echoed in Stanley Kubrick's (1928–1999) film *2001: A Space Odyssey*, in which six astronauts encounter an alien artifact on the Moon and six crew members travel in a spaceship to Jupiter.

The smallest four perfect numbers, 6, 28, 496 and 8128, were discovered over two thousand years ago. Since the time of the Pythagoreans, there has been great interest in finding larger and larger perfect numbers. The search for perfect numbers is intimately connected to the search for large primes (which we will investigate in the next chapter). We have shown (à la Euler) that even perfect numbers are of the form $2^{p-1}(2^p - 1)$, where p and $2^p - 1$ are prime. Euler proved that if n is an odd perfect number, then

$$n = p_1^{4m+1} \prod_{i=2}^{k} p_i^{2e_i},$$

where $p_1 \equiv 1 \pmod 4$ and all the p_i are distinct. Since Euler's time a few further criteria have been found for odd perfect numbers, but so far no one has found an odd perfect number or shown that they do not exist.

A last word: Augustine (354–430) was clearly awed by perfect numbers, as this assertion from his *The City of God* shows:

> Six is a number perfect in itself, and not because God created the world in six days; rather the contrary is true. God created the world in six days because this number is perfect, and it would remain perfect, even if the work of the six days did not exist.

Pioneers of integer partitions

Integer partitions were first investigated by Gottfried Wilhelm Leibniz (1646–1716). Leibniz counted the number of partitions of n for small values of n but found no general method of computation. Euler solved the problem of counting partitions, in principle, by using generating functions. His calculation is performed via our Corollary 6.56. Recall that the basis for this corollary is Euler's Pentagonal Number Theorem. The proof we showed is due to Fabian Franklin (1853–1939), a Hungarian–American mathematician and student of James Sylvester (1814–1897). Fabian's proof has been called the first major achievement of American mathematics. At the midpoint of his life, Fabian turned his energies to writing about politics and economics. His works include "Plain Talk On Economics" and "What Prohibition Has Done to America." Fabian was married to Christine Ladd-Franklin (1847–1930), an aspiring mathematician who found it difficult to obtain lecturing positions because of her gender.

Sylvester can be said to have started the modern study of partitions. In his treatise "A Constructive Theory of Partitions, Arranged in Three Acts, An Interact, and an Exodion" (1882), he wrote, "It is essential to consider a partition as a definite thing, which end is attained by regularization of the succession of its parts according to some prescribed law. The simplest law for the purpose is that the arrangement of parts shall be according to their order of magnitude." Sylvester is proposing that a partition be represented by what we call a Ferrers Diagram. Ferrers had corresponded with Sylvester about bijective proofs for certain partition identities. Sylvester collaborated with many other mathematicians, including students, colleagues, and researchers at other universities. He was at the focal point of American mathematics while he was at Johns Hopkins University (from 1877 to 1883), founding the *American Journal of Mathematics*, the first mathematical journal in the United States. Among many diverse contributions to mathematics, Sylvester coined the term "matrix" (in 1850). Sylvester's obituary in the *London Times* stated, "As Professor and as editor of the *American Journal of Mathematics*, he practically founded the study of higher mathematics in the United States."

In 1918 G. H. Hardy (1877–1947) and Ramanujan obtained an asymptotic estimate for the partition function:

$$p(n) \sim \frac{1}{4n\sqrt{3}} e^{\pi\sqrt{2n/3}}.$$

(We say that two functions $f(n)$ and $g(n)$ are asymptotic, and write $f(n) \sim g(n)$, if

$$\lim_{n\to\infty} f(n)/g(n) = 1.)$$

More surprisingly, Hardy and Ramanujan found an exact formula for $p(n)$. The formula is very complicated and uses advanced methods of generating functions. Ramanujan also discovered many identities concerning partition numbers, for example, that $p(5n + 4)$ is always divisible by 5. Another of Ramanujan's observations is that $p(n \mid \text{parts} \equiv \pm 1 \pmod 5) = p(n \mid \text{parts differ by at least } 2)$. This identity had been discovered earlier by Leonard James Rogers (1862–1933).

Chapter 7

Large Primes

> $2^{30}(2^{31} - 1)$ is the greatest perfect number that will ever be discovered, for, as they are merely curious without being useful, it is not likely that any person will attempt to find a number beyond it.
>
> PETER BARLOW (1776–1862)

The above prognostication was short-sighted. In fact, 44 perfect numbers have been discovered so far; the one mentioned above is only the eighth, and the prime indicated, $2^{31}-1$, is not particularly large. As we saw in Chapter 6, even perfect numbers are in one-to-one correspondence with primes of the form $2^p - 1$. We also saw (in Chapter 3) that the question of which regular n-gons are constructible has to do with prime numbers of a certain form. Prime numbers are interesting in part because of their connections with famous problems such as these. In this chapter we discuss the ongoing search for large primes. How do we find large primes? Computers will be very useful in our calculations.

7.1 Prime listing, primality testing, and prime factorization

In his 1801 *Disquisitiones Arithmeticae*, Gauss noted the central importance in number theory of the processes of primality testing and prime factorization:

> The problem of distinguishing prime numbers from composite numbers and of resolving the latter into their prime factors is known to be one of the most important and useful in arithmetic. It has engaged the industry and wisdom of ancient and modern geometers to such an extent that it would be superfluous to discuss the problem at length.... Further, the dignity of the science itself seems to require that every possible means be explored for the solution of a problem so elegant and so celebrated.

We begin our search for large primes by addressing three basic tasks: (1) listing the prime numbers up to a given number, (2) testing a given number to determine whether it is a prime, and (3) determining the prime factorization of a given number.

The following simple observation, first recorded in Chapter 2, is useful in devising our algorithms.

PROPOSITION 7.1
If a number n has a proper divisor, then it has one less than or equal to $\sqrt{n}$.

Prime listing

Recall the Sieve of Eratosthenes from Chapter 2.

ALGORITHM 7.2 (Sieve of Eratosthenes)
Given n, the prime numbers up to n are identified.

(1) Let n be a positive integer; set $S = \{2, 3, \ldots, n\}$.

(2) For $2 \le i \le \lfloor \sqrt{n} \rfloor$, do:
If i has not been crossed out in S, then cross out all proper multiples of i in S.

(3) The elements of S that are not crossed out are the prime numbers up to n.

We implement Algorithm 7.2 in *Mathematica.*

Mathematica

```
primelist[n_] := Module[{candidatelist, i, propermultiple},
candidatelist = Range[1, n];
candidatelist[[1]] = 0;
Do[
  If[candidatelist[[i]] != 0,
  Do[
    candidatelist[[propermultiple]] = 0,
    {propermultiple, 2i, n, i}
  ]],
  {i, 2, Floor[Sqrt[n]]}
];
DeleteCases[candidatelist, 0]
]
```

We test our program, finding all primes less than 100.

Mathematica

```
primelist[100]
{2, 3, 5, 7, 11, 13, 17, 19, 23, 29, 31, 37, 41, 43, 47, 53,
59, 61, 67, 71, 73, 79, 83, 89, 97}
```

Here are the corresponding *Maple* calculations.

Maple

```
primelist:=proc(n)
local candidatelist, i, propermultiple;
candidatelist:=array([seq(i,i=1..n)]);
candidatelist[1]:=0;
for i from 2 to floor(sqrt(n)) do
  if candidatelist[i] <> 0 then
  for propermultiple from 2*i to n by i do
    candidatelist[propermultiple]:=0
  end do;
  end if;
end do;
{seq(candidatelist[i],i=1..n)} minus {0};
end proc:

primelist(100);
{2, 3, 5, 7, 11, 13, 17, 19, 23, 29, 31, 37, 41, 43,
47, 53, 59, 61, 67, 71, 73, 79, 83, 89, 97}
```

We confirm the values of $\pi(n)$ in Table 2.1, using the *Mathematica* command

```
Table[Length[primelist[10^k]], {k, 1, 6}] // TableForm
```

or the *Maple* command

```
map(x->[nops(primelist(x))],[seq(10^k,k=1..6)]);
```

(`Length` (*Mathematica*) and `nops` (*Maple*) give the size of a list).

```
4
25
168
1229
9592
78498
```

Therefore, $\pi(10) = 4$, $\pi(10^2) = 25$, $\pi(10^3) = 168$, $\pi(10^4) = 1229$, $\pi(10^5) = 9592$, and $\pi(10^6) = 78498$.

How many steps does Algorithm 7.2 take? The algorithm has an outer loop and an inner loop. In the outer loop, the index i goes from 2 to $\lfloor\sqrt{n}\rfloor$. The inner loop is executed only when i is a prime, in which case there are $\lfloor n/i\rfloor - 1$ steps (striking out proper multiples of i). Hence, the total number of steps is

$$\sum_{p\leq\sqrt{n}}\left(\left\lfloor\frac{n}{p}\right\rfloor - 1\right) \sim n\sum_{p\leq\sqrt{n}}\frac{1}{p} \sim n\log\log\sqrt{n} \sim n\log\log n.$$

The asymptotic estimate of $\sum_{p\leq\sqrt{n}} p^{-1}$ is due to Franz Mertens (1840–1927).

We refine the implementation of Algorithm 7.2, to make it faster.

1. Array values are 0 for composite and 1 for prime.
2. Even numbers greater than 2 are composite, so don't need to be tested. We consider only odd numbers $i = 2j+1$, where $1 \leq j \leq \lfloor (n-1)/2\rfloor$. In the inner loop, $(2j+1)^2 \leq n$; hence, $j \leq \lfloor(\sqrt{n}-1)/2\rfloor$.
3. Proper multiples of a prime i are dealt with starting at i^2, since array values corresponding to any smaller multiples have already been set to 0. This means starting at $2j(j+1)$ and increasing by steps of size $2j+1$.

Mathematica

```
newprimelist[n_] := Module[{m, x},
  x = Table[1, {m = Floor[(n - 1)/2]}];
  If[x[[#]] == 1, x[[Range[2#(# + 1), m, 2# + 1]]] = 0] &
  /@ Range[Floor[(Sqrt[n] - 1)/2]];
  Prepend[Flatten[2 Position[x, 1] + 1], 2]
]
```

Maple

```
newprimelist:=proc(n)
local m, x, j, i;
m:=floor((n-1)/2);
x:=array([seq(1,i=1..m)]);
for j from 1 to floor(evalf((sqrt(n)-1)/2)) do
  if x[j]=1 then
  for i from 2*j*(j+1) to m by 2*j+1 do
    x[i]:=0;
  end do;
  end if;
end do;
{seq(x[i]*(2*i+1),i=1..m)} minus {0} union {2};
end proc:
```

In order to facilitate our later computations, we store the set of primes up to 10^7. Thus, if we want, say, the 500,000th prime, we can produce it easily.

Mathematica

```
ourlist = newprimelist[10^7];

ourlist[[500000]]

7368787
```

Maple

```
ourlist:=convert(newprimelist(10^7),'list'):

ourlist[500000];

7368787
```

Primality testing

Our first prime testing procedure uses hardly any memory. We merely check each integer k, with $2 \leq k \leq \lfloor\sqrt{n}\rfloor$, to see if k divides n.

ALGORITHM 7.3 (Test for primality)
Given an integer $n > 1$, it is reported whether n is prime or composite.
For $2 \leq k \leq \lfloor\sqrt{n}\rfloor$, check whether k divides n. If some k divides n, then stop and report that n is not a prime. If none of these numbers is a divisor, then report that n is a prime.

Here is our algorithm in *Mathematica*.

Mathematica

```
primetest[n_] := Module[{upperlimit, trialdivisor},
upperlimit = Floor[Sqrt[n]];
For[trialdivisor = 2, trialdivisor <= upperlimit,
trialdivisor++,
  If[Mod[n, trialdivisor] == 0, Return[False]]
];
Return[True]
]
```

We test our algorithm on the prime mentioned at the beginning of the chapter and on a composite number.

Note: In these and later calculations, we include `Timing` (*Mathematica*) and `time` (*Maple*) commands to get a sense of how much time computations take. This information is for comparison and to get an overall idea; we do not give details about what computer is used.

Mathematica

```
primetest[2^31-1] // Timing
{0.711 Second, True}
primetest[Prime[500000] Prime[500001]] // Timing
{102.677 Second, False}
```

Here are the corresponding *Maple* commands.

Maple

```
primetest:=proc(n)
local upperlimit,trialdivisor;
upperlimit:=floor(evalf(sqrt(n)));
for trialdivisor from 2 to upperlimit do
  if modp(n,trialdivisor)=0 then
  return(False);
  end if;
end do;
return(True);
end proc:

starttime:=time():
primetest(2^31-1);
time()-starttime;
True
0.130
primetest(ithprime(500000)*ithprime(500001));
time()-starttime;
False
time = 15.753
```

Although our `primetest` procedure uses almost no memory, it has a shortcoming in that $\sqrt{n}$ is a large number when n is very large, so there are many potential divisors to test. However, if we are willing to store a list of primes (such as `ourlist`), we can create a procedure that is much faster.

ALGORITHM 7.4 (New test for primality)
Given an integer $n > 1$, it is reported whether n is prime or composite.

For primes $p \le \lfloor \sqrt{n} \rfloor$, check whether p divides n. If some p divides n, then stop and report that n is not a prime. If none of these numbers is a divisor, then report that n is a prime.

Mathematica

```
newprimetest[n_] := Module[{upperlimit, i, trialdivisor},
  upperlimit = Floor[Sqrt[n]];
  For[i=1, (trialdivisor = ourlist[[i]]) <= upperlimit, i++,
  If[Mod[n, trialdivisor] == 0, Return[False]]
  ];
  Return[True]
]

newprimetest[2^31-1] // Timing
{0.1 Second, True}
newprimetest[Prime[500000] Prime[500001]] // Timing
{10.265 Second, False}
```

Maple

```
newprimetest:=proc(n)
local upperlimit, i;
upperlimit:=floor(evalf(sqrt(n)));
for i from 1 while ourlist[i]<=upperlimit do
  if modp(n,ourlist[i])= 0 then
  return(False);
  end if;
end do;
return(True);
end proc:

starttime:=time():
newprimetest(2^31-1);
time()-starttime;
True
0.020
newprimetest(ithprime(500000)*ithprime(500001));
time()-starttime;
False
1.783
```

Note: We can use `newprimetest` only when n has a prime factor less than 10^7 (recall that `ourlist` is based on this bound) or n itself is a prime less than 10^{14} (i.e., $(10^7)^2$).

Prime factorization

Here is a simple procedure for factoring integers.

ALGORITHM 7.5 (Prime factorization)
Given an integer $n > 1$, the prime factors of n, together with their multiplicities, are reported.

Let n be given and set $n^* = n$. For $2 \leq k \leq \lfloor\sqrt{n^*}\rfloor$, check whether k divides n^*; if k divides n^*, then report k as a prime factor and set $n^* = n^*/k$. Continue until the condition $k \leq \lfloor\sqrt{n^*}\rfloor$ fails, and report n^* as a prime factor.

Here is our procedure in *Mathematica*.

Mathematica

```
primefactorization[n_] :=
  Module[{primefactors, nstar, upperlimit, trialdivisor},
  primefactors = {};
  nstar = n;
  upperlimit = Floor[Sqrt[nstar]];
  trialdivisor = 2;
  While[trialdivisor <= upperlimit,
    If[Mod[nstar, trialdivisor] == 0,
    AppendTo[primefactors, trialdivisor];
    nstar = nstar/trialdivisor;
    upperlimit = Floor[Sqrt[nstar]],
    trialdivisor++]
  ];
  AppendTo[primefactors, nstar];
  {#, Count[primefactors, #]} & /@ Union[primefactors]
]
```

We test our program.

Mathematica

```
primefactorization[297308799018312231 5334] // Timing

{128.665 Second,
{{2,1},{3,2},{1999,1},{7368787,1},{11213119751,1}}}
```

Thus,

$$297308799018312231533 4 = 2 \cdot 3^2 \cdot 1999 \cdot 7368787 \cdot 11213119751.$$

Here are the computations in *Maple*.

Maple

```
primefactorization:=proc(n)
local primefactors, nstar, upperlimit, trialdivisor;
primefactors:=[];
nstar:=n;
upperlimit:=floor(evalf(sqrt(nstar)));
trialdivisor:=2;
while trialdivisor <= upperlimit do
  if modp(nstar,trialdivisor)=0 then
  primefactors:=[op(primefactors),trialdivisor];
  nstar:=nstar/trialdivisor;
  upperlimit:=floor(evalf(sqrt(nstar)));
  else
  trialdivisor:=trialdivisor+1;
  end if;
end do;
primefactors:=[op(primefactors),nstar];
{seq([x,ListTools[Occurrences](x,primefactors)],x in
primefactors)};
end proc:

starttime:=time():
primefactorization(2973087990183122315334);
time()-starttime;
{[1999, 1], [7368787, 1], [3, 2], [11213119751, 1], [2, 1]}
24.904
```

Our `primefactorization` procedure has the virtue of using almost no memory, but it is slow. We can make a much faster algorithm by testing only primes as potential divisors.

***ALGORITHM 7.6* (New prime factorization)**

Given an integer $n > 1$, the prime factors of n, together with their multiplicities, are reported.

Let n be given and set $n^* = n$. For primes $p \leq \lfloor \sqrt{n^*} \rfloor$, check whether p divides n^*; if p divides n^*, then report p as a prime factor and set $n^* = n^*/p$. Continue until the condition $p \leq \lfloor \sqrt{n^*} \rfloor$ fails, and report n^* as a prime factor.

Here is our procedure in *Mathematica*.

Mathematica

```
newprimefactorization[n_] :=
  Module[{primefactors, nstar, upperlimit, i, trialdivisor},
  primefactors = {};
  nstar = n;
  upperlimit = Floor[Sqrt[nstar]];
  trialdivisor = ourlist[[i=1]];
  While[trialdivisor <= upperlimit,
    If[Mod[nstar, trialdivisor] == 0,
    AppendTo[primefactors, trialdivisor];
    nstar = nstar/trialdivisor;
    upperlimit = Floor[Sqrt[nstar]],
    trialdivisor = ourlist[[++i]]]
  ];
  AppendTo[primefactors, nstar];
  {#, Count[primefactors, #]} & /@ Union[primefactors]
]
```

We ask our program to factor 2973087990183122315334.

Mathematica

```
newprimefactorization[2973087990183122315334] // Timing

{11.627 Second,
{{2,1},{3,2},{1999,1},{7368787,1},{11213119751,1}}}
```

Notice that this computation is much faster than with the previous factorization program.

This program factors numbers in which all prime factors are less than 10^7 (the bound on `ourlist`), with the possible exception of one additional prime factor less than $10^{14} = (10^7)^2$. In all other cases, *Mathematica* eventually reports an error (the size of the array is exceeded), but shows the factors found up to that point. For example, with the command

```
newprimefactorization[9199838343019266106321698] // Timing,
```

Mathematica reports an error but still shows the following partial results:

```
{20.72 Second,{{2,1},{3,2},{1999,1},{255678904536136571239,1}}}.
```

Thus, we know that part of the prime factorization here is $2 \cdot 3^2 \cdot 1999$ and that 255678904536136571239 is a prime or composite factor. In fact,

$$255678904536136571239 = 11213119751 \cdot 22801763489$$

(as determined by `FactorInteger`).

Here are the corresponding *Maple* computations.

Maple

```
newprimefactorization:=proc(n)
local primefactors, nstar, upperlimit, i, trialdivisor;
primefactors:=[];
nstar:=n;
upperlimit:=floor(evalf(sqrt(n)));
i:=1;
trialdivisor:=ourlist[i];
while trialdivisor <= upperlimit do
  if modp(nstar,trialdivisor)=0 then
  primefactors:=[op(primefactors),trialdivisor];
  nstar:=nstar/trialdivisor;
  upperlimit:=floor(evalf(sqrt(nstar)));
  else
  i:=i+1;
  trialdivisor:=ourlist[i];
  end if;
end do;
primefactors:=[op(primefactors),nstar];
{seq([x,ListTools[Occurrences](x,primefactors)],x in
primefactors)};
end proc:

starttime:=time():
newprimefactorization(297308799018312231533 4);
time()-starttime;
{[7368787, 1], [11213119751, 1], [1999, 1], [3, 2], [2, 1]}
2.083
newprimefactorization(9199838343019266106321698);
Error, "invalid subscript selector"
ifactor(255678904536136571239);
(22801763489) (11213119751)
```

We will see methods for prime factorization using continued fractions in Chapter 8 and elliptic curves in Chapter 11.

Exercises

◇**7.1** Use an implementation of Algorithm 7.2 (i.e., `newprimelist`) to determine $\pi(10^7)$. What is the relative error of the estimate for this number provided by the Prime Number Theorem? How many primes in this set are of the form $4n+1$? How many are of the form $4n+3$?

7.2 How many steps does `newprimelist[n]` take?

◇7.3 Recall from p. 45 that "twin primes" are pairs p and $p+2$ with both numbers prime. Use `newprimelist` to count the number of twin prime pairs up to 10^7.

7.4 A "smooth number" is a number all of whose prime factors are small. Specifically, a k-smooth number has all prime factors less than or equal to k. Which prime factorization algorithm works better for smooth numbers, `primefactorization` or `newprimefactorization`?

◇7.5 Modify the `primefactorization` procedure so that, beyond 2, only odd trial divisors are tested. Compare the speed of this new version to that of the original by finding the factorization indicated on p. 212.

◇7.6 The "lucky numbers" are defined by a sieving process as follows. Start with the counting numbers:

$$\{1, 2, 3, 4, 5, 6, 7, 8, 9, 10, 11, 12, 13, 14, 15, 16, 17, 18, 19, 20, \ldots\}.$$

Strike out every second number:

$$\{1, \not{2}, 3, \not{4}, 5, \not{6}, 7, \not{8}, 9, \not{10}, 11, \not{12}, 13, \not{14}, 15, \not{16}, 17, \not{18}, 19, \not{20}, \ldots\}.$$

The first number left after 1 is 3; strike out every third number remaining (beginning the count with 1):

$$\{1, \not{2}, 3, \not{4}, \not{5}, \not{6}, 7, \not{8}, 9, \not{10}, \not{11}, \not{12}, 13, \not{14}, 15, \not{16}, \not{17}, \not{18}, 19, \not{20}, \ldots\}.$$

The first number left after 3 is 7; strike out every seventh number remaining (beginning the count with 1):

$$\{1, \not{2}, 3, \not{4}, \not{5}, \not{6}, 7, \not{8}, 9, \not{10}, \not{11}, \not{12}, 13, \not{14}, 15, \not{16}, \not{17}, \not{18}, \not{19}, \not{20}, \ldots\}.$$

The numbers left by this process are the lucky numbers. Create a computer program to produce the set of lucky numbers up to n. Make a conjecture as to the proportion of numbers up to n that are lucky numbers. How does this proportion compare with the limiting proportion in the Prime Number Theorem (see p. 46)?

7.7 Explain why Wilson's theorem (Theorem 3.35) does not provide an effective method for testing whether a number is prime.

◇7.8 Use `newprimefactorization` to create a procedure that emulates *Mathematica*'s `DivisorSigma` command or *Maple*'s `numtheory[sigma]` command (see Section 6.1).

7.2 Fermat numbers

In this section and the next one, we discuss some special numbers that have a celebrated place in number theory. As it happens, some of these numbers are very large primes. In the discussion, we will develop special techniques for dealing with these numbers.

Recall the "Fermat numbers" introduced in Section 3.9.

DEFINITION 7.7 *The* Fermat number F_n *is given by the formula*

$$F_n = 2^{2^n} + 1, \quad n \geq 0.$$

If a Fermat number is a prime, it is called a Fermat prime.

(A little joke: It's the format that makes it a Fermat.)

We have: $F_0 = 3$, $F_1 = 5$, $F_2 = 17$, $F_3 = 257$, $F_4 = 65537$. These numbers are all primes, and in fact they are the only known Fermat primes. Recall from Section 3.9 that a regular n-gon is constructible via straightedge and compass if and only if n is a product of distinct Fermat primes and/or a power of 2. Construction of a regular (equilateral) 3-gon (triangle) is learned in beginning geometry classes. We outlined the steps for constructing a regular 5-gon and a regular 17-gon. Using the same method, one could construct a regular 257-gon and 65537-gon. In fact, Johann Gustav Hermes (1846–1912) carried out the calculations for constructing a regular 65537-gon (taking about twelve years to do so!).

We can use *Mathematica* to verify that the first five Fermat numbers are primes.

Mathematica

```
fermatnumber[n_] := 2^(2^n) + 1
Table[fermatnumber[n], {n, 0, 4}]
{3, 5, 17, 257, 65537}

Table[primetest[fermatnumber[n]], {n, 0, 4}]
{True, True, True, True, True}
```

Here are the corresponding *Maple* computations.

Maple

```
fermatnumber:=n->2^(2^n)+1:
seq(fermatnumber(n),n=0..4);
3, 5, 17, 257, 65537
seq(primetest(fermatnumber(n)),n=0..4);
True, True, True, True, True
```

Fermat conjectured that all Fermat numbers are primes, but this was disproven by Euler, who factored F_5 in 1732. We can verify that F_5 is composite with *Mathematica.*

Mathematica

```
fermatnumber[5]
4294967297
primetest[%]
False
```

Therefore, F_5 is composite. We factor F_5.

Mathematica

```
newprimefactorization[fermatnumber[5]] //Timing
{0.01 Second, {{641,1}, {6700417,1}}}
```

Hence,

$$F_5 = 4294967297 = 641 \cdot 6700417. \tag{7.1}$$

Here are the corresponding *Maple* computations.

Maple

```
fermatnumber(5);
4294967297
primetest(%);
False
starttime:=time():
newprimefactorization(fermatnumber(5));
time()-starttime;
{[641, 1], [6700417, 1]}
0.050
```

How did Euler find this factorization? We will explore this question in a moment. First, we prove an interesting fact about Fermat numbers. Although not all Fermat numbers are primes, they are either primes or pseudoprimes.

PROPOSITION 7.8

The numbers F_n are either pseudoprimes for the base 2 or are prime numbers. That is, $F_n \mid 2^{F_n} - 2$, for all $n \geq 0$.

PROOF Note that $n + 1 \leq 2^n$, for $n \geq 0$, and hence 2^{n+1} divides 2^{2^n}. It follows that

$$\begin{aligned}
2^{F_n} - 2 &= 2^{2^{2^n}+1} - 2 \\
&= 2(2^{2^{2^n}} - 1) \\
&= 2(2^{2^{n+1}k} - 1) \quad \text{(for some integer } k\text{)} \\
&= 2(2^{2^{n+1}} - 1)m \quad \text{(for some integer } m\text{)} \\
&= 2((2^{2^n})^2 - 1)m \\
&= 2(2^{2^n} + 1)(2^{2^n} - 1)m \\
&= 2F_n(2^{2^n} - 1)m.
\end{aligned}$$

(The fourth equality holds because, in general, $x - 1$ divides $x^k - 1$.) ∎

We have shown that the Fermat numbers F_n satisfy the congruence

$$2^{F_n} \equiv 2 \pmod{F_n}. \tag{7.2}$$

Now we investigate the process of factoring a Fermat number. First, we record a theorem of Euler.

THEOREM 7.9 (Euler)

Every divisor of F_n is of the form $2^{n+1}k + 1$.

PROOF Let p be a prime factor of F_n. The order of 2 modulo p is 2^{n+1}, since $2^{2^n} \equiv -1 \pmod{p}$ and $2^{2^{n+1}} \equiv 1 \pmod{p}$. It follows by Fermat's little theorem (Theorem 3.18) and Proposition 3.42 that $2^{n+1} \mid p-1$, and therefore $p = 2^{n+1}k + 1$. The fact that all divisors are of this form is an immediate consequence. ∎

Let's use Theorem 7.9 to factor F_5. According to the theorem, the only possible factors are: $1 \cdot 64 + 1 = 65$, $2 \cdot 64 + 1 = 129$, $3 \cdot 64 + 1 = 193$, $4 \cdot 64 + 1 = 257$, $5 \cdot 64 + 1 = 321$, $6 \cdot 64 + 1 = 385$, $7 \cdot 64 + 1 = 449$, $8 \cdot 64 + 1 = 513$, $9 \cdot 64 + 1 = 577$, $10 \cdot 64 + 1 = 641$, etc. Some of the numbers shown in our list are composite and hence do not have to be tested. The primes are 193, 257, 449, and 641. We do not need to check 257 as it is a Fermat number and distinct Fermat numbers are coprime (Exercise 7.11). It is easy to verify that 641 divides F_5 and so obtain the factorization in (7.1).

Later, Eduard Lucas (1842–1891) improved on Euler's result. We state and prove Lucas' theorem.

n	no. of digits in F_n	no. of digits in prime factors	discoverer
5	10	3, 7	L. Euler, 1732
6	20	6, 14	F. Landry, 1880
7	39	17, 22	M. A. Morrison and J. Brillhart, 1970
8	78	16, 62	R. Brent and J. M. Pollard, 1981
9	155	7, 49, 99	M. S. Manasse, A. K. Lenstra, and H. W. Lenstra 1993
10	309	8, 10, 40, 252	R. Brent, 1995
11	617	6, 6, 21, 22, 564	R. Brent, 1988

TABLE 7.1: Fermat numbers and their prime factors.

THEOREM 7.10 (Lucas)
For $n > 1$, every divisor of F_n is of the form $2^{n+2}k + 1$.

PROOF As in the proof of Theorem 7.9, the order of 2 modulo p is 2^{n+1}. By Fermat's little theorem and Proposition 3.42, $8 \mid p - 1$ (for $n \geq 2$). It follows by Theorem 5.12 that $\left(\frac{2}{p}\right) = 1$. Hence, by Proposition 5.8, we have $2^{(p-1)/2} \equiv 1 \pmod{p}$, and so $2^{n+1} \mid (p-1)/2$. Therefore, $p = 2^{n+2}k + 1$. As in the proof of Euler's theorem, it follows that all divisors are of this form. ▯

The only Fermat numbers F_n, with $n > 4$, that have been completely factored are F_5, F_6, F_7, F_8, F_9, F_{10}, and F_{11}. Table 7.1 shows the current state of knowledge (see [33]).

Let's go on to investigate some of these results. Factoring F_6 is within our reach, even without Lucas' theorem.

Mathematica

```
newprimefactorization[fermatnumber[6]] //Timing
{12.8 Second, {{274177,1}, {67280421210721,1}}}
```

Maple

```
starttime:=time():
newprimefactorization(fermatnumber(6));
time()-starttime;
{[67280421310721, 1], [274177, 1]}
2.173
```

Hence,

$$F_6 = 18446744073709551617 = 274177 \cdot 67280421210721. \tag{7.3}$$

We now use Lucas' theorem to investigate the Fermat number F_{10} (a 309-digit number). The theorem tells us that the factors of this number are of the form $2^{12}k + 1 = 4096k + 1$. We write a simple *Mathematica* procedure (a modified version of `primefactorization`) to check such numbers in turn as possible divisors. When a divisor is found, it is reported and the procedure terminates.

Note: Of course, searches based on Lucas' theorem are guaranteed to stop because F_n has *some* factor. But can we expect the computations to come to a conclusion in a reasonable amount of time?

Mathematica

```
(f = fermatnumber[10];
trialdivisor = 4097;
While[Mod[f, trialdivisor] != 0,
trialdivisor = trialdivisor + 4096];
trialdivisor) // Timing

{0.121 Second, 45592577}
```

The 8-digit prime factor 45592577 is found instantly. Let's proceed to find the next factor.

Mathematica

```
(fprime = f / trialdivisor;
While[Mod[fprime, trialdivisor] != 0,
trialdivisor = trialdivisor + 4096];
trialdivisor) // Timing

{21.07 Second, 6487031809}
```

Hence, the 10-digit prime 6487031809 is also a factor (and this is found in about 21 seconds).

Here are the corresponding *Maple* computations.

Maple

```
starttime:=time():
f:=fermatnumber(10):
trialdivisor:=4097:
while modp(f,trialdivisor) <> 0 do
  trialdivisor:=trialdivisor+4096:
end do:
trialdivisor;
time() - starttime;
45592577
0.050
fprime:=f/trialdivisor:
while modp(fprime,trialdivisor) <> 0 do
  trialdivisor:=trialdivisor+4096:
end do:
trialdivisor;
time() - starttime;
6487031809
10.165
```

Despite these good results, let's not think that things are always this simple. Consider the problem of finding the 17-digit and 22-digit prime factors of F_7 (or even the other prime factors of F_{10}). A straightforward program, again based on Lucas' theorem, could do the search. For example, in *Mathematica* we would use the following.

```
(f = fermatnumber[7];
trialdivisor = 513;
While[Mod[f, trialdivisor] != 0,
    trialdivisor = trialdivisor + 512];
trialdivisor)
```

But the search would take years on a personal computer. Better methods and/or faster computers are needed. Michael A. Morrison and John Brillhart factored F_7 using continued fractions, and Richard Brent factored F_{10} and F_{11} using the elliptic curve method. See Chapters 8 and 11 for discussions of these methods.

Pepin's test (invented by P. Pepin) gives a method for determining whether F_n is prime.

THEOREM 7.11 (Pepin's test)
The Fermat number F_n, where $n > 0$, is a prime if and only if F_n divides $3^{(F_n-1)/2} + 1$.

We defer the proof of Pepin's test until after an illustration of its use.

Pepin's test is fairly easy to apply. Here is the main idea. We must test the congruence

$$3^{(F_n-1)/2} \equiv -1 \pmod{F_n}.$$

Although the numbers involved are potentially huge, we can devise a simple algorithm to do the check, as the quantity $3^{(F_n-1)/2}$ may be arrived at by repeated squaring. We have

$$3^{(F_n-1)/2} = 3^{2^{2^n-1}},$$

and the power of 3 on the right may be computed by successive uses of the identity

$$3^{2^k} = \left(3^{2^{k-1}}\right)^2.$$

Let's consider, for example, how to use Pepin's test to show that F_7 is composite. We have

$$F_7 = 2^{128} + 1.$$

Hence, $(F_7 - 1)/2 = 2^{127}$ and the congruence to test is

$$3^{2^{127}} \equiv -1 \pmod{\left(2^{128} + 1\right)}.$$

Thus, we start with 3 and successively take squares modulo $2^{128} + 1$ until we arrive at the expression on the left (this takes 127 steps).

Here is a *Mathematica* procedure that performs Pepin's test on F_n.

Mathematica

```
pepinstest[n_] := Module[{m, a},
  m = fermatnumber[n];
  a = 3;
  Do[a = Mod[a^2, m], {2^n - 1}];
  If[Mod[a, m] == m-1, Return["prime"], Return["composite"]]
]
```

We use our procedure to test F_7.

Mathematica

```
pepinstest[7] // Timing
{0.01 Second, "composite"}
```

Here are the *Maple* computations.

Maple

```
pepintest:= proc(n)
local m, a;
m:=fermatnumber(n);
a:=3;
to 2^n-1 do
  a:=modp(a^2,m);
end do;
if modp(a,m) = m-1 then
return("prime")
else
return("composite");
end if;
end proc:
starttime:=time();
pepintest(7);
time()-starttime;
"composite"
0.70
```

Thus, we have shown that F_7 is composite.

Pepin's test works well to show that Fermat numbers are composite, but it does not identify the factors of these numbers.

PROOF (of Theorem 7.11)

Assume that F_n is prime. By Theorem 5.13, since $F_n \equiv 1 \pmod 4$ and $F_n \equiv 2 \pmod 3$, we have

$$\left(\frac{3}{F_n}\right) = \left(\frac{F_n}{3}\right) = \left(\frac{2}{3}\right) = -1$$

(i.e., 3 is a quadratic nonresidue modulo F_n). Hence, by Euler's criterion (Theorem 5.5),

$$3^{(F_n-1)/2} \equiv -1 \pmod{F_n}.$$

Now assume that the congruence

$$3^{(F_n-1)/2} \equiv -1 \pmod{F_n} \tag{7.4}$$

holds. Squaring both sides, we obtain

$$3^{F_n-1} \equiv 1 \pmod{F_n}.$$

Letting p be some prime that divides F_n, we have

$$3^{F_n-1} \equiv 1 \pmod{p}.$$

It follows that the order of 3 modulo p is a power of 2. In fact, the order must be $2^{2^n} = F_n - 1$ (or else repeated squaring modulo p would yield a contradiction to (7.4)). Since, by Fermat's theorem, this number is at most $p - 1$, we have $F_n \leq p$, and therefore $F_n = p$, i.e., F_n is a prime. □

Note: Pepin's test also follows from the forthcoming Lucas' theorem (Theorem 7.14). A proof of Pepin's test was called for in Exercise 5.22.

One of the consequences of our analysis of Fermat numbers is the inescapable observation that most Fermat numbers seem to be composite. Indeed, there is a conjecture that all Fermat numbers but the first five are composite. So our search for large primes has not netted any new exhibits (except, perhaps, the factors of some of the Fermat numbers). However, we will see in the next section that Mersenne numbers offer a richer opportunity for finding large primes.

Exercises

7.9 What form does a Fermat number have in base 2?

7.10 Prove the following facts about Fermat numbers:

(a) No Fermat number ends in a 1 (in base 10).

(b) No Fermat numbers except 3 and 5 are divisible by 3 or 5.

7.11 (a) Prove that the Fermat numbers F_n satisfy the recurrence relation

$$F_n = F_0 F_1 \ldots F_{n-1} + 2, \quad n \geq 1.$$

(b) Prove that every two different Fermat numbers are relatively prime.

(c) Show that the result (b) implies that there are infinitely many prime numbers (thus giving another proof of Theorem 2.36).

7.12 Give an alternate proof of Proposition 7.8 by raising the congruence

$$2^{2^n} \equiv -1 \pmod{F_n}$$

to the power $2^{2^n - n}$ and multiplying by 2.

◇**7.13** Use Pepin's test (Theorem 7.11) to show that the Fermat number F_8 is composite.

◇**7.14** (a) Use Lucas' theorem (Theorem 7.10) and a computer to find prime factors of the Fermat numbers F_9 and F_{11}.

(b) Similarly, find factors of F_{12} and F_{18} (thereby showing that these numbers are composite).

7.3 Mersenne numbers

Recall from Chapter 6 that if $2^n - 1$ is a prime, then $2^{n-1}(2^n - 1)$ is a perfect number. Numbers of the form $2^n - 1$ are called "Mersenne numbers." Thus, Mersenne numbers that are primes yield perfect numbers.

DEFINITION 7.12 *The* Mersenne number M_n *is given by the formula*

$$M_n = 2^n - 1, \quad n \geq 1.$$

If a Mersenne number is a prime it is called a Mersenne prime.

The first few Mersenne numbers are

$$1, 3, 7, 15, 31, 63, 127, 255, 511, 1023, 2047, 4095.$$

We have already shown (Proposition 6.7) that a necessary condition for a Mersenne number M_n to be prime is that n is prime. We therefore focus our attention on Mersenne numbers $M_p = 2^p - 1$, where p is prime.

Marin Mersenne (1588–1648) studied these numbers and made an assertion about which ones are prime. He claimed that M_p is prime for

$$p = 2, 3, 5, 7, 13, 17, 19, 31, 67, 127, 257$$

and composite for all other primes $p < 257$. We check Mersenne's assertion using *Mathematica*.

Mathematica

```
mersennenumber[n_] := 2^n-1
Select[Range[257], (PrimeQ[#] && PrimeQ[mersennenumber[#]])&]
{2, 3, 5, 7, 13, 17, 19, 31, 61, 89, 107, 127}
```

Here are the *Maple* computations.

Maple

```
mersennenumber:= n->2^n-1:
select(n->isprime(mersennenumber(n)),
select(isprime,[seq(i,i=1..257)]));
[2, 3, 5, 7, 13, 17, 19, 31, 61, 89, 107, 127]
```

We see that Mersenne made some errors of commission (M_{67} and M_{257} are not primes) and some errors of omission (M_{61}, M_{89}, and M_{107} are primes). Even so, Mersenne's claim is remarkable for the number of correct determinations made.

Mersenne primes are examples of very large prime numbers (for example, M_{127} has 39 digits). As such, they have been eagerly sought by mathematicians hoping to set "prime records." Table 7.2 shows the known Mersenne primes as of 2007 (see [33]).

How do we determine some of the entries in Table 7.2? Let's start by ruling out some numbers as Mersenne primes. A straightforward way to do this is by a pseudoprime test. However, we cannot use a base 2 test, because, like Fermat numbers, Mersenne numbers are either primes or base 2 pseudoprimes.

PROPOSITION 7.13

The numbers M_p (p prime) are either pseudoprimes for the base 2 or are prime numbers. That is, $M_p \mid 2^{M_p} - 2$, for all primes p.

The proof is called for in the exercises.

The proposition shows that Mersenne numbers M_p (p prime) satisfy the congruence

$$2^{M_p} \equiv 2 \pmod{M_p}. \tag{7.5}$$

Although we cannot rule out Mersenne numbers as primes via a base 2 pseudoprime test, we can use a base 3 pseudoprime test. If k is a prime greater than 3, then by Fermat's little theorem,

$$3^{k-1} \equiv 1 \pmod{k}.$$

If this congruence does not hold, then k is not prime. We can apply this test to the Mersenne numbers M_p. We must test

$$3^{2^p-2} \equiv 1 \pmod{2^p - 1}.$$

We implement the test in *Mathematica*.

Mathematica

```
pseudoprimetest[n_] := If[PowerMod[3, n - 1, n] == 1,
Return["test fails"], Return["composite"]]
```

Let's check that M_{23} is composite.

Mathematica

```
pseudoprimetest[mersennenumber[23]]
"composite"
```

Similarly, we can disprove Mersenne's claim that M_{67} is prime.

number	p	year	discoverer
1	2	antiquity	
2	3	antiquity	
3	5	antiquity	
4	7	antiquity	
5	13	1461	?
6	17	1588	P. Cataldi
7	19	1588	P. Cataldi
8	31	1750	L. Euler
9	61	1883	I. M. Pervouchine
10	89	1911	R. E. Powers
11	107	1913	R. E. Powers
12	127	1876	E. Lucas
13	521	1952	D. H. Lehmer, R. M. Robinson
14	607	1952	D. H. Lehmer, R. M. Robinson
15	1279	1952	D. H. Lehmer, R. M. Robinson
16	2203	1952	D. H. Lehmer, R. M. Robinson
17	2281	1952	D. H. Lehmer, R. M. Robinson
18	3217	1957	H. Riesel
19	4253	1961	A. Hurwitz
20	4423	1961	A. Hurwitz
21	9689	1963	D. Gillies
22	9941	1963	D. Gillies
23	11213	1963	D. Gillies
24	19937	1971	B. Tuckerman
25	21701	1978	L. Noll, L. Nickel
26	23209	1979	L. Noll
27	44497	1979	H. Nelson, D. Slowinski
28	86243	1982	D. Slowinski
29	110503	1988	W. Colquitt, L. Welsh
30	132049	1983	D. Slowinski
31	216091	1985	D. Slowinski
32	756839	1992	P. Gage, D. Slowinski
33	859433	1994	P. Gage, D. Slowinski
34	1257787	1996	D. Slowinski, P. Gage
35	1398269	1996	J. Armengaud, GIMPS*
36	2976221	1997	G. Spence, GIMPS*
37	3021377	1998	R. Clarkson, GIMPS*
38	6972593	1999	N. Hajratwala, GIMPS*
39	13466917	2001	M. Cameron, GIMPS*
40	20996011	2003	M. Shafer, GIMPS*
41	24036583	2004	J. Findley, GIMPS*
42	25964951	2005	M. Nowak, GIMPS*
43	30402457	2006	C. Cooper, S. Boone, GIMPS*
44	32582657	2006	C. Cooper, S. Boone, GIMPS*

TABLE 7.2: Exponents p of known Mersenne primes $2^p - 1$.

* GIMPS = Great Internet Mersenne Prime Search

Mathematica

```
pseudoprimetest[mersennenumber[67]]
"composite"
```

Of course, the test will fail whenever M_p is prime. It has been conjectured that the test will not fail for composite values.

Here are the *Maple* computations.

Maple

```
pseudoprimetest:=proc(n)
if modp(3&^(n-1),n)=1
then return("test fails")
else return("composite")
end if;
end proc:

pseudoprimetest(mersennenumber(23));
"composite"
pseudoprimetest(mersennenumber(67));
"composite"
```

We could expand our test by checking the congruence $a^{n-1} \equiv 1 \pmod{n}$ for several primes a. A number n that satisfies many such tests is with high probability a prime. However, absolute pseudoprimes (see p. 78) satisfy all of these tests (with $\gcd(a, n) = 1$).

The Lucas–Lehmer test (discovered by Lucas and D. H. Lehmer (1905–1991)) is a powerful if and only if criterion for primality.

THEOREM (Lucas–Lehmer test)

Let p be an odd prime. Define

$$r_1 = 4 \qquad \text{and} \quad r_k = r_{k-1}^2 - 2, \quad \text{for} \quad 2 \leq k \leq p-1.$$

Then M_p is prime if and only if M_p divides r_{p-1}.

The proof (see, for example, [26]) is a *tour-de-force* that showcases Lucas sequences, Euler's criterion, and other central topics of number theory.

Mathematica

```
lucaslehmertest[p_] := Module[{m, r},
  m = mersennenumber[p];
  r = 4;
  Do[r = Mod[r^2 - 2, m], {p - 2}];
  If [r == 0, Return["prime"], Return["composite"]]
]
```

Maple

```
lucaslehmertest:=proc(p)
local m, r;
m:=mersennenumber(p);
r:=4;
to p-2 do
r:=modp(r^2-2,m)
end do;
if r=0
then return("prime")
else return("composite")
end if;
end proc:
```

Let's show that M_{521} is a prime (see Table 7.2).

Mathematica

```
lucaslehmertest[521]
"prime"
```

Maple

```
lucaslehmertest(521);
"prime"
```

We can check the entries of Table 7.2 over a fairly large range, with the calculations taking a couple of minutes.

Mathematica

```
Select[Select[Range[5000],PrimeQ],
(lucaslehmertest[#] == "prime") &]
{3, 5, 7, 13, 17, 19, 31, 61, 89, 107,
127, 521, 607, 1279, 2203, 2281, 3217, 4253, 4423}
```

Maple

```
select(n->evalb(lucaslehmertest(n)="prime"),
select(isprime,[seq(i,i=1..5000)]));
[3, 5, 7, 13, 17, 19, 31, 61, 89, 107, 127, 521, 607, 1279,
2203, 2281, 3217, 4253, 4423]
```

We have tested Mersenne numbers for primality but we have not commented upon the problem of factoring composite Mersenne numbers. To convey an idea of the difficulty of factoring Mersenne numbers without modern computers, we mention that Frank N. Cole (1861–1926) worked over a twenty-year period to find the factors of M_{67}, completing the search in 1903:

$$M_{67} = 193707721 \cdot 761838257287. \tag{7.6}$$

Our `primefactorization` procedure takes less than an hour to factor M_{67}. Of course, the Mersenne number M_{67} is only "moderately large."

Exercises

7.15 What form does a Mersenne number have in base 2?

7.16 Prove that 3 is the only number that is both a Fermat number and a Mersenne number.

7.17 Prove that if $a^b - 1$ is a prime and $b > 1$, then $a = 2$ and b is a prime.

7.18 Prove Proposition 7.13.

7.19 Suppose that p and q are odd primes, and q is a divisor of the Mersenne number M_p. Prove that $q \equiv 1 \pmod{p}$ and $q \equiv \pm 1 \pmod{8}$.

◇**7.20** Use the Lucas–Lehmer test to show that M_{5003} is composite.

◇**7.21** List the base 2 pseudoprimes less than 5000. Find the smallest base 2 pseudoprime divisible by a perfect square.

7.4 Prime certificates

Suppose that we believe that a certain large number is a prime. How can we convince other people that it is? In light of Proposition 7.1, we could prove that p is a prime by showing that p has no divisor k, with $2 \leq k \leq \sqrt{p}$. But this entails an unreasonable number of checks to make. A "prime certificate" is a set of information that yields a short proof that a specific number is prime. Simple "composite certificates" exist for composite numbers: to prove that n is composite, just furnish a proper divisor of n.

A useful prime certification procedure is based on a theorem of Lucas.

THEOREM 7.14 (Lucas)

A number n is prime if and only if there exists an integer a such that $a^{n-1} \equiv 1 \pmod{n}$ and $a^{(n-1)/q} \not\equiv 1 \pmod{n}$, for every prime divisor q of $n-1$.

Note: Lucas' theorem (with $a = 3$) applied to Fermat numbers (Definition 7.7) yields what is essentially Pepin's test (Theorem 7.11).

PROOF If n is a prime, then there exists a primitive root a modulo n (Theorem 3.53). It is clear that a will have the properties called for in the theorem.

Now suppose that there exists a number a as required in the theorem. We will show that the order of a modulo n is $n-1$. This means that $\phi(n) = n-1$. And since $\phi(n) = n-1$ if and only if n is prime, it follows that n is prime.

Let k be the order of a modulo n. Then $k \mid n-1$, so that $n-1 = kl$, for some integer l. If $l > 1$, then l is divisible by some prime q, i.e., $l = l'q$. Now $n-1 = kl'q$, and

$$1 \not\equiv a^{(n-1)/q} \equiv a^{kl'} \equiv \left(a^k\right)^{l'} \equiv 1 \pmod{n},$$

a contraction. Hence, $l = 1$ and $k = n-1$. ∎

Here are computer procedures to search for an a value (as well as the prime factors of $n-1$) called for in Theorem 7.14.

Mathematica

```
primecertificate[n_] := Module[{primedivisors, exponents, a},
  primedivisors = Part[FactorInteger[n - 1], All, 1];
  exponents = (n - 1)/primedivisors;
  a = 2;
  While[MemberQ[PowerMod[a, exponents, n], 1], a++];
  {n, a, primedivisors}
]
```

Maple

```
primecertificate:=proc(n)
local primedivisors, exponents, a;
primedivisors:={seq(ifactors(n-1)[2][i][1],
i=1..nops(ifactors(n-1)[2]))};
exponents:=map(x->(n-1)/x,primedivisors);
a:=2;
while member(1,map(x->modp(a&^x,n),exponents)) do
  a:=a+1
end do;
return[n,a,primedivisors];
end proc:
```

Example 7.15 Let's work with the 10,000th prime.

Mathematica

```
n = Prime[10^4]
104729
primecertificate[n]
{104729, 12, {2, 13, 19, 53}}
```

Maple

```
n := ithprime(10^4);
n := 104729
primecertificate(n);
[104729, 12, {2, 13, 19, 53}]
```

Thus, a prime certificate for 104729 is the value $a = 12$ and the set of prime divisors of 104728: $\{2, 13, 19, 53\}$. To convince someone that 104729 is prime, we can say, "Use 12 in Lucas' theorem, where the appropriate prime divisors are 2, 13, 19, and 53." □

There are a couple of technicalities related to our prime certificate procedure to consider. First, we must be able to show that the list of prime divisors of $n-1$ is correct. This can be verified by repeatedly dividing $n-1$ by the primes in the list (checking for multiplicities). Second, the listed primes might be so large that it isn't obvious that they are primes. (In the example above, all the prime divisors are less than 100, so it is clear that they really are primes.) Large prime divisors should themselves be given certificates. In practice, let's assume the primality of primes less than 10^7 (the bound on `ourlist`).

We modify our code to recursively provide prime certificates for those prime divisors of $n-1$ that are larger than 10^7.

Note: As a recursive procedure, `newprimecertificate` calls itself with n replaced by the set of all prime divisors of $n-1$. This means that the operation of the program "branches" at each step of the recursion. This branching can be seen in the output of our next example.

Mathematica

```
newprimecertificate[n_] := n /; n < 10^7

newprimecertificate[n_] := Module[{primedivisors, exponents,
a},
  primedivisors = Part[FactorInteger[n - 1], All, 1];
  exponents = (n - 1)/primedivisors;
  a = 2;
  While[MemberQ[PowerMod[a, exponents, n], 1], a++];
  {n, a, Map[newprimecertificate, primedivisors]}
] /; n > 10^7
```

Maple

```
newprimecertificate:=proc(n)
local primedivisors, exponents, a;
if n<10^7
then return(n)
else
primedivisors:={seq(ifactors(n-1)[2][i][1],
i=1..nops(ifactors(n-1)[2]))};
exponents:=map(x->(n-1)/x,primedivisors);
a:=2;
while member(1,map(x->modp(a&^x,n),exponents)) do
  a:=a+1
end do;
return([n,a,map(newprimecertificate,primedivisors)])
end if;
end proc:
```

Conciseness of certificates becomes more evident as we take larger prime numbers. We now show an example with a very large prime number.

Example 7.16 Let's generate a certificate for the Mersenne number M_{521} (a prime with 157 digits).

Using the *Mathematica* command

```
newprimecertificate[mersennenumber[521]]
```

or the *Maple* command

```
newprimecertificate(mersennenumber(521));
```

we obtain the certificate

```
{68647976601306097149819007990813932172694353001433054093944
6345918554318339765605212255964066145455497729631139148085803
712198799971664381257402829111505715l, 3, {2, 3, 5, 11, 17,
31, 41, 53, 131, 157, 521, 1613, 2731, 8191, 42641, 51481, 61681,
409891, 858001, 5746001, 7623851, {34110701, 15, {2, 5, 13, 19,
1381}}, {308761441, 17, {2, 3, 5, 13, 49481}}, {2400573761, 3,
{2, 5, 13, 347, 1663}}, {65427463921, 17, {2, 3, 5, 7, 13,
2995763}}, {108140989558681, 17, {2, 3, 5, 13, 683, 1433, 23609}},
{145295143558111, 7, {2, 3, 5, 7, 13, {2534364967, 3, {2, 3, 7,
8620289}}}}, {173308343918874810521923841, 3, {2, 5, 13, 28793,
{3617255895172730l7, 17, {2, 3, 7, 3191, {674750394557, 2, {2, 7,
{24098228377, 7, {2, 3, 109, 9211861}}}}}}}}}}.
```

The above prime certificate for M_{521} consists of the value $a = 3$ and the prime divisors of $M_{521} - 1$, together with prime certificates for the prime divisors greater than 10^7. Given this certificate, one can quickly check (using Lucas' theorem), that M_{521} is indeed a prime.

□

More intricate theorems yield more advanced prime certification procedures. For example, in the following generalization of Lucas' theorem, a different value of a may be used for each prime divisor of $n - 1$. See [27] for a proof.

THEOREM (Lucas)

A number n is prime if and only if, for each prime divisor q of $n - 1$, there exists a number a, with $2 \le a \le n - 1$, such that $a^{n-1} \equiv 1 \pmod{n}$ and $a^{(n-1)/q} \not\equiv 1 \pmod{n}$.

Exercises

◇**7.22** Use `primecertificate` to find prime certificates for 1999, 7368787, and 2038074743.

◇**7.23** Use `newprimecertificate` to find a prime certificate for 313062201143.

◇**7.24** (a) Use Theorem 7.14 to show that the 626-digit number $278^{2^8} + 1$ is prime.

(b) Use Theorem 7.14 to show that the 2986-digit number $824^{2^{10}} + 1$ is prime.

7.5 Finding large primes

In Chapter 4, we examined several techniques that require prime numbers with one hundred or more digits to guarantee security. How can we find such large primes in a reasonable amount of time? One possibility is to pick a number (at random if we like) with the desired number of digits. We could check to see whether our number is prime or not by using a procedure such as the one employed in our `primetest` command. If it is not prime, we increase our number by 1 and test this new number, and so on.

Mathematica

```
nextprime[n_]:=Module[{p},
  p=n;
  While[primetest[p]==False,p++];
p]
```

Maple

```
ournextprime:=proc(n)
  local p;
  p:=n;
  while primetest(p)=False
    do
    p:=p+1
    end do:
  return(p);
end proc:
```

For example, our procedure will find that the next prime after 24 is 29.

One practical problem with this procedure is that it may spend a large amount of time analyzing composite numbers before it arrives at a prime number. For instance, if we begin with a number that is a product of two large prime numbers, it will take a long time to complete the `primetest` command for that number. One way to modify this procedure is to test whether the conclusion of Fermat's theorem holds for each given number. That is, check whether

$$a^{n-1} \equiv 1 \pmod{n}$$

for our current value of n and some $a > 1$. We know that for any choice of a, pseudoprimes will exist that "pass" this test, but such pseudoprimes are few and far between. We can use our previously defined `pseudoprimetest`, but to really cut down on the composite numbers that pass the test, we will check three different bases.

Mathematica

```
newpseudoprimetest[n_] :=
  If[PowerMod[2, n - 1, n] == 1 &&
  PowerMod[3, n - 1, n] == 1 &&
  PowerMod[5, n - 1, n] == 1,
  Return["test fails"], Return["composite"]
]
```

We now use this command in place of **primetest** in our prime finding procedure.

Mathematica

```
nextprimecandidate[n_]:=Module[{p},
  p=n;
  While[newpseudoprimetest[p]=="composite",p++];
  p
]

nextprimecandidate[10^100]
10000000000000000000000000000000000000000000000000
000000000000000000000000000000000000000000000000267
```

Here are the corresponding *Maple* commands.

Maple

```
newpseudoprimetest:=proc(n)
  if modp(2&^(n-1),n)=1 and
  modp(3&^(n-1),n)=1 and
  modp(5&^(n-1),n)=1
  then
  return("test fails")
  else
  return("composite")
  end if;
end proc:

nextprimecandidate:=proc(n)
  local p;
  p:=n;
  while newpseudoprimetest(p)="composite" do
    p:=p+1
  end do;
  return(p);
end proc:

nextprimecandidate(10^100);
10000000000000000000000000000000000000000000000000
000000000000000000000000000000000000000000000000267
```

Let's compare the speed of this method with our earlier one.

Mathematica

```
(nextprime[10^12])//Timing

{18.216 Second,1000000000039}

(nextprimecandidate[10^12])//Timing

{0.02 Second,1000000000039}
```

Maple

```
starttime:=time():
ournextprime(10^12);
time()-starttime;
1000000000039
10.84
starttime:=time();
nextprimecandidate(10^12);
time()-starttime();
1000000000039
0.3
```

We emphasize that this new method may return a number that is not prime. Composite numbers that are pseudoprimes to the bases 2, 3, and 5 will cause the procedure to end and will be returned as the result. While there are infinitely many such numbers, the probability of one turning up in our prime search is low. In fact, there are only 11 such numbers less than 100000. If we wish to be sure our number is prime, we can employ the slower `primetest` on the result of the search.

Another approach to the problem of finding a large prime number is to build up to the number in such a way that it is easy to check whether the result is prime or not. Recall that Lucas' theorem (Theorem 7.14) tells us that a number n is prime if and only if there exists an integer a such that $a^{n-1} \equiv 1 \pmod{n}$ and $a^{(n-1)/q} \not\equiv 1 \pmod{n}$, for every prime divisor q of $n-1$. In order to use this theorem to check whether n is prime, we must be in possession of the prime factorization of $n-1$. Suppose that we have a prime p. We choose a number k at random and set $n = kp+1$. If k is small enough for us to quickly factor, then we can easily obtain the prime factorization of $n-1$. At this point we can begin to look for a value of a that satisfies the conditions of Lucas' theorem. If we find cannot find one, we choose a new value of k and try again. If we do find one, then we know n is prime. Thus, starting with a prime p we obtain a larger prime n. We will have increased the number of digits in our prime by roughly the number of digits of k. Employing this procedure repeatedly, we can quickly obtain primes with several hundred digits. We summarize the algorithm below.

ALGORITHM 7.17 (Finding a large prime using Lucas' theorem)
Given d, a prime number with at least d digits is returned.

(1) Let d be a positive integer. Choose a small prime number p.

(2) If $p > 10^d$, then return p.

(3) Choose a small random number k, let $p' = kp + 1$, and let $a = 2$.

(4) If $a^{p'-1} \not\equiv 1 \pmod{p'}$, then go to step (3).

(5) If $a^{(p'-1)/q} \equiv 1 \pmod{p'}$ for any prime q dividing $p' - 1$, then increase a by 1 and go to step (4).

(6) Set $p = p'$ and go to (2).

Note: By Dirichlet's theorem (see Chapter 10), there are infinitely many primes of the form $kp + 1$, so with probability 1 we will find such a prime eventually (as long as the bound on k is sufficiently large).

To implement this algorithm, we first define a function that, given a number n and prime divisor p of $n-1$, will search for an a that satisfies the conditions of Lucas' theorem. If such an a is found, the procedure reports that n is prime. Otherwise, it reports that a is composite.

Mathematica

```
primecertificatetest[n_, p_] := Module[{primedivisors,
exponents, a},
  primedivisors = Union[Part[FactorInteger[(n - 1)/p], All,
1], {p}];
  exponents = (n - 1)/primedivisors;
  a = 2;
  While[PowerMod[a, n - 1, n] == 1,
  If[MemberQ[PowerMod[a, exponents, n], 1], a++,
Return["prime"]]
  ];
  Return["composite"]
]
```

Maple

```
primecertificatetest:=proc(n,p)
  local primedivisors, exponents, a;
  primedivisors:={seq(ifactors((n-1)/p)[2][i][1],
  i=1..nops(ifactors((n-1)/p)[2]))} union {p};
  exponents:=map(x->(n-1)/x,primedivisors);
  a:=2;
  while modp(a&^(n-1),n)=1 do
  if member(1,map(x->modp(a&^x,n),exponents))
  then a:=a+1
  else return("prime")
  end if;
  end do;
  return("composite");
end proc:
```

We now implement Algorithm 7.17.

Mathematica

```
largeprime[d_] :=
  Module[{p,pprime},
  p = Prime[Random[Integer, {1, 10^4}]];
  While[p < 10^d,
  pprime = Random[Integer, {10^6, 10^7}]*p + 1;
  While[primecertificatetest[pprime, p] == "composite",
  pprime = Random[Integer, {10^6, 10^7}]*p + 1];
  p=pprime];
  p
]
```

Maple

```
largeprime:=proc(d)
  local p, pprime, i;
  i:=1;
  p:=ithprime(RandomTools[Generate]
  (integer(range=1..10^4)));
  while p<10^d do
  pprime:=RandomTools[Generate]
  (integer(range=10^6..10^7))*p+1;
  while primecertificatetest(pprime,p)="composite" do
  pprime:=RandomTools[Generate]
  (integer(range=10^6..10^7))*p+1;
  end do;
  p:=pprime;
  end do;
  return(p);
end proc:
```

Our first prime is the jth prime number, where j is a randomly chosen number between 1 and 10000. Thus, this first prime is at most 104729. Also, each time through the `While` loop, our randomly chosen value for k is a number between 10^6 and 10^7. Thus, k (and hence $p' - 1$) is easily factored, and p' has roughly 7 more digits than p. We now use the algorithm to find a prime with at least 100 digits.

Mathematica

```
largeprime[100]//Timing
{3.816 Second,
577747426606634775799359289659322132026179345211854246530383б
200274484458986042980901988517259041591974989}
```

Maple

```
starttime:=time();
largeprime(100);
8592230864084622643898930254827963465937151478363342490б
78265625820018587222977982409457646619543994116З
time()-starttime;
31.48
```

Exercises

◇**7.25** Write a computer program to list the composite numbers less than 100,000 for which the `newpseudoprimetest` procedure fails (i.e., numbers that are pseudoprimes to the bases 2, 3, and 5).

◇**7.26** Use Algorithm 7.17 (or `largeprime`) to find a prime number with at least 200 digits.

◇**7.27** Modify the `largeprime` procedure to output a certificate of the prime produced.

◇**7.28** Write a computer program, similar to `largeprime`, based on the generalized Lucas theorem (Section 7.4), to find large prime numbers.

7.6 Notes

Eratosthenes

Eratosthenes (276–194 B.C.E) was a scientist and mathematician who made significant contributions in diverse fields. Perhaps his most celebrated achievement was finding a way to measure the Earth's circumference using elementary geometry (his result was fairly accurate). Also in geometry/astronomy, he estimated the distances between Earth and Sun and between Earth and Moon. In geography, he mapped the route of the Nile, hypothesized that lakes are the source of this river, and gave an explanation for its flooding behavior. In mechanics, he developed a device for "duplicating the cube." In the humanities, he wrote about the theater and ethics. Befitting his "renaissance" abilities and erudition, Eratosthenes served as librarian of the famous Library at Alexandra.

Chapter 8

Continued Fractions

"Say," he grunted, "you've worked on this, I'm sure. Did you try continued fractions?

[The devil speaking to a mathematician about Fermat's Last Theorem, in "The Devil and Simon Flagg."]

Arthur Porges (1915–2006)

On many college and high school mathematics competitions, one frequently encounters problems of the following sort:

Determine the real number represented by the expression below.

$$1 + \cfrac{1}{1 + \cfrac{1}{1 + \cfrac{1}{1 + \cfrac{1}{1 + \ddots}}}} \tag{8.1}$$

The "trick" is to observe that if we take the reciprocal of this expression and add 1, then we get the same expression back. That is, if we denote the expression by x, then

$$x = 1 + \frac{1}{x}. \tag{8.2}$$

Now we have an equation that we can easily solve. Multiplying equation (8.2) through by x, and rearranging, we get

$$x^2 - x - 1 = 0.$$

Using the quadratic formula, we find the roots to be $(1 \pm \sqrt{5})/2$. Since $x > 0$, we must have

$$x = \frac{1 + \sqrt{5}}{2}.$$

The expression given in (8.1) is an example of a *continued fraction*. These objects provide us with an alternative to decimal expansion for approximating irrational numbers and have remarkable number-theoretic applications.

8.1 Finite continued fractions

We begin with a definition.

DEFINITION 8.1 *A* finite continued fraction *is an expression of the form*

$$a_0 + \cfrac{1}{a_1 + \cfrac{1}{a_2 + \ddots + \cfrac{1}{a_{n-1} + \cfrac{1}{a_n}}}} \tag{8.3}$$

where $a_0, a_1, \ldots, a_n$ *are real numbers and* $a_1, \ldots, a_n$ *are positive. The* a_i *are the* partial quotients *of the continued fraction. If the partial quotients are all integers, the continued fraction is* simple. *We use the notation*

$$[a_0; a_1, a_2, \ldots, a_n]$$

to represent the continued fraction given in (8.3). When $n = 0$ *we write* $[a_0;]$.

Example 8.2 Let's write 125/54 as a finite simple continued fraction. We first observe that

$$125 = 54 \cdot 2 + 17.$$

Dividing the equation by 54, we have

$$\frac{125}{54} = 2 + \frac{17}{54}.$$

We can rewrite the right side as

$$2 + \frac{1}{\frac{54}{17}}.$$

We next apply the division algorithm to 54 and 17 and repeat the process. At this point one might notice that these first two division algorithm computations are the first two steps in the Euclidean algorithm applied to 125 and 54. Before proceeding, let's complete this Euclidean algorithm computation.

$$\begin{aligned}
125 &= 54 \cdot 2 + 17 \\
54 &= 17 \cdot 3 + 3 \\
17 &= 3 \cdot 5 + 2 \\
3 &= 2 \cdot 1 + 1 \\
2 &= 1 \cdot 2
\end{aligned}$$

We now rewrite each row in the algorithm as a division equality.

$$\frac{125}{54} = 2 + \frac{17}{54} \tag{8.4}$$

$$\frac{54}{17} = 3 + \frac{3}{17} \tag{8.5}$$

$$\frac{17}{3} = 5 + \frac{2}{3} \tag{8.6}$$

$$\frac{3}{2} = 1 + \frac{1}{2} \tag{8.7}$$

We can rewrite (8.4) as

$$\frac{125}{54} = 2 + \frac{1}{\frac{54}{17}}.$$

We now use (8.5) to rewrite the right side above as

$$2 + \frac{1}{3 + \frac{3}{17}}.$$

With equation (8.6) this becomes

$$2 + \frac{1}{3 + \frac{1}{5 + \frac{2}{3}}}.$$

Finally, making use of (8.7), we obtain

$$\frac{125}{54} = 2 + \frac{1}{3 + \frac{1}{5 + \frac{1}{1 + \frac{1}{2}}}},$$

and so $125/54 = [2; 3, 5, 1, 2]$.

□

The procedure employed in Example 8.2 can be applied to any rational number to obtain a finite simple continued fraction. Not surprisingly, it is also the case that any finite simple continued fraction represents a rational number.

THEOREM 8.3

Any finite simple continued fraction represents a rational number. Conversely, any rational number can be represented by a finite simple continued fraction.

PROOF The first statement is easily proved by induction on the length of the continued fraction. First note that $[a_0;] = a_0$. Now observe that

$$[a_0; a_1] = a_0 + \frac{1}{a_1} = \frac{a_0 a_1 + 1}{a_1},$$

and so the statement is true for $n = 1$. For $n > 1$, we have

$$[a_0; a_1, \ldots, a_n] = a_0 + \frac{1}{[a_1; a_2, \ldots, a_n]}.$$

By inductive hypothesis, we can write

$$[a_0; a_1, \ldots, a_n] = a_0 + \frac{1}{p/q}$$

for some rational number p/q. Since the quantity on the right can be expressed as $(pa_0 + q)/p$, we have proved that $[a_0; a_1, \ldots, a_n]$ is rational.

Now suppose that a/b is a rational number with $b > 0$. We will prove that a/b can be written as a finite continued fraction using induction on b. If $b = 1$, then $a/b = [a;]$. Now suppose that any rational number with denominator less than b can be written as a finite simple continued fraction. Using the division algorithm (Theorem 2.9), we can write

$$a = bq + r$$

for some integers q and r with $0 \leq r < b$. Dividing through by b we have

$$\frac{a}{b} = q + \frac{r}{b}. \tag{8.8}$$

If $r = 0$, then $a/b = [q;]$. Otherwise, we can rewrite (8.8) as

$$\frac{a}{b} = q + \frac{1}{b/r}.$$

We may apply our induction hypothesis to b/r, and so

$$b/r = [a_1; a_2, \ldots, a_n]$$

for some integers a_1, a_2, ..., a_n. Since $b/r > 1$, we know that a_1 is positive. (See Exercise 8.6.) It follows that

$$\frac{a}{b} = [q; a_1, \ldots, a_n].$$

□

We can use ideas from the proof of the theorem to write *Mathematica* and *Maple* procedures to convert rational numbers to finite simple continued fractions and vice versa. (*Mathematica* and *Maple* have built-in functions to accomplish this.)

Mathematica

```
ourcontinuedfraction[x_]:=Module[{xprime, a},
  xprime = x;
  a = {};
  While[!IntegerQ[xprime],
    AppendTo[a, IntegerPart[xprime]];
      xprime = 1/(xprime - IntegerPart[xprime])];
    AppendTo[a, IntegerPart[xprime]]
    ]

ourcontinuedfraction[125/54]

{2,3,5,1,2}

ourfromcontinuedfraction[a_]:=Module[{c},
  c=a;
  While[Length[c]>1,
    c=Delete[ReplacePart[c,c[[-2]]+1/c[[-1]],-2],-1]];
  c[[1]]
  ]

ourfromcontinuedfraction[{2,3,5,1,2}]

125/54
```

Notice that we do not make use of the semi-colon after the initial partial quotient.

Maple

```
ourcontinuedfraction:=proc(x)
  local xprime, a;
  xprime:=x;
  a:=[];
  while not type(xprime,integer) do
    a:=[op(a),floor(xprime)];
    xprime:=1/(xprime-floor(xprime));
  end do;
  a:=[op(a),floor(xprime)];
end proc:

ourcontinuedfraction(125/54);

[2, 3, 5, 1, 2]

ourfromcontinuedfraction:=proc(a)
  local c;
  c:=a;
  while nops(c)>1 do
    c[-2]:=c[-2]+1/c[-1];
    c:=c[1..-2];
  end do;
  c[1];
end proc:

ourfromcontinuedfraction([2,3,5,1,2]);

125/54
```

In Example 8.2, we used the Euclidean algorithm to write a rational number as a simple continued fraction. The proof of Theorem 8.3 shows that this works in general. That is, if we perform the Euclidean algorithm on the pair (a, b), and obtain

$$\begin{aligned} a &= bq_1 + r_1 \\ b &= r_1q_2 + r_2 \\ r_1 &= r_2q_3 + r_3 \\ &\vdots \\ r_{n-2} &= r_{n-1}q_n + r_n \\ r_{n-1} &= r_nq_{n+1} + 0, \end{aligned}$$

then $a/b = [q_1; q_2, \ldots, q_{n+1}]$. A formal proof of this fact is called for in the exercises.

We observe that most rational numbers will have more than one simple continued fraction expansion. For instance,

$$5 + \frac{1}{3} = 5 + \cfrac{1}{2 + \cfrac{1}{1}}.$$

Thus, $16/3 = [5; 3] = [5; 2, 1]$. Notice that in general,

$$[a_0; a_1, \ldots, a_{n-1}, a_n, 1] = [a_0; a_1, \ldots, a_{n-1}, a_n + 1],$$

and so any rational number (other than 1) has a simple continued fraction expansion whose last partial quotient is larger than 1. Such a representation is unique.

PROPOSITION 8.4
Suppose that

$$[a_0; a_1, \ldots, a_m] = [b_0; b_1, \ldots, b_n],$$

for two simple continued fractions such that a_m and b_n are both larger than 1. Then $m = n$ and $a_i = b_i$ for $0 \le i \le n$.

The proof is called for in Exercise 8.7.

DEFINITION 8.5 *For $1 \le k \le n$, the kth* convergent*, C_k, of a continued fraction $[a_0; a_1, \ldots, a_n]$ is the continued fraction*

$$C_k = [a_0; a_1, \ldots, a_k].$$

We extend this definition to include $k = 0$, and so we set

$$C_0 = a_0.$$

Example 8.6 Recall from Example 8.2 that the continued fraction expansion of $125/54$ is $[2; 3, 5, 1, 2]$. Thus, the convergents of this continued fraction are

$$\begin{aligned}
C_0 &= 2\\
C_1 &= [2; 3] = 7/3\\
C_2 &= [2; 3, 5] = 37/16\\
C_3 &= [2; 3, 5, 1] = 44/19\\
C_4 &= [2; 3, 5, 1, 2] = 125/54.
\end{aligned}$$

□

If a finite continued fraction is simple, then Theorem 8.3 implies that all of its convergents represent rational numbers. The numerator and denominator of the kth convergent can be expressed recursively in terms of the numerators and denominators of the previous convergents. This recursive formula will hold for non-simple continued fractions as well.

PROPOSITION 8.7

Let a_0, a_1, ..., a_n be real numbers with a_1, ..., a_n positive. For the continued fraction $[a_0; a_1, \ldots, a_n]$, define

$$\begin{aligned} p_0 &= a_0 & q_0 &= 1 \\ p_1 &= a_1 p_0 + 1 & q_1 &= a_1 \\ p_k &= a_k p_{k-1} + p_{k-2} & q_k &= a_k q_{k-1} + q_{k-2} & (2 \le k \le n). \end{aligned}$$

Then $C_k = p_k/q_k$, for $0 \le k \le n$.

PROOF Since $C_0 = a_0$ and $C_1 = a_0 + 1/a_1 = (a_0 a_1 + 1)/a_1$, the result holds for $k = 0$ and $k = 1$. By way of induction, suppose that the result is true for $k = m \ge 1$. Observe that

$$[a_0; a_1, \ldots, a_{m-1}, a_m, a_{m+1}] = \left[a_0; a_1, \ldots, a_{m-1}, a_m + \frac{1}{a_{m+1}}\right]. \tag{8.9}$$

Since the first m partial quotients (i.e., a_0, ..., a_{m-1}) are the same in both continued fractions, our inductive hypothesis tells us that the continued fraction on the right side of (8.9) is equal to

$$\frac{\left(a_m + \dfrac{1}{a_{m+1}}\right) p_{m-1} + p_{m-2}}{\left(a_m + \dfrac{1}{a_{m+1}}\right) q_{m-1} + q_{m-2}}.$$

Hence,

$$\begin{aligned} C_{m+1} &= \frac{a_m p_{m-1} + p_{m-2} + \dfrac{p_{m-1}}{a_{m+1}}}{a_m q_{m-1} + q_{m-2} + \dfrac{q_{m-1}}{a_{m+1}}} \\ &= \frac{a_{m+1} p_m + p_{m-1}}{a_{m+1} q_m + q_{m-1}} \\ &= \frac{p_{m+1}}{q_{m+1}}. \end{aligned}$$

□

We summarize the proposition by presenting a table similar to the one given in Section 2.5 for writing $\gcd(a, b)$ as a linear combination of a and b.

		a_0	a_1	$\cdots$	a_{k-2}	a_{k-1}	a_k	$\cdots$	a_n
0	1	p_0	p_1	$\cdots$	p_{k-2}	p_{k-1}	p_k	$\cdots$	p_n
1	0	q_0	q_1	$\cdots$	q_{k-2}	q_{k-1}	q_k	$\cdots$	q_n

The two leftmost columns of the table are always set to $\{0, 1\}$ and $\{1, 0\}$. For each $k = 0, \ldots, n$, we compute p_k by multiplying the entry to the left of p_k by a_k, and adding the product to the entry two columns to the left of p_k. We compute q_k similarly. For the continued fraction from Example 8.2, the table is completed as below.

		2	3	5	1	2
0	1	2	7	37	44	125
1	0	1	3	16	19	54

Using the theorem, we can easily construct *Mathematica* and *Maple* procedures to compute the list of convergents of a finite simple continued fraction.

Mathematica

```
ourconvergents[x_]:=Module[{p,q,a,c},
  If[ListQ[x],a=x,
    a=ourcontinuedfraction[x]];
  p[0]=a[[1]];
  q[0]=1;
  c[0]=p[0]/q[0];
  If[Length[a]>0,
    p[1]=a[[1]]a[[2]]+1;
    q[1]=a[[2]];
    c[1]=p[1]/q[1]];
  Do[
    p[i]=a[[i+1]]p[i-1]+p[i-2];
    q[i]=a[[i+1]]q[i-1]+q[i-2];
    c[i]=p[i]/q[i],
    {i,2,Length[a]-1}];
  Table[c[i],{i,0,Length[a]-1}]]
```

This command will accept both a rational number and a finite simple continued fraction as input.

Mathematica

```
ourconvergents[125/54]

{2, 7/3, 37/16, 44/19, 125/54}

ourconvergents[{2,3,5,1,2}]

{2, 7/3, 37/16, 44/19, 125/54}
```

Here are the corresponding *Maple* commands.

Maple

```
ourconvergents:=proc(x)
  local p, q, a, c, i;
  if type(x,list)
    then
      a:=x;
    else
      a:=ourcontinuedfraction(x);
  end if;
  p[0]:=a[1];
  q[0]:= 1;
  c[0]:= p[0]/q[0];
  if nops(a) > 0 then
    p[1]:=a[1]*a[2] + 1;
    q[1]:=a[2];
    c[1]:=p[1]/q[1];
    for i from 2 to nops(a)-1 do
      p[i]:=a[i+1]*p[i-1]+p[i-2];
      q[i]:=a[i+1]*q[i-1]+q[i-2];
      c[i]:=p[i]/q[i];
    end do;
  end if;
  [seq(c[i],i=0..nops(a)-1)];
end proc:

ourconvergents(125/54);
[2, 7/3, 37/16, 44/19, 125/54]

ourconvergents([2,3,5,1,2]);
[2, 7/3, 37/16, 44/19, 125/54]
```

Observe in our example that p_k/q_k gives the convergent C_k in lowest terms. This will always be the case for a simple continued fraction and will follow from the next theorem.

THEOREM 8.8
Let p_k and q_k be as defined in Proposition 8.7. Then for $1 \leq k \leq n$,

$$p_k q_{k-1} - p_{k-1} q_k = (-1)^{k-1}.$$

PROOF We proceed by induction. For $k = 1$, we have

$$p_1 q_0 - p_0 q_1 = (a_0 a_1 + 1) \cdot 1 - a_0 a_1 = (-1)^0.$$

Now suppose that

$$p_m q_{m-1} - p_{m-1} q_m = (-1)^{m-1},$$

for some $m \geq 1$, and consider the case $k = m + 1$. We have

$$\begin{aligned} p_{m+1} q_m - p_m q_{m+1} &= (a_{m+1} p_m + p_{m-1}) q_m - p_m (a_{m+1} q_m + q_{m-1}) \\ &= -(p_m q_{m-1} - p_{m-1} q_m) \\ &= -(-1)^{m-1}. \end{aligned}$$

Therefore, $p_{m+1} q_m - p_m q_{m+1} = (-1)^m$. ∎

COROLLARY 8.9
Let $[a_0, a_1, \ldots, a_n]$ be a simple continued fraction. Then $\gcd(p_k, q_k) = 1$ for $0 \leq k \leq n$.

PROOF Since $q_0 = 1$, we have $\gcd(p_0, q_0) = 1$. For $1 \leq k \leq n$, the result follows from Theorem 8.8 and Theorem 2.16. ∎

COROLLARY 8.10
With notation as in Proposition 8.7,

$$C_k - C_{k-1} = \frac{(-1)^{k-1}}{q_k q_{k-1}} \tag{8.10}$$

for $1 \leq k \leq n$, and

$$C_k - C_{k-2} = \frac{a_k (-1)^k}{q_k q_{k-2}} \tag{8.11}$$

for $1 < k \leq n$.

PROOF By Proposition 8.7,

$$C_k - C_{k-1} = \frac{p_k}{q_k} - \frac{p_{k-1}}{q_{k-1}} = \frac{p_k q_{k-1} - p_{k-1} q_k}{q_k q_{k-1}}.$$

Equation (8.10) now follows from Theorem 8.8. The proof of (8.11) is called for in the exercises. ∎

COROLLARY 8.11

Let $C_0, C_1, C_2, \ldots$ be the convergents of a finite continued fraction. Then

$$C_0 < C_2 < C_4 < \cdots < C_5 < C_3 < C_1.$$

That is, whenever $k \geq 0$ is even and $l \geq 1$ is odd, we have

$$C_k < C_{k+2}, \tag{8.12}$$

$$C_{l+2} < C_l, \tag{8.13}$$

$$C_k < C_l. \tag{8.14}$$

PROOF Observe that q_i is positive for all i, and so (8.12) and (8.13) follow from equation (8.11) in Corollary 8.10. Equation (8.10) in Corollary 8.10 tells us that for even $k \geq 2$,

$$C_k < C_{k-1}.$$

It follows that

$$C_k < C_{k-1} < C_{k-3} < \cdots < C_1.$$

Thus, if l is odd and $k > l$, then $C_k < C_l$. Similarly, for odd l,

$$C_{l-1} < C_l,$$

and so

$$C_0 < C_2 < \cdots < C_{l-1} < C_l.$$

It follows that $C_k < C_l$ when $k < l$. ∎

Exercises

8.1 Write

$$1 + \cfrac{1}{1 + \cfrac{1}{1 + \cfrac{1}{1 + \cfrac{1}{1}}}}$$

in the form p/q for some integers p and q.

8.2 Write

$$3+\cfrac{1}{1+\cfrac{1}{8+\cfrac{1}{1+\cfrac{1}{2}}}}$$

in the form p/q for some integers p and q.

8.3 (a) Write $21/13$ as a simple continued fraction.

(b) Write $133/29$ as a simple continued fraction.

◇**8.4** Use a computer to find the list of convergents for $1997/473$.

†**8.5** Let a and b be integers with $b > 0$. Suppose that we perform the Euclidean algorithm on the pair (a, b), and obtain

$$\begin{aligned} a &= bq_1 + r_1 \\ b &= r_1 q_2 + r_2 \\ r_1 &= r_2 q_3 + r_3 \\ &\vdots \\ r_{n-2} &= r_{n-1} q_n + r_n \\ r_{n-1} &= r_n q_{n+1} + 0. \end{aligned}$$

Show that $a/b = [q_1; q_2, \ldots, q_{n+1}]$.

Hint: Use induction on the number of steps in the Euclidean algorithm, and follow the proof of Theorem 8.3.

8.6 Suppose that $\alpha = [a_0; a_1, \ldots, a_n]$, where $a_1, \ldots, a_n$ are positive integers and a_0 is any integer and $a_n > 1$.

(a) Show that $\alpha > 1$ if and only if $a_0 \geq 1$.

(b) Show that $\lfloor \alpha \rfloor = a_0$.

8.7 Prove Proposition 8.4.

Hint: Compare with the forthcoming Theorem 8.21.

8.8 Prove that equation (8.11) from Corollary 8.10 holds, that is, show that

$$C_k - C_{k-2} = \frac{a_k(-1)^k}{q_k q_{k-2}}.$$

8.9 With notation as in Proposition 8.7, show that $\gcd(p_k, p_{k-1}) = 1$ and $\gcd(q_k, q_{k-1}) = 1$, for $1 \leq k \leq n$.

8.2 Infinite continued fractions

We know from Theorem 8.3 that an irrational number cannot be represented by a finite simple continued fraction. Similarly, an irrational number cannot be expressed as a decimal number with a finite number of digits. One can always approximate an irrational number with a finite decimal expansion. Let's attempt to do the same with continued fractions. We can mimic the procedure we used to expand a rational number into a simple continued fraction. If α is a non-integral real number, we can write

$$\alpha = \lfloor \alpha \rfloor + \frac{1}{1/(\alpha - \lfloor \alpha \rfloor)}, \tag{8.15}$$

where $\lfloor \alpha \rfloor$ is the greatest integer less than or equal to α, and so

$$\alpha = [\lfloor \alpha \rfloor; 1/(\alpha - \lfloor \alpha \rfloor)].$$

If α is irrational, then $1/(\alpha - \lfloor \alpha \rfloor)$ will also be irrational and greater than 1. Hence, if α is irrational, we can expand $1/(\alpha - \lfloor \alpha \rfloor)$ as we did α in (8.15) and obtain a continued fraction expansion of the form

$$\alpha = [a_0; a_1, \alpha_2],$$

where a_0 and a_1 are integers and α_2 is an irrational number greater than 1. Let's show that we can repeat this process to obtain, for any $n \geq 1$, a continued fraction expansion

$$\alpha = [a_0; a_1, \ldots, a_{n-1}, \alpha_n],$$

where $a_0, \ldots, a_{n-1}$ are integers and α_n is an irrational number greater than 1. If we have an expansion of the form

$$\alpha = [a_0; a_1, \ldots, a_{n-2}, \alpha_{n-1}]$$

with α_{n-1} an irrational number greater than 1, then

$$\begin{aligned} \alpha &= \left[a_0; a_1, \ldots, a_{n-2}, [\alpha_{n-1}] + \frac{1}{1/(\alpha_{n-1} - \lfloor \alpha_{n-1} \rfloor)} \right] \\ &= \left[a_0; a_1, \ldots, a_{n-2}, [\alpha_{n-1}], 1/(\alpha_{n-1} - \lfloor \alpha_{n-1} \rfloor) \right]. \end{aligned}$$

Note that $[\alpha_{n-1}]$ is a positive integer and $1/(\alpha_{n-1} - [\alpha_{n-1}])$ is an irrational number greater than 1. We record the results of our discussion as a theorem below.

THEOREM 8.12
If α is an irrational number, then α has a finite continued fraction expansion of the form

$$[a_0; a_1, \ldots, a_{n-1}, \alpha_n],$$

where a_0, ..., a_{n-1} are integers, a_1, ..., a_{n-1} are positive, and α_n is an irrational number greater than 1. One can obtain such a representation by taking

$$\begin{aligned} a_0 &= \lfloor \alpha \rfloor, \\ \alpha_1 &= 1/(\alpha - \lfloor \alpha \rfloor), \end{aligned}$$

and by recursively defining

$$\begin{aligned} a_{n-1} &= \lfloor \alpha_{n-1} \rfloor, \\ \alpha_n &= 1/(\alpha_{n-1} - \lfloor \alpha_{n-1} \rfloor), \end{aligned}$$

for $n > 1$.

Example 8.13 Let's apply this algorithm to $\sqrt{3}$. Since $\sqrt{3} \doteq 1.73$, we write

$$\sqrt{3} = 1 + \frac{1}{1/(\sqrt{3}-1)}.$$

We simplify $\dfrac{1}{\sqrt{3}-1}$ by rationalizing the denominator:

$$\frac{1}{\sqrt{3}-1} = \frac{\sqrt{3}+1}{2}.$$

Thus,

$$\sqrt{3} = \left[1; \frac{\sqrt{3}+1}{2}\right]. \tag{8.16}$$

Since $\dfrac{\sqrt{3}+1}{2} \doteq 1.37$, we write

$$\frac{\sqrt{3}+1}{2} = 1 + \frac{1}{1/((\sqrt{3}+1)/2-1)}.$$

With a little algebra, we see that

$$\frac{1}{(\sqrt{3}+1)/2-1} = \frac{2}{\sqrt{3}-1} = 1+\sqrt{3}.$$

Our new continued fraction expansion is

$$\sqrt{3} = [1; 1, \sqrt{3}+1]. \tag{8.17}$$

Moving on, since $\sqrt{3}+1 \doteq 2.73$, we write

$$\sqrt{3}+1 = 2 + \frac{1}{1/(\sqrt{3}+1-2)}.$$

We simplify

$$1/(\sqrt{3}+1-2) = \frac{\sqrt{3}+1}{2},$$

and so

$$\sqrt{3} = \left[1; 1, 2, \frac{\sqrt{3}+1}{2}\right]. \tag{8.18}$$

Notice that each step of the algorithm uses only the last partial quotient of the continued fraction from the previous step. The last partial quotient we have in (8.18) is the same as the last partial quotient in (8.16). Our algorithm has entered an infinite loop. We see that the next few steps will produce the continued fractions

$$\left[1; 1, 2, 1, \sqrt{3}+1\right] \tag{8.19}$$

$$\left[1; 1, 2, 1, 2, \frac{\sqrt{3}+1}{2}\right] \tag{8.20}$$

$$\left[1; 1, 2, 1, 2, 1, \sqrt{3}+1\right]. \tag{8.21}$$

Let's compute the convergents of the continued fraction given in (8.21).

k	kth convergent
0	1
1	2
2	$5/3 \doteq 1.66667$
3	$7/4 = 1.75$
4	$19/11 \doteq 1.72727$
5	$26/15 \doteq 1.73333$

One can easily produce the results of the table with our *Mathematica* command `ourconvergents[{1, 1, 2, 1, 2, 1}]` or with our *Maple* command `ourconvergents([1, 1, 2, 1, 2, 1])`.

Since $\sqrt{3} \doteq 1.7321$, our convergents appear to be approaching $\sqrt{3}$ (hence the name "convergents"). □

When we say that a real number α has a decimal expansion

$$d_0.d_1d_2d_3\ldots,$$

what we mean is that the sequence

$$\left\{\sum_{i=0}^{n} 10^{-i} d_i\right\}_{n=0}^{\infty}$$

converges to the real number α. In light of our the previous, it appears reasonable to attempt to describe real numbers in a similar way using continued fractions. The following theorem indicates that this will be feasible.

THEOREM 8.14
Let $a_0, a_1, a_2, \ldots$ be an infinite sequence of integers with a_i positive for $i \geq 1$. Let

$$C_n = [a_0; a_1, a_2, \ldots, a_n].$$

Then the sequence $C_0, C_1, C_2, \ldots$ converges.

PROOF We will show that, given $\varepsilon > 0$, there exists an N such that

$$|C_l - C_k| < \varepsilon,$$

whenever $N \leq k \leq l$. (A sequence satisfying this condition is called a *Cauchy sequence* and will necessarily converge to some real number.) Observe that for $n \leq l$, the term C_n is the nth convergent of C_l. If k is odd, then Corollary 8.11 implies that

$$C_{k-1} < C_l \leq C_k,$$

while if f k is even, then the corollary implies that

$$C_k \leq C_l < C_{k-1}.$$

In either case,

$$|C_l - C_k| \leq |C_k - C_{k-1}|.$$

Let $p_0, q_0, p_1, q_1, \ldots, p_l, q_l$ be the quantities defined in Proposition 8.7 for the continued fraction $[a_0; a_1, \ldots, a_l]$. We may conclude, using Corollary 8.10, that

$$|C_l - C_k| \leq \frac{1}{q_k q_{k-1}}. \tag{8.22}$$

Since the q_i grow arbitrarily large, we can find N such that $q_N > 1/\varepsilon$. Now for $N \leq k \leq l$, we have

$$|C_l - C_k| \leq \frac{1}{q_k q_{k-1}} \leq \frac{1}{q_N q_{N-1}} < \varepsilon.$$

Hence, the sequence is Cauchy. ∎

DEFINITION 8.15 *Let $a_0, a_1, a_2, \ldots$ be an infinite sequence of integers with a_i positive for $i \geq 1$. The* infinite simple continued fraction $[a_0; a_1, a_2, \ldots]$ *is the real number*

$$\lim_{n \to \infty} [a_0; a_1, a_2, \ldots, a_n].$$

We again refer to each a_i as a partial quotient and define the nth convergent of $[a_0; a_1, a_2, \ldots]$ to be $C_n = [a_0; a_1, a_2, \ldots, a_n]$.

In the next section we will show that an infinite simple continued fraction always represents an irrational number.

The infinite simple continued fraction $[1; 1, 2, 1, 2, 1, \ldots]$ represents a real number, which our calculations in Example 8.13 suggest is $\sqrt{3}$. The next theorem verifies this.

THEOREM 8.16

Let α be an irrational number, and suppose that

$$[a_0; a_1, \ldots, a_{n-1}, \alpha_n]$$

is the continued fraction expansion of α given in Theorem 8.12. Then

$$\alpha = [a_0; a_1, a_2, \ldots].$$

PROOF Let $C_n = [a_0; a_1, \ldots, a_n]$ be the nth convergent of the infinite simple continued fraction $[a_0; a_1, \ldots]$. We will show that $\lim_{n\to\infty} |\alpha - C_n| = 0$. Let p_0/q_0, $\ldots$, p_n/q_n, $\hat{p}_{n+1}/\hat{q}_{n+1}$ be the convergents of the continued fraction

$$[a_0; a_1, \ldots, a_n, \alpha_{n+1}]$$

as given in Proposition 8.7. That is,

$$\begin{aligned} p_0 &= a_0 & q_0 &= 1 \\ p_1 &= a_0 a_1 + 1 & q_1 &= a_1 \\ p_k &= a_k p_{k-1} + p_{k-2} & q_k &= a_k q_{k-1} + q_{k-2} \qquad (2 \le k \le n) \\ \hat{p}_{n+1} &= \alpha_{n+1} p_n + p_{n-1} & \hat{q}_{n+1} &= \alpha_{n+1} q_n + q_{n-1}. \end{aligned}$$

We observe that

$$\begin{aligned} p_n/q_n &= C_n, \\ \hat{p}_{n+1}/\hat{q}_{n+1} &= \alpha. \end{aligned}$$

Applying Corollary 8.10, we have

$$|\alpha - C_n| = \frac{1}{\hat{q}_{n+1} q_n}. \tag{8.23}$$

Since a_1, $\ldots$, a_n are all positive integers, it follows that q_n is at least as big as the nth Fibonacci number. We know that α_{n+1} is greater than 1, and so

$$\hat{q}_{n+1} > q_n + q_{n-1}.$$

We conclude that $\hat{q}_{n+1}$ is greater than the $(n+1)$st Fibonacci number. Hence, $1/(\hat{q}_{n+1}q_n)$ goes to zero as n goes to infinity. ▯

We can easily modify the `ourcontinuedfraction` command defined above to compute the first n partial quotients of an infinite simple continued fraction expansion of any real number. The command below computes the first n partial quotients for the expansion of x.

Mathematica

```
ourcontinuedfraction[x_, n_] := Module[{xprime, a, i},
  xprime = x;
  a = {IntegerPart[xprime]};
  i=1;
  While[i<n && !IntegerQ[xprime],
    xprime = 1/(xprime - IntegerPart[xprime]);
    AppendTo[a, IntegerPart[xprime]];
    i++];
  a
  ]

ourcontinuedfraction[Sqrt[3], 6]

{1,1,2,1,2,1}
```

This is written to accept rational number inputs, but the procedure will terminate if the expansion is complete before i reaches n.

Mathematica

```
ourcontinuedfraction[125/54,8]

{2,3,5,1,2}
```

Here are the corresponding *Maple* commands.

Maple

```
ourcontinuedfraction2:=proc(x,n)
  local xprime, a, i;
  xprime:=x;
  a:=[floor(xprime)];
  i:=1;
  while i<n and not type(xprime,integer) do
    xprime:=1/(xprime-floor(xprime));
    a:=[op(a),floor(xprime)];
    i:=i+1;
  end do;
  a;
end proc:

ourcontinuedfraction2(sqrt(3),6);

[1, 1, 2, 1, 2, 1]

ourcontinuedfraction2(125/54,8);

[2, 3, 5, 1, 2]
```

We can also extend the `ourconvergents` command in a similar way. In *Mathematica* we compose the command `ourconvergents` with the command `ourcontinuedfraction`.

```
ourconvergents[x_,n_] :=
  ourconvergents[ourcontinuedfraction[x,n]]
```

The corresponding code for *Maple* is similar.

```
ourconvergents2:=(x,n)->ourconvergents
  (ourcontinuedfraction2(x,n)):
```

For the rational number 125/54 with $n = 8$, we obtain the list

```
2, 7/3, 37/16, 44/19, 125/54.
```

Notice that we get only five convergents, as the continued fraction expansion of 125/54 has only five partial quotients. For the irrational number $\sqrt{3}$ with $n = 8$, the following list is reported.

```
1, 2, 5/3, 7/4, 19/11, 26/15, 71/41, 97/56
```

Example 8.17 Let's take a second look at the example from the beginning of the chapter, i.e.,

$$\frac{1+\sqrt{5}}{2} = [1; 1, 1, \ldots].$$

We compute

$$\begin{aligned}
\frac{1+\sqrt{5}}{2} &= 1 + \frac{-1+\sqrt{5}}{2} \\
&= 1 + \frac{1}{2/(-1+\sqrt{5})} \\
&= 1 + \frac{1}{\left(\frac{2}{-1+\sqrt{5}}\right)\left(\frac{1+\sqrt{5}}{1+\sqrt{5}}\right)} \\
&= 1 + \frac{1}{(1+\sqrt{5})/2}.
\end{aligned}$$

Thus, $(1+\sqrt{5})/2 = [1; (1+\sqrt{5})/2]$. Following the procedure from Theorem 8.12, we see that

$$(1+\sqrt{5})/2 = [1; 1, 1, \ldots, 1, (1+\sqrt{5})/2],$$

and so by Theorem 8.16, we have

$$(1+\sqrt{5})/2 = [1; 1, 1, \ldots].$$

□

Exercises

8.10 Write $\sqrt{5}$ as a simple continued fraction.

8.11 Write $1+\sqrt{3}$ as a simple continued fraction.

8.12 Let a be a positive integer. Show that

$$[a; a, a, \ldots] = \frac{a+\sqrt{a^2+4}}{2}.$$

8.13 Use the fact that $[1; 1, 1, \ldots] = (1+\sqrt{5})/2$ to show that

$$\lim_{n\to\infty} \frac{f_n}{f_{n-1}} = \frac{1+\sqrt{5}}{2},$$

where f_n is the nth Fibonacci number.

8.14 Let a_0 be any real number and a_1, b_1, a_2, b_2, ... nonzero real numbers. Suppose that

$$\mathcal{C}_n = a_0 + \cfrac{b_1}{a_1 + \cfrac{b_2}{a_2 + \ddots + \cfrac{b_{n-1}}{a_{n-1} + \cfrac{b_n}{a_n}}}}.$$

(a) Show that $\mathcal{C}_n = \mathcal{P}_n/\mathcal{Q}_n$, where

$$\begin{array}{lll} \mathcal{P}_0 = a_0 & \mathcal{Q}_0 = 1 & \\ \mathcal{P}_1 = a_1\mathcal{P}_0 + b_1 & \mathcal{Q}_1 = a_1 & \\ \mathcal{P}_n = a_n\mathcal{P}_{n-1} + b_n\mathcal{P}_{n-2} & \mathcal{Q}_n = a_n\mathcal{Q}_{n-1} + b_n\mathcal{Q}_{n-2} & (n \geq 2). \end{array}$$

(b) Show that for $n \geq 1$,

$$\mathcal{P}_n\mathcal{Q}_{n-1} - \mathcal{P}_{n-1}\mathcal{Q}_n = (-1)^{n-1}b_1b_2\ldots b_n.$$

(c) If t_1, t_2, ..., t_n are nonzero real numbers, show that

$$\mathcal{C}_n = a_0 + \cfrac{b_1t_1}{a_1t_1 + \cfrac{b_2t_1t_2}{a_2t_2 + \cfrac{b_3t_2t_3}{a_3t_3 + \ddots + \cfrac{b_{n-1}t_{n-2}t_{n-1}}{a_{n-1}t_{n-1} + \cfrac{b_nt_{n-1}t_n}{a_nt_n}}}}}.$$

8.15 Let c_1, c_2, ..., c_n be nonzero real numbers. Show that

$$\sum_{k=1}^{n} \frac{1}{c_k} = \cfrac{1}{c_1 - \cfrac{c_1^2}{c_1 + c_2 - \cfrac{c_2^2}{c_2 + c_3 - \ddots - \cfrac{c_{n-1}^2}{c_{n-1} + c_n}}}}$$

8.3 Rational approximation of real numbers

As we saw in the preceding section, we can regard an infinite simple continued fraction expansion

$$\alpha = [a_0; a_1, a_2, \ldots]$$

as an analog of an infinite decimal expansion of a real number. Truncating an infinite decimal expansion of an irrational number produces a rational approximation of the number. The same is true for continued fraction expansions. That is, $[a_0; a_1, \ldots, a_n]$ is a rational approximation of α. The following theorem and its corollary allow us to determine how good the approximation is.

THEOREM 8.18
Suppose that α has an infinite simple continued fraction expansion

$$\alpha = [a_0; a_1, a_2, \ldots].$$

Let C_n be the nth convergent of this continued fraction. Then

$$C_0 < C_2 < C_4 < \cdots < \alpha < \cdots < C_5 < C_3 < C_1.$$

PROOF Note that C_0, C_1, ..., C_n are also the convergents of the finite continued fraction $[a_0; a_1, \ldots, a_n]$. Thus, in light of Corollary 8.11, we only need to show that

$$C_k < \alpha < C_l$$

for all even k and odd l. Since the sequence C_0, C_1, C_2, ... converges to α, any subsequence also converges to α. In particular

$$\lim_{n\to\infty} C_{2n} = \alpha \text{ and} \tag{8.24}$$

$$\lim_{n\to\infty} C_{2n+1} = \alpha. \tag{8.25}$$

Since $C_0 < C_2 < C_4 < \cdots$, it follows from (8.24) that $C_k < \alpha$ for all even k. Similarly, equation (8.25) implies that $\alpha < C_l$ for all odd l. ∎

COROLLARY 8.19
Let C_n be the nth convergent of an infinite simple continued fraction expansion of a real number α, and let p_n and q_n have their usual meanings for this continued fraction. Then

$$|\alpha - C_n| < \frac{1}{q_n q_{n+1}}. \tag{8.26}$$

PROOF By the theorem,

$$|\alpha - C_n| < |C_{n+1} - C_n|.$$

Equation (8.26) now follows from Corollary 8.10. ∎

COROLLARY 8.20
An infinite simple continued fraction represents an irrational number.

PROOF By way of contradiction, suppose that

$$\frac{a}{b} = [a_0; a_1, \ldots]$$

for some relatively prime integers a and b. Let p_n/q_n be the nth convergent for this continued fraction, and choose n large enough so that $|b| < q_n$. Since a/b and p_n/q_n are expressed in lowest terms (by Corollary 8.9), they are different numbers, and consequently

$$aq_n - bp_n \neq 0.$$

Using Corollary 8.19 we have

$$\left|\frac{a}{b} - \frac{p_n}{q_n}\right| < \frac{1}{q_n q_{n+1}}.$$

Since $|b| < q_n < q_{n+1}$, we conclude that

$$0 < |aq_n - bp_n| < \frac{|b|}{q_{n+1}} < 1.$$

The number $aq_n - bp_n$ is an integer, and so we have a contradiction. ∎

Unlike a decimal expansion, a continued fraction expansion of an irrational number is unique.

THEOREM 8.21
Let α be an irrational number, and suppose that

$$\alpha = [a_0; a_1, \ldots] = [b_0; b_1, \ldots],$$

where a_n and b_n are integers for all $n \geq 0$. Then $a_n = b_n$ for all $n \geq 0$.

PROOF By Theorem 8.18,

$$a_0 < \alpha < a_0 + \frac{1}{a_1}.$$

Consequently, $a_0 = \lfloor \alpha \rfloor$. Similarly, $b_0 = \lfloor \alpha \rfloor$, and therefore $a_0 = b_0$.

Now,

$$\begin{aligned}\alpha &= \lim_{m\to\infty} [a_0; a_1, \ldots, a_m] \\ &= a_0 + \frac{1}{\lim_{m\to\infty} [a_1; a_2, \ldots, a_m]} \\ &= a_0 + \frac{1}{[a_1; a_2, \ldots]}.\end{aligned}$$

Similarly,

$$\alpha = b_0 + \frac{1}{[b_1; b_2, \ldots]}.$$

Hence,

$$[a_1; a_2, \ldots] = [b_1; b_2, \ldots].$$

Our argument in the first paragraph now applies to give us

$$a_1 = b_1.$$

Arguing inductively, we can show that

$$[a_n; a_{n+1}, \ldots] = [b_n; b_{n+1}, \ldots],$$

for all $n \geq 0$, and so $a_n = b_n$ for all $n \geq 0$. $\square$

In light of this uniqueness, we may talk about *the* infinite simple continued fraction expansion of an irrational number. If we have an infinite simple continued fraction representation

$$\alpha = [a_0; a_1, \ldots],$$

then

$$a_n = \lfloor \alpha_n \rfloor,$$

where α_n is defined as in Theorem 8.12.

Example 8.22 We enlist the help of *Mathematica* or *Maple* to compute the first few convergents of π.

```
3, 22/7, 333/106, 355/113, 103993/33102
```

Let's compare these convergents to π to see how good an approximation each one provides.

n	C_n	$\lvert\pi - C_n\rvert$
0	3	0.141593
1	22/7	0.00126449
2	333/106	0.0000832196
3	355/113	$2.66764 \cdot 10^{-7}$
4	103993/33102	$5.77891 \cdot 10^{-10}$

Notice, for instance, that the error in the approximation given by 333/106 is within the bound of

$$\frac{1}{106 \cdot 113} \doteq 0.0000834864$$

given in Corollary 8.19. In Theorem 8.18 we showed, in general, that the convergent p_{n+2}/q_{n+2} provides a better approximation than p_n/q_n. In the above example, each *successive* convergent is a better approximation than the previous one. This also is true in general. □

THEOREM 8.23

Let α be an irrational number, and let p_n and q_n be the usual quantities associated to the infinite simple continued fraction expansion of α. Then for $n \geq 1$,

$$\left|\alpha - \frac{p_{n+1}}{q_{n+1}}\right| < \left|\alpha - \frac{p_n}{q_n}\right|. \tag{8.27}$$

Moreover,

$$|q_{n+1}\alpha - p_{n+1}| < |q_n\alpha - p_n|. \tag{8.28}$$

PROOF We first observe that the first inequality follows from the second. Since $q_{n+1} > q_n$ when $n \geq 1$, we have

$$\frac{1}{q_{n+1}} < \frac{1}{q_n}.$$

Now multiplying the left side of (8.28) by $1/q_{n+1}$ and the right side by $1/q_n$, we obtain (8.27).

Now we prove that the inequality in (8.28) holds. We use the notation from Theorems 8.12 and 8.16, where

$$\alpha = [a_0; a_1, \ldots, a_{n-1}, \alpha_n],$$

and the convergents for this continued fraction are

$$\frac{p_0}{q_0}, \frac{p_1}{q_1}, \ldots, \frac{p_{n-1}}{q_{n-1}}, \frac{\hat{p}_n}{\hat{q}_n}.$$

In Theorem 8.12, a_n is defined to be the greatest integer less than or equal to α_n, and so

$$\alpha_n < a_n + 1. \tag{8.29}$$

From equation (8.23) in the proof of Theorem 8.16, we have

$$\left|\alpha - \frac{p_n}{q_n}\right| = \frac{1}{\hat{q}_{n+1}q_n} = \frac{1}{(\alpha_{n+1}q_n + q_{n-1})q_n}.$$

From (8.29) we have

$$\begin{aligned}\left|\alpha - \frac{p_n}{q_n}\right| &> \frac{1}{((a_{n+1}+1)q_n + q_{n-1})q_n} \\ &= \frac{1}{(q_{n+1}+q_n)q_n} \\ &\geq \frac{1}{q_{n+2}q_n}.\end{aligned}$$

We conclude that

$$|q_n\alpha - p_n| > \frac{1}{q_{n+2}}. \tag{8.30}$$

On the other hand, by Corollary 8.19,

$$\left|\alpha - \frac{p_{n+1}}{q_{n+1}}\right| < \frac{1}{q_{n+1}q_{n+2}}.$$

It follows that

$$|q_{n+1}\alpha - p_{n+1}| < \frac{1}{q_{n+2}}.$$

Comparing this inequality with that in (8.30) yields the inequality in (8.28).

□

In our computation of the convergents of π, the familiar rational approximation 22/7 turns up. How does this approximation stack up against others with reasonably small denominators? The *Mathematica* procedure below will search for rational approximations of π with denominators at most 106 (the denominator of the next convergent of π) that provide a better approximation than that of 22/7. For each such fraction a/b, the quantity $|b\pi - a|$ is also recorded in the third column.

Mathematica

```
Do[
  a = Round[b*Pi];
  If[Abs[Pi - a/b] < Abs[Pi - 22/7],
    Print[{a/b, N[Abs[Pi - a/b]], N[Abs[b*Pi - a]]}]],
  {b,1,106}]
```

Here are the corresponding *Maple* commands.

Maple

```
Digits:=20;
for b from 1 to 106 do
  a:=round(b*Pi):
  if abs(evalf(Pi)-a/b) < abs(evalf(Pi)-22/7) then
    print([a/b, evalf(abs(Pi-a/b)), evalf(abs(b*Pi-a))]);
  end if
end do:
```

The results of the procedure are presented in the following table.

a/b	$\lvert\pi - a/b\rvert$	$\lvert b\pi - a\rvert$
179/57	0.0012418	0.07078
201/64	0.0009677	0.06193
223/71	0.0007476	0.05308
245/78	0.0005670	0.04423
267/85	0.0004162	0.03538
289/92	0.0002883	0.02652
311/99	0.0001785	0.01767
333/106	0.0000832	0.00882

We observe that 333/106, a convergent of the continued fraction expansion, is the best approximation by fractions with denominators less than 107. Note also that

$$|7\pi - 22| \doteq 0.00885,$$

which is smaller than the corresponding value for any other fraction in the table apart from the subsequent convergent 333/106. The following theorem and corollary show that such statements can be made in general.

THEOREM 8.24
Let α be an irrational number, and let p_n/q_n be the nth convergent of the infinite simple continued fraction expansion of α. Suppose that a and b are integers with $0 < b < q_n$. Then

$$|q_{n-1}\alpha - p_{n-1}| \le |b\alpha - a|.$$

PROOF Suppose that $a/b = p_{n-1}/q_{n-1}$, and so $a = tp_{n-1}$ and $b = tq_{n-1}$ for some integer t. In this case, we have

$$|q_{n-1}\alpha - p_{n-1}| \le |t||q_{n-1}\alpha - p_{n-1}| = |b\alpha - a|.$$

We may now suppose that $a/b \ne p_{n-1}/q_{n-1}$.

We begin with consideration of the linear system of equations

$$p_{n-1}x + p_n y = a \tag{8.31}$$

$$q_{n-1}x + q_n y = b. \tag{8.32}$$

Using Theorem 8.8, we can solve this system (see Exercise 8.17) to obtain

$$x = (-1)^{n-1}(bp_n - aq_n) \tag{8.33}$$

$$y = (-1)^{n-1}(aq_{n-1} - bp_{n-1}). \tag{8.34}$$

We immediately see that x and y are integers. We claim that they are both nonzero. Observe that $y \neq 0$ since we are assuming $a/b \neq p_{n-1}/q_{n-1}$. If $x = 0$, then we would have $bp_n = aq_n$. By Corollary 8.9, $\gcd(p_n, q_n) = 1$. Thus, we would have $q_n \mid b$, which contradicts our supposition that $b < q_n$.

We further claim that x and y have opposite signs. Since q_n, q_{n-1}, and b are all positive, (8.32) implies that x and y cannot both be negative. On the other hand, if x and y were both positive, then we would have

$$b = q_{n-1}x + q_n y > q_n,$$

which is not the case.

We also claim that the quantities $(q_{n-1}\alpha - p_{n-1})$ and $(q_n\alpha - p_n)$ have opposite signs. This follows from Theorem 8.18, which tells us that either

$$\frac{p_n}{q_n} < \alpha < \frac{p_{n-1}}{q_{n-1}} \quad \text{or} \quad \frac{p_{n-1}}{q_{n-1}} < \alpha < \frac{p_n}{q_n}.$$

It now follows that the numbers $x(q_{n-1}\alpha - p_{n-1})$ and $y(q_n\alpha - p_n)$ have the same sign. Consequently,

$$|x(q_{n-1}\alpha - p_{n-1}) + y(q_n\alpha - p_n)| = |x(q_{n-1}\alpha - p_{n-1})| + |y(q_n\alpha - p_n)|. \tag{8.35}$$

From (8.31) and (8.32), we have

$$\begin{aligned} b\alpha - a &= (q_{n-1}x + q_n y)\alpha - (p_{n-1}x + p_n y) \\ &= x(q_{n-1}\alpha - p_{n-1}) + y(q_n\alpha - p_n). \end{aligned}$$

From (8.35), we conclude that

$$|b\alpha - a| = |x(q_{n-1}\alpha - p_{n-1})| + |y(q_n\alpha - p_n)|.$$

Since x is nonzero, we have

$$|b\alpha - a| \geq |q_{n-1}\alpha - p_{n-1}|.$$

$\square$

COROLLARY 8.25
Let α be an irrational number, and let p_n/q_n be the nth convergent of the infinite simple continued fraction expansion of α. Suppose that a and b are integers with $0 < b < q_n$. Then

$$|\alpha - p_n/q_n| < |\alpha - a/b|.$$

PROOF By Theorem 8.23,

$$|q_n\alpha - p_n| < |q_{n-1}\alpha - p_{n-1}| \leq |b\alpha - a|.$$

By hypothesis, $0 < 1/q_n < 1/b$, and so

$$|1/q_n||q_n\alpha - p_n| < |1/b||b\alpha - a|.$$

The result now follows. □

The theorem tells us that if we want a good rational approximation of an irrational number, then we should look at convergents of the continued fraction expansion. We can use the theorem to show that if a rational approximation is *too good*, then it has to be a convergent.

THEOREM 8.26
Let α be an irrational number, and let a and b be integers with $b > 0$. If

$$\left|\alpha - \frac{a}{b}\right| < \frac{1}{2b^2},$$

then $a/b = p_n/q_n$ for some convergent p_n/q_n in the infinite simple continued fraction expansion of α.

PROOF We first observe that our assumption implies

$$|b\alpha - a| < \frac{1}{2b}. \tag{8.36}$$

Since the sequence $\{q_n\}$ grows without bound, there exists a positive integer n such that

$$q_n \leq b < q_{n+1}.$$

By Theorem 8.24, we have

$$|q_n\alpha - p_n| \leq |b\alpha - a|,$$

and so by (8.36), we obtain

$$\left|\alpha - \frac{p_n}{q_n}\right| < \frac{1}{2bq_n}. \tag{8.37}$$

Now, by way of contradiction, suppose that $a/b \neq p_n/q_n$, and consequently $aq_n - bp_n \neq 0$. It follows that

$$\begin{aligned}\frac{1}{bq_n} &\leq \frac{|aq_n - bp_n|}{bq_n} \\ &= \left|\frac{a}{b} - \frac{p_n}{q_n}\right| \\ &= \left|\frac{a}{b} - \alpha + \alpha - \frac{p_n}{q_n}\right| \\ &\leq \left|\alpha - \frac{a}{b}\right| + \left|\alpha - \frac{p_n}{q_n}\right|.\end{aligned}$$

Putting equation (8.37) together with our assumption in the statement of the theorem, we have

$$\frac{1}{bq_n} < \frac{1}{2b^2} + \frac{1}{2bq_n}.$$

Multiplying the inequality through by $b^2 q_n$ yields

$$b < \frac{q_n}{2} + \frac{b}{2},$$

which contradicts the fact that $b \geq q_n$. Thus, we must have $a/b = p_n/q_n$. $\square$

The previous theorem asserts that *if there exists* a rational number that is a sufficiently good approximation of an irrational number, then it must be a convergent of the continued fraction. It does not assert that all (or even any) convergents will satisfy the given inequality. In fact, not all convergents will, but an infinite number of them will.

THEOREM 8.27
If α is an irrational number, then there are an infinite number of rational numbers a/b such that

$$\left|\alpha - \frac{a}{b}\right| < \frac{1}{2b^2}. \tag{8.38}$$

PROOF We will show that, for any n, either p_n/q_n or p_{n+1}/q_{n+1} satisfies the inequality in (8.38). Suppose, by way of contradiction, that this is not the case, and so

$$\left|\alpha - \frac{p_n}{q_n}\right| + \left|\alpha - \frac{p_{n+1}}{q_{n+1}}\right| > \frac{1}{2q_n^2} + \frac{1}{2q_{n+1}^2}. \tag{8.39}$$

(Since α is irrational, equality is not possible in (8.38), and so our strict inequality in (8.39) is justified.) By Theorem 8.18, we can rewrite the left side as

$$\left|\frac{p_n}{q_n} - \frac{p_{n+1}}{q_{n+1}}\right|,$$

which, by Corollary 8.10, is equal to

$$\frac{1}{q_{n+1}q_n}.$$

Thus, (8.39) becomes

$$\frac{1}{q_{n+1}q_n} > \frac{1}{2q_n^2} + \frac{1}{2q_{n+1}^2}.$$

Multiplying both sides of the inequality by $2q_n^2 q_{n+1}^2$, we have

$$2q_n q_{n+1} > q_{n+1}^2 + q_n^2.$$

This inequality is equivalent to

$$(q_{n+1} - q_n)^2 < 0,$$

and so we have reached a contradiction. □

We can actually even do a little better.

THEOREM 8.28 (Hurwitz)
If α is an irrational number, then there are an infinite number of rational numbers a/b such that

$$\left|\alpha - \frac{a}{b}\right| < \frac{1}{\sqrt{5}b^2}. \tag{8.40}$$

PROOF We will show that, for any n, at least one of p_n/q_n, p_{n+1}/q_{n+1}, or p_{n+2}/q_{n+2} satisfies the inequality in (8.40). Let's assume that this is not the case.

As in the proof of Theorem 8.27,

$$\left|\alpha - \frac{p_n}{q_n}\right| + \left|\alpha - \frac{p_{n+1}}{q_{n+1}}\right| = \frac{1}{q_n q_{n+1}}.$$

Our hypothesis that (8.40) does not hold for p_n/q_n and p_{n+1}/q_{n+1} implies that

$$\frac{1}{q_n q_{n+1}} > \frac{1}{\sqrt{5}q_n^2} + \frac{1}{\sqrt{5}q_{n+1}^2}.$$

(Again, the strict inequality is justified due to the irrationality of α.) Multiplying through by $\sqrt{5}q_n q_{n+1}$ yields the inequality

$$\sqrt{5} > \frac{q_{n+1}}{q_n} + \frac{q_n}{q_{n+1}}.$$

If we set $b_n = q_{n+1}/q_n$, then this inequality becomes

$$\sqrt{5} > b_n + \frac{1}{b_n},$$

from which it follows that

$$b_n^2 - \sqrt{5}b_n + 1 < 0.$$

Completing the square, we obtain

$$\left(b_n - \frac{\sqrt{5}}{2}\right)^2 < \frac{1}{4},$$

and so

$$b_n < \frac{1+\sqrt{5}}{2}. \tag{8.41}$$

Since we are also assuming that p_{n+2}/q_{n+2} does not satisfy (8.40), we also have

$$b_{n+1} < \frac{1+\sqrt{5}}{2}, \tag{8.42}$$

where $b_{n+1} = q_{n+2}/q_{n+1}$. Now $q_{n+2} \geq q_{n+1} + q_n$, and so

$$b_{n+1} \geq 1 + \frac{1}{b_n}. \tag{8.43}$$

Since $b_n > 0$, (8.41) implies that

$$\frac{1}{b_n} > \frac{2}{1+\sqrt{5}}.$$

Putting this together with (8.42) and (8.43), we conclude that

$$\frac{1+\sqrt{5}}{2} > 1 + \frac{2}{1+\sqrt{5}} = \frac{1+\sqrt{5}}{2}.$$

Thus, we have obtained our contradiction. ∎

This turns out to be the best that we can do. It is not true in general that there are an infinite number of rational numbers satisfying

$$\left|\alpha - \frac{a}{b}\right| < \frac{1}{cb^2},$$

unless $c \le \sqrt{5}$. To see why, let's try to determine what would be a worst-case scenario. We have the following crude estimate:

$$\begin{aligned}\left|\alpha - \frac{p_n}{q_n}\right| &\le \frac{1}{q_{n+1}q_n} \\ &= \frac{1}{(a_{n+1}q_n + q_{n-1})q_n} \\ &\le \frac{1}{a_{n+1}q_n^2}.\end{aligned}$$

This suggests that we look at

$$\alpha = [1; 1, 1, \ldots],$$

which is exactly the continued fraction example from the introduction to this chapter. We proved that this is the continued fraction expansion of the golden ratio,

$$\alpha = \frac{1+\sqrt{5}}{2}.$$

In fact, in Example 8.17, we showed that

$$\alpha = \left[1; 1, 1, \ldots, 1, \frac{1+\sqrt{5}}{2}\right].$$

If p_n/q_n is the nth convergent for α, then by Corollary 8.10,

$$\left|\alpha - \frac{p_n}{q_n}\right| = \frac{1}{\left(\left(\frac{1+\sqrt{5}}{2}\right)q_n + q_{n-1}\right)q_n} = \frac{1}{\left(\frac{1+\sqrt{5}}{2} + \frac{q_{n-1}}{q_n}\right)q_n^2}. \tag{8.44}$$

Since $a_n = 1$ for all n, we have $p_n = f_{n+2}$ and $q_n = f_{n+1} = p_{n-1}$. Thus,

$$\lim_{n\to\infty} \frac{q_n}{q_{n-1}} = \lim_{n\to\infty} \frac{p_{n-1}}{q_{n-1}} = \frac{1+\sqrt{5}}{2},$$

and so the coefficient of q_n^2 in the last expression in (8.44) is approaching

$$\frac{1+\sqrt{5}}{2} + \frac{2}{1+\sqrt{5}} = \sqrt{5}.$$

Consequently, if $c > \sqrt{5}$, then

$$\left|\alpha - \frac{p_n}{q_n}\right| > \frac{1}{cq_n^2}$$

for sufficiently large n.

One can show (see Exercise 8.18) that the set of continued fractions for which Hurwitz's theorem provides the best bound is exactly the collection of continued fractions whose partial quotients are eventually all equal to 1. An irrational number with such a continued fraction expansion is called a *noble number*.

Exercises

◇8.16 Find the best rational approximation a/b of the following numbers with $b < 100$.

(a) $\sqrt{2}$

(b) $\sqrt[3]{2}$

(c) e

8.17 In this exercise we will solve the system of equations (8.31) and (8.32) from the proof of Theorem 8.24.

(a) Multiply (8.31) by q_n and (8.32) by p_n and subtract. Now use Theorem 8.8 to obtain (8.33).

(b) Use a similar method to deduce (8.34).

8.18 Let $\alpha = [a_0; a_1, \ldots]$, an infinite simple continued fraction, and suppose that the partial quotients are not all eventually 1 (i.e., $a_n \geq 2$ for an infinite number of n). Show that there exists a constant $c > \sqrt{5}$ such that there exist an infinite number of rational numbers a/b with

$$\left|\alpha - \frac{a}{b}\right| < \frac{1}{cb^2}.$$

8.4 Periodic continued fractions

We have already discussed the parallel between decimal expansions and continued fraction expansions of real numbers. We know that numbers with finite decimal expansion and finite continued fraction expansion must be rational. As we have seen, only numbers with finite continued fraction expansion are rational. However, decimal number expansions of rational numbers can be infinite. We know that an infinite decimal expansion of a rational number must be periodic, i.e., its terms become repeating at some point. In fact, any periodic decimal expansion represents a rational number. What is the corresponding result for continued fractions?

DEFINITION 8.29 *An infinite continued fraction $\alpha = [a_0; a_1, \ldots]$ is* periodic *if there exist a positive integer s and an integer N such that $a_k = a_{k+s}$ for all $k \geq N$. The smallest such s is called the* period *of the continued fraction. For such a continued fraction, we will write*

$$\alpha = [a_0; a_1, \ldots, a_{N-1}, \overline{a_N, \ldots, a_{N+s-1}}].$$

When we may take $N = 0$, we call the continued fraction purely periodic, *and we will simply write*

$$\alpha = [\overline{a_0; a_1, \ldots, a_{s-1}}].$$

We observe that periodic infinite simple continued fractions necessarily represent irrational numbers. We have already encountered two irrational numbers with periodic continued fractions:

$$[1; \overline{1, 2}] = [1; 1, 2, 1, 2, 1, \ldots] = \sqrt{3},$$

$$[\overline{1};] = [1; 1, 1, 1, \ldots] = \frac{1 + \sqrt{5}}{2}.$$

We could say that these numbers are, in some sense, barely irrational.

DEFINITION 8.30 *A real number is called a* quadratic irrational *if it is a root of an irreducible degree 2 polynomial with integer coefficients.*

Using the quadratic formula, α is a quadratic irrational if and only if it can be expressed in the form

$$\alpha = q + r\sqrt{d}$$

where q and r are rational numbers, $r \neq 0$, and d is a positive integer that is not a perfect square (Exercise 8.19). In this section we will show that a number is a quadratic irrational if and only if its continued fraction expansion is periodic.

To begin, let's recall how to obtain the rational number represented by an infinite periodic decimal expansion. Suppose that $\alpha = 2.1\overline{345}$. Let $\beta = 0.\overline{345}$, and so

$$\alpha = 2.1 + \frac{1}{10}\beta. \tag{8.45}$$

Observe that

$$1000\beta = 345 + \beta,$$

and so we obtain

$$\beta = \frac{345}{999}.$$

Inserting this into (8.45), we get

$$\alpha = 2.1 + \left(\frac{1}{10}\right)\left(\frac{345}{999}\right) = \frac{21324}{9990}.$$

The proof that any periodic infinite simple continued fraction is a quadratic irrational is not altogether different. We will need a simple result about a transformation of a quadratic irrational.

LEMMA 8.31
Let β be a quadratic irrational, and let

$$\alpha = \frac{a\beta + b}{c\beta + f} \tag{8.46}$$

where a, b, c and f are integers with c and f not both zero. Then α is a quadratic irrational if and only if $af - bc \neq 0$.

PROOF By rationalizing the denominator of the right side of (8.46), we see that α is either a quadratic irrational or a rational number. We need to show that α is irrational if and only if $af - bc \neq 0$.

If $c \neq 0$, then

$$\alpha = \frac{a\beta + b}{c\beta + f} = \frac{a}{c} + \frac{bc - af}{c(c\beta + f)}.$$

The expression on the right is rational if and only if $bc - af = 0$.

If $c = 0$ and $f \neq 0$, then

$$\alpha = \frac{a\beta}{f} + \frac{b}{f}.$$

The expression in the right is rational if and only if $a/f = 0$. Since $f \neq 0$, this is equivalent to $af = 0$. ∎

THEOREM 8.32
If α is a real number with periodic infinite simple continued fraction expansion, then α is a quadratic irrational.

PROOF First, we deal with the special case in which the continued fraction is purely periodic. In this case we have $\alpha = [\overline{a_0; a_1, \ldots, a_{s-1}}]$, and so

$$\alpha = [a_0; a_1, \ldots, a_{s-1}, \alpha]. \tag{8.47}$$

Let $p_0/q_0, \ldots, p_{s-1}/q_{s-1}, \hat{p}_s/\hat{q}_s$ be the convergents of the continued fraction in (8.47). By Proposition 8.7, we have

$$\alpha = \frac{\hat{p}_s}{\hat{q}_s} = \frac{\alpha p_{s-1} + p_{s-2}}{\alpha q_{s-1} + q_{s-2}}.$$

(This expression for α still makes sense and is valid in the $s = 1$ case if we follow the convention $p_{-1} = 1$ and $q_{-1} = 0$.) Rearranging, we find that

$$q_{s-1}\alpha^2 + (q_{s-2} - p_{s-1})\alpha - p_{s-2} = 0.$$

Thus, α satisfies a degree 2 polynomial with integer coefficients. Since α has an infinite simple continued fraction expansion, it must be irrational and hence a quadratic irrational.

Now suppose that $\alpha = [a_0; a_1, \ldots, a_{N-1}, \overline{a_N, \ldots, a_{N+s-1}}]$ is a periodic infinite simple continued fraction with $N > 0$. Let $\beta = [\overline{a_N; a_{N+1}, \ldots, a_{N+s-1}}]$, and so

$$\alpha = [a_0; a_1, \ldots, a_{N-1}, \beta]. \tag{8.48}$$

Let p_0/q_0, ..., p_{N-1}/q_{N-1}, $\hat{p}_N/\hat{q}_N$ be the convergents of the continued fraction in (8.48). Again, by Proposition 8.7, we have

$$\alpha = \frac{\hat{p}_N}{\hat{q}_N} = \frac{\beta p_{N-1} + p_{N-2}}{\beta q_{N-1} + q_{N-2}}.$$

Since β is purely periodic, we know that it is a quadratic irrational. The fact that α is a quadratic irrational follows from Lemma 8.31 and Theorem 8.8. ☐

Example 8.33 Let's determine the quadratic irrational represented by the continued fraction

$$\beta = [\overline{1; 2, 3}].$$

We have

$$\beta = [1; 2, 3, \beta],$$

and if we let p_0/q_0, p_1/q_1, and p_2/q_2 be the first three convergents of this continued fraction, then

$$\beta = \frac{p_2\beta + p_1}{q_2\beta + q_1}. \tag{8.49}$$

We record the necessary values of p_n and q_n in the table below.

		1	2	3
0	1	1	3	10
1	0	1	2	7

Equation (8.49) now becomes

$$\beta = \frac{10\beta + 3}{7\beta + 2}.$$

Rearranging we get

$$7\beta^2 - 8\beta - 3 = 0.$$

Using the quadratic formula, we conclude that

$$\beta = \frac{4 - \sqrt{37}}{7} \text{ or } \frac{4 + \sqrt{37}}{7}.$$

Since β is clearly positive, it must be the case that

$$\beta = \frac{4 + \sqrt{37}}{7}.$$

□

Now we begin to work toward the proof of the converse of Theorem 8.32. By Exercise 8.20, a real quadratic irrational α can be written in the form

$$\alpha = \frac{P + \sqrt{d}}{Q} \tag{8.50}$$

where P, Q, and d are integers, d is a positive number that is not a perfect square, and Q divides $d^2 - P$. Let's explore the infinite simple continued fraction expansion of such an expression. By Theorem 8.12, we have

$$\begin{aligned} a_0 &= \lfloor \alpha \rfloor \\ a_1 &= \lfloor \alpha_1 \rfloor = \left\lfloor \frac{1}{\alpha - a_0} \right\rfloor. \end{aligned}$$

The number α_1 will also be a quadratic irrational (since it has a periodic continued fraction expansion), and so we should be able to write it in the form

$$\alpha_1 = \frac{P_1 + \sqrt{d}}{Q_1}. \tag{8.51}$$

We have

$$\begin{aligned} \alpha_1 &= \frac{1}{\dfrac{P + \sqrt{d}}{Q} - a_0} \\ &= \frac{Q}{(P - a_0 Q) + \sqrt{d}}. \end{aligned}$$

Multiplying the top and bottom by $\sqrt{d} - (P - a_0 Q)$, this becomes

$$\frac{Q(a_0 Q - P + \sqrt{d})}{d - (P - a_0 Q)^2},$$

which can be rewritten as

$$\frac{(a_0 Q - P) + \sqrt{d}}{(d - (a_0 Q - P)^2)/Q}.$$

Thus, if we take

$$\begin{aligned} P_1 &= a_0 Q - P, \\ Q_1 &= \frac{d - P_1^2}{Q}, \end{aligned}$$

then equation (8.51) will be satisfied. We note that P_1 is an integer. Since Q divides $d - P^2$, and

$$Q_1 = \frac{d - P^2}{Q} + 2a_0P - a_0^2Q,$$

Q_1 is also an integer. Repeating these calculations will lead to a recursive formula for the partial quotients in the continued fraction expansion of α.

LEMMA 8.34

Let α be a real quadratic irrational written in the form

$$\alpha = \frac{P_0 + \sqrt{d}}{Q_0}$$

where P_0, Q_0 and d are integers, $d \neq 0$ is not a perfect square, and Q_0 divides $d^2 - P_0$. For $n \geq 0$, define

$$\alpha_n = \frac{P_n + \sqrt{d}}{Q_n} \tag{8.52}$$

$$a_n = \lfloor \alpha_n \rfloor \tag{8.53}$$

$$P_{n+1} = a_nQ_n - P_n \tag{8.54}$$

$$Q_{n+1} = \frac{d - P_{n+1}^2}{Q_n}. \tag{8.55}$$

Then P_n and Q_n are integers for all n, and

$$\alpha = [a_0; a_1, a_2, \ldots].$$

PROOF We proceed by induction. The $n = 0$ case holds by the hypotheses of the lemma, and the argument preceding the lemma proves the $n = 1$ case. Now suppose that P_0, Q_0, P_1, Q_1, $\ldots$, P_n, Q_n are all integers. From (8.54) it follows that P_{n+1} is an integer. We can write

$$\begin{aligned} Q_{n+1} &= \frac{d - (a_nQ_n - P_n)^2}{Q_n} \\ &= \frac{d - P_n^2}{Q_n} + 2a_nP_n - a_n^2Q_n \\ &= \frac{Q_nQ_{n-1}}{Q_n} + 2a_nP_n - a_n^2Q_n \\ &= Q_{n-1} + 2a_nP_n - a_n^2Q_n. \end{aligned}$$

From our inductive hypothesis we conclude that Q_{n+1} is also an integer.

To show that $\alpha = [a_0; a_1, a_2, \ldots]$, it suffices (by Theorem 8.12) to prove that

$$\frac{P_{n+1} + \sqrt{d}}{Q_{n+1}} = \frac{1}{(P_n + \sqrt{d})/Q_n - a_n}.$$

This verification is similar to the one carried out prior to the statement of the lemma, and so we leave it as an exercise. □

LEMMA 8.35
With notation as in Lemma 8.34, suppose that $P_N = P_M$ and $Q_N = Q_M$ for some $N < M$. Let $s = M - N$. Then $a_{k+s} = a_k$ for all $k \geq N$, and consequently the infinite simple continued fraction expansion of α is periodic.

PROOF We will prove inductively that

$$P_{k+s} = P_k \tag{8.56}$$

$$Q_{k+s} = Q_k \tag{8.57}$$

$$\alpha_{k+s} = \alpha_k \tag{8.58}$$

$$a_{k+s} = a_k. \tag{8.59}$$

In our base case, $k = N$, we see (8.56) and (8.57) hold by hypothesis. Equation (8.58) follows from (8.56) and (8.57), and equation (8.59) follows from (8.58). Now suppose that (8.56) through (8.59) hold for $k = n$. Then

$$P_{n+1+s} = a_{n+s}Q_{n+s} - P_{n+s} = a_nQ_n - P_n = P_{n+1},$$

and

$$Q_{n+1+s} = \frac{d - P_{n+s+1}^2}{Q_{n+s}} = \frac{d - P_{n+1}^2}{Q_n} = Q_{n+1}.$$

With $k = n + 1$, (8.58) and (8.59) once again follow from (8.56) and (8.57). □

We have already shown that periodic infinite simple continued fractions represent quadratic irrationals (see Theorem 8.32). Lemma 8.35 will allow us to prove the converse of Theorem 8.32. Before we give that proof, we need one more ingredient.

DEFINITION 8.36 *Let $\alpha = q + r\sqrt{d}$ where q and r are rational numbers and d is a positive integer that is not a perfect square.* The conjugate *of α is*

$$\alpha' = q - r\sqrt{d}.$$

Assuming that $r \neq 0$, then α is a root of an irreducible degree 2 polynomial $f(x)$ with integer coefficients. Thus, the quadratic formula implies that the conjugate of α is the second root of $f(x)$.

LEMMA 8.37
Let $\alpha_1 = q_1 + r_1\sqrt{d}$ *and* $\alpha_2 = q_2 + r_2\sqrt{d}$ *be real quadratic irrationals. Then*

1. $(\alpha_1 + \alpha_2)' = \alpha_1' + \alpha_2'$
2. $(\alpha_1 - \alpha_2)' = \alpha_1' - \alpha_2'$
3. $(\alpha_1\alpha_2)' = \alpha_1'\alpha_2'$
4. $(\alpha_1/\alpha_2)' = \alpha_1'/\alpha_2'$.

The proofs of these identities are routine, and we leave them as exercises.

THEOREM 8.38
A real number α *has a periodic infinite simple continued fraction expansion if and only if it is a quadratic irrational.*

PROOF As we noted above, we have already proved half of this statement in the proof of Theorem 8.32. It remains to show that quadratic irrationals have periodic infinite simple continued fractions. Let α be a quadratic irrational, and write

$$\alpha = \frac{P_0 + \sqrt{d}}{Q_0}$$

for some integers P_0, Q_0, and d, with $Q_0 \mid d - P_0^2$. We again use the notation from Lemma 8.34.

As usual, let $p_0/q_0, \ldots, p_{n-1}/q_{n-1}, \hat{p}_n/\hat{q}_n$ be the convergents for the continued fraction $\alpha = [a_0; a_1, \ldots, a_{n-1}, \alpha_n]$. We have

$$\alpha = \frac{\hat{p}_n}{\hat{q}_n} = \frac{\alpha_n p_{n-1} + p_{n-2}}{\alpha_n q_{n-1} + q_{n-2}}.$$

Taking the conjugate of α and applying Lemma 8.37, we get

$$\alpha' = \frac{\alpha_n' p_{n-1} + p_{n-2}}{\alpha_n' q_{n-1} + q_{n-2}}.$$

Solving this equation for α_n', we find that

$$\alpha_n' = \frac{p_{n-2} - q_{n-2}\alpha'}{-p_{n-1} + q_{n-1}\alpha'} = \frac{q_{n-2}\left(\frac{p_{n-2}}{q_{n-2}} - \alpha'\right)}{-q_{n-1}\left(\frac{p_{n-1}}{q_{n-1}} - \alpha'\right)}. \tag{8.60}$$

Since p_n/q_n approaches α as n goes to infinity,

$$\lim_{n\to\infty} \frac{\left(\frac{p_{n-2}}{q_{n-2}} - \alpha'\right)}{\left(\frac{p_{n-1}}{q_{n-1}} - \alpha'\right)} = 1.$$

It now follows from (8.60) that $\alpha'_n < 0$ for all $n \geq N$, for some integer N. Since $\alpha_n > 0$ for $n \geq 1$, we have

$$\frac{2\sqrt{d}}{Q_n} = \alpha_n - \alpha'_n > 0$$

for $n \geq N$. It follows that $Q_n > 0$, and so from the definition of α_n we have $-\sqrt{d} < P_n$ for such n. On the other hand, when $\alpha'_n < 0$ we have $P_n < \sqrt{d}$, and so

$$-\sqrt{d} < P_n < \sqrt{d},$$

for $n \geq N$. Thus, there are a finite number of possible values of P_n.

Now from (8.55), we have

$$Q_n \leq Q_n Q_{n+1} = d - P_{n+1}^2 \leq d,$$

for $n \geq N$. Since Q_n is positive for such n, there are a finite number of possible values for Q_n as well. Consequently, there are only a finite number of possibilities for the ordered pairs (P_n, Q_n), and in particular, $(P_n, Q_n) = (P_m, Q_m)$, for some $n < m$. Lemma 8.35 now implies that the continued fraction expansion of α is periodic. ∎

Example 8.39 Using Lemma 8.35, we compute the simple continued fraction expansion of $(10 + \sqrt{11})/7$. The function `quadirrcontinuedfraction` defined below will report the continued fraction expansion of $(x + \sqrt{d})/y$ with the first period enclosed in brackets.

Mathematica

```
quadirrcontinuedfraction[x_, y_, d_] :=
  Module[{n, alpha, a, P, Q, ourlist},
  n = 0;
  P[n] = x;
  Q[n] = y;
  While[! MemberQ[ourlist =
      Table[{P[k], Q[k]}, k, 0, n - 1], {P[n], Q[n]}],
    alpha[n] = (P[n] + Sqrt[d])/Q[n];
    a[n] = Floor[alpha[n]];
    P[n + 1] = a[n] Q[n] - P[n];
    Q[n + 1] = (d - (P[n + 1])^2)/Q[n];
    n++];
  s = n - Flatten[Position[ourlist, {P[n], Q[n]}]][[1]] + 1;
  Append[Table[a[k], {k, 0, n - s - 1}],
  Table[a[k], {k, n - s, n - 1}]]
  ]
```

Maple

```
quadirrcontinuedfraction:=proc(x,y,d)
  local n, alpha, a, P, Q, ourlist, position, s, k;
  n:=0;
  P[n]:=x;
  Q[n]:=y;
  ourlist:=[];
  while not member([P[n],Q[n]],ourlist,'position') do
    alpha[n]:=(P[n]+sqrt(d))/Q[n];
    ourlist:=[op(ourlist),[P[n],Q[n]]];
    a[n]:=floor(alpha[n]);
    P[n+1]:=a[n]*Q[n]-P[n];
    Q[n+1]:=(d-(P[n+1])^2)/Q[n];
    n:=n+1;
    end do;
  s:=n-position+1;
  [seq(a[k],k=0..position-2),
    [seq(a[k],k=position-1..position+s-2)]];
end proc:
```

Computing `quadirrcontinuedfraction` on the triple $\{10, 7, 11\}$, we obtain the following.

```
[1, 1, [9, 4, 9, 23]]
```

We conclude that

$$\frac{10+\sqrt{11}}{7} = [1; 1, \overline{9, 4, 9, 23}].$$

□

Recall that a continued fraction is *purely periodic* if it has the form

$$[\overline{a_0; a_1, \ldots, a_{s-1}}].$$

In Example 8.33, we found that

$$\frac{4+\sqrt{37}}{7} = [\overline{1; 2, 3}],$$

and so the continued fraction expansion of $(4+\sqrt{37})/7$ is purely periodic. On the other hand, our calculation in Example 8.39 shows that the continued fraction expansion of $(10+\sqrt{11})/7$ is not purely periodic. We close this section with a result that classifies the collection of quadratic irrationals that have purely periodic continued fraction expansions.

THEOREM 8.40

Let α be a real quadratic irrational. The infinite simple continued fraction expansion of α is purely periodic if and only if $\alpha > 1$ and $-1 < \alpha' < 0$.

Note: Quadratic irrationals that satisfy the hypotheses of the theorem are called *reduced*.

PROOF Suppose that α is a reduced quadratic irrational. Recall from Theorem 8.12 that we obtain the quantities a_n and α_n by first setting $\alpha_0 = \alpha$ and recursively defining

$$a_n = \lfloor \alpha_n \rfloor \tag{8.61}$$

$$\alpha_{n+1} = \frac{1}{\alpha_n - a_n}. \tag{8.62}$$

We claim that

$$-1 < \alpha_n' < 0, \tag{8.63}$$

for all $n \geq 0$. Since α is reduced, this certainly holds for $n = 0$. Now by way of induction, suppose that $-1 < \alpha_n' < 0$. Rearranging (8.62), we obtain

$$\alpha_n = \frac{1}{\alpha_{n+1}} + a_n. \tag{8.64}$$

Taking conjugates of both sides and using Lemma 8.37, this becomes

$$\alpha_n' = \frac{1}{\alpha_{n+1}'} + a_n. \tag{8.65}$$

Employing our inductive hypothesis, we conclude that

$$\frac{1}{\alpha_{n+1}'} + a_n < 0,$$

or equivalently

$$\frac{1}{\alpha_{n+1}'} < -a_n. \tag{8.66}$$

By definition $a_n \geq 1$ for all $n \geq 1$, and since $\alpha > 1$, we also have $a_0 \geq 1$. Thus, (8.66) implies that

$$\frac{1}{\alpha_{n+1}'} < -1,$$

and so $-1 < \alpha_{n+1}' < 0$.

Putting equations (8.63) and (8.65) together we have

$$-1 < \frac{1}{\alpha_{n+1}'} + a_n < 0.$$

We may conclude from this that

$$1 + \left\lfloor \frac{1}{\alpha'_{n+1}} \right\rfloor = -a_n. \tag{8.67}$$

The proofs of Lemma 8.35 and Theorem 8.38 imply that there exist integers $N < M$ such that $\alpha_N = \alpha_M$. For such N and M, we also have $\alpha'_N = \alpha'_M$, and so by (8.67),

$$a_{N-1} = a_{M-1}.$$

Using equation (8.64), we conclude that

$$\alpha_{N-1} = \alpha_{M-1}.$$

Repeating this argument $N-1$ times, we eventually are led to the conclusion that $\alpha_0 = \alpha_{M-N}$. We now have

$$\begin{aligned} \alpha &= [a_0; a_1, \ldots, a_{M-N-1}, \alpha_{M-N}] \\ &= [a_0; a_1, \ldots, a_{M-N-1}, \alpha_0] \\ &= [a_0; a_1, \ldots, a_{M-N-1}, \alpha] \\ &= [\overline{a_0; a_1, \ldots, a_{M-N-1}}], \end{aligned}$$

and so the infinite simple continued fraction expansion of α is purely periodic.

We now prove the converse. We assume that the infinite simple continued fraction expansion of α is purely periodic, and so we can write

$$\alpha = [\overline{a_0; a_1, \ldots, a_k}]$$

for some integer $k \geq 1$. (Note that even when the period of the continued fraction is 1, we may take $k = 1$ by writing $\alpha = [\overline{a_0; a_1}]$.) Since

$$\lfloor \alpha \rfloor = a_0 = a_{k+1} \geq 1,$$

we see that $\alpha > 1$. We may also write

$$\alpha = [a_0; a_1, \ldots, a_k, \alpha].$$

If p_0/q_0, p_1/q_1, $\ldots$, p_k/q_k, $\hat{p}_{k+1}/\hat{q}_{k+1}$ are the convergents for this continued fraction, then by Proposition 8.7 we have

$$\alpha = \frac{\hat{p}_{k+1}}{\hat{q}_{k+1}} = \frac{\alpha p_k + p_{k-1}}{\alpha q_k + q_{k-1}}.$$

(The second equality is valid because $k \geq 1$.) Rearranging, we find that

$$q_k \alpha^2 + (q_{k-1} - p_k)\alpha - p_{k-1} = 0.$$

Thus, the polynomial

$$f(x) = q_k x^2 + (q_{k-1} - p_k)x - p_{k-1}$$

has α as a root. We know that α' must be the second root of $f(x)$. Since $\alpha > 1$, we need to show only that $f(x)$ has a root between -1 and 0 to complete the proof that α is reduced. We observe that

$$f(-1) = (q_k - q_{k-1}) + (p_k - p_{k-1})$$
$$f(0) = -p_{k-1}.$$

Since $a_0 > 0$, we see from the definitions of p_{k-1}, p_k, q_{k-1}, and q_k that $q_k - q_{k-1} \geq 0$ and $p_k - p_{k-1} > 0$, and so $f(-1) > 0$. On the other hand, $p_{k-1} > 0$ (since $a_0 > 0$), and so $f(0) < 0$. Thus, the Intermediate Value Theorem tells us that $f(x)$ has a root between -1 and 0. Since $\alpha > 1$, this root must be α'. $\square$

Referring back to Examples 8.33 and 8.39, we observe that $\frac{4+\sqrt{37}}{7} \doteq 1.4$ and $\frac{4-\sqrt{37}}{7} \doteq -0.3$, but $\frac{10-\sqrt{11}}{7} \doteq 0.7$.

Exercises

8.19 Show that α is a quadratic irrational if and only if it can be expressed in the form

$$\alpha = q + r\sqrt{d},$$

where q and r are rational numbers and d is an integer that is not a perfect square.

8.20 Let α be a quadratic irrational. Show that α can be written in the form

$$\alpha = \frac{P+\sqrt{d}}{Q},$$

where P, Q, and d are integers, d is not a perfect square, and Q divides $d - P^2$.

8.21 Write the following periodic continued fractions in the form $r + q\sqrt{d}$ where r and q are rational numbers and d is an integer with no perfect square divisors greater than 1.

(a) $[\overline{3; 1, 5, 7}]$

(b) $[4; 2, \overline{1, 2, 3}]$

8.22 Prove Lemma 8.37.

8.23 Let d be an integer that is not a perfect square.

(a) Show that $\sqrt{d}$ is irrational. That is, show that $d \neq (a/b)^2$ for any integers a and b.
Hint: Use the Fundamental Theorem of Arithmetic.

(b) Let x and y be rational numbers. Show that

$$x + y\sqrt{d} = 0,$$

if and only if $x = 0$ and $y = 0$.

(c) Let x_1, x_2, y_1, and y_2 be rational numbers. Show that

$$x_1 + y_1\sqrt{d} = x_2 + y_2\sqrt{d},$$

if and only if $x_1 = x_2$ and $y_1 = y_2$.

8.24 Let $F_n = 2^{2^n} + 1$ be the nth Fermat number. Show that

$$\sqrt{F_n} = [2^{2^{n-1}}; \overline{2^{2^{n-1}+1}}].$$

8.25 Let $\alpha = P + \sqrt{d}$ where P and $d > 0$ are integers and d is not a perfect square. Show that α is reduced if and only if $P = \lfloor\sqrt{d}\rfloor$.

8.26 Let $\alpha = (P + \sqrt{d})/Q$ be a reduced quadratic irrational and define Q_n as in Lemma 8.34. Show that

$$0 < Q_n < 2\sqrt{d}.$$

8.5 Continued fraction factorization

In this section, we will describe an algorithm for factoring integers that makes use of the theory that we developed in the preceding sections. An implementation of this algorithm was used by Morrison and Brillhart [25] to obtain the first factorization of the Fermat number $F_7 = 2^{2^7} + 1$.

Suppose that an odd number d can be represented in the form $x^2 - y^2 = d$. Factoring, we have

$$(x - y)(x + y) = d.$$

Conversely, given a factorization $d = ab$, the system of equations

$$a = x - y$$
$$b = x + y$$

has the unique solution

$$x = \frac{a + b}{2}$$
$$y = \frac{-a + b}{2}.$$

Since d is odd, a and b are necessarily odd, and so x and y are indeed integers. Thus, we have proved the following statement.

PROPOSITION 8.41

Let d be an odd positive integer. There is a one-to-one correspondence between factorizations

$$d = ab$$

and solutions to the equation

$$x^2 - y^2 = d. \tag{8.68}$$

We can use this observation to create an algorithm (called Fermat factorization) for factoring integers. Beginning with $x = \lceil \sqrt{d} \rceil$ (the least integer greater than or equal to $\sqrt{d}$), we check whether $x^2 - d$ is a perfect square. If so, we obtain the factorization

$$d = (x - y)(x + y),$$

where $y = \sqrt{x^2 - d}$. If the quantity $x^2 - d$ is not a perfect square, we increase x by 1 and try again.

Mathematica

```
fermatfactor[d_] :=
  Module[{x},
    x = Ceiling[Sqrt[d]];
    While[Sqrt[x^2 - d] != Floor[Sqrt[x^2 - d]], x++];
    {x + Sqrt[x^2 - d], x - Sqrt[x^2 - d]}]
```

We test our procedure with $d = 320813$.

Mathematica

```
fermatfactor[320813]

{593, 541}
```

Here are the corresponding *Maple* commands.

Maple

```
fermatfactor:=proc(d)
  local x;
  x:=ceil(sqrt(d));
  while not sqrt(x^2-d)=floor(sqrt(x^2-d)) do
    x:=x+1;
  end do;
  [x+sqrt(x^2-d),x-sqrt(x^2-d)];
end proc:

fermatfactor(320813);

[593, 541]
```

In light of Proposition 8.41, the procedure necessarily terminates when d is odd, but in the event that d is prime, it will not terminate until it reaches the factorization $d = d \cdot 1$.

If we happen to know, for instance, that d is a product of two odd primes that are about the same size, this procedure will find the factorization much faster than simply checking each possible factor between 2 and $\sqrt{n}$. However, for a randomly chosen odd number d, there is no reason to expect that our `fermatfactor` procedure will be faster than checking every possible factor. We can improve our luck by instead trying to find a solution to the congruence

$$x^2 - y^2 \equiv 0 \pmod{d}.$$

The following proposition tells us that such a solution, while weaker than the equality $x^2 - y^2 = d$, is still likely to produce a factorization of d.

PROPOSITION 8.42
Suppose that x and y are integers such that $x \not\equiv \pm y \pmod{d}$. If

$$x^2 - y^2 \equiv 0 \pmod{d},$$

then $\gcd(x + y, d)$ *and* $\gcd(x - y, d)$ *are proper divisors of* d.

PROOF Let $g = \gcd(x + y, d)$. Observe that $g \neq d$ since $x \not\equiv -y \pmod{d}$. We have

$$d \mid (x + y)(x - y).$$

If $g = 1$, then d must divide $x - y$. This contradicts our assumption that $x \not\equiv y \pmod{d}$. Hence, g is a proper divisor of d. A similar argument shows that $\gcd(x - y, d)$ is also a proper divisor of d. ∎

We now can try the following strategy for factoring d.

(1) Construct a list of congruences

$$\begin{aligned} x_1^2 &\equiv z_1 \pmod{d} \\ x_2^2 &\equiv z_2 \pmod{d} \\ &\vdots \\ x_r^2 &\equiv z_r \pmod{d}. \end{aligned}$$

(2) Find a subsequence $z_{i_1}, \ldots, z_{i_k}$ such that $z_{i_1} \ldots z_{i_k}$ is a square (or even just a square mod d).

(3) Set $x = x_{i_1} \ldots x_{i_k}$ and $y = \sqrt{z_{i_1} \ldots z_{i_k}}$. If $x \not\equiv \pm y \pmod{d}$, then Proposition 8.42 tells us that $\gcd(x+y, d)$ and $\gcd(x-y, d)$ are proper factors of d.

To complete step (2), we will need to factor the z_i, and so we would like to generate these congruences in such a way as to have these numbers be as small as possible. The following theorem will provide a list of congruences as in (1) through the computation of the continued fraction expansion $\sqrt{d}$.

THEOREM 8.43
Let d be a positive integer that is not a perfect square. Let p_n/q_n be the nth convergent of the continued fraction expansion of $\sqrt{d}$, and define Q_n as in Lemma 8.34 for $\alpha = \sqrt{d}$. Then

$$p_n^2 - dq_n^2 = (-1)^{n-1} Q_{n+1} \tag{8.69}$$

for $n \geq 0$.

PROOF We will also define a_n, α_n, and P_n as in Lemma 8.34. We first check the $n = 0$ case. We have

$$\begin{aligned} p_0 &= a_0 \\ q_0 &= 1 \\ P_0 &= 0 \\ Q_0 &= 1 \\ P_1 &= a_0 Q_0 - P_0 = a_0 \\ Q_1 &= \frac{d - P_1^2}{Q_0} = d - a_0^2. \end{aligned}$$

Hence,

$$p_0^2 - dq_0^2 = a_0^2 - d = (-1)^{-1} Q_1,$$

and so (8.69) holds for $n = 0$.

By Lemma 8.34, we have

$$\sqrt{d} = [a_0; a_1, \ldots, a_n, \alpha_{n+1}],$$

where

$$\alpha_{n+1} = \frac{P_{n+1} + \sqrt{d}}{Q_{n+1}}.$$

For $n \geq 1$, we may apply Proposition 8.7 to obtain

$$\sqrt{d} = \frac{\alpha_{n+1}p_n - p_{n-1}}{\alpha_{n+1}q_n + q_{n-1}} = \frac{\left(\dfrac{P_{n+1} + \sqrt{d}}{Q_{n+1}}\right) p_n + p_{n-1}}{\left(\dfrac{P_{n+1} + \sqrt{d}}{Q_{n+1}}\right) q_n + q_{n-1}}.$$

From this equality, we deduce that

$$dq_n + (P_{n+1}q_n + Q_{n+1}q_{n-1})\sqrt{d} = (P_{n+1}p_n + Q_{n+1}p_{n-1}) + p_n\sqrt{d}.$$

Using Exercise 8.23, we obtain the pair of relations

$$P_{n+1}q_n + Q_{n+1}q_{n-1} = p_n, \tag{8.70}$$

$$P_{n+1}p_n + Q_{n+1}p_{n-1} = dq_n. \tag{8.71}$$

Multiplying (8.70) by p_n and (8.71) by q_n, we find that

$$p_n^2 - dq_n^2 = (p_n q_{n-1} - p_{n-1}q_n)Q_{n+1}.$$

Equation (8.69) now follows with the help of Theorem 8.8.

□

The theorem tells us that, for every $n \geq 0$, we have

$$p_n^2 \equiv (-1)^{n-1}Q_{n+1} \pmod{d}.$$

Using Exercise 8.26 and the proof of the forthcoming Lemma 9.59, one can show that $Q_n < 2\sqrt{d}$, and hence it is small relative to the size of d.

Example 8.44 We attempt to factor the number $d = 320813$ by examining the congruences provided by Theorem 8.43. The *Mathematica* function defined below will provide us with the quantities p_n (reduced mod d) and $(-1)^{n-1}Q_{n+1}$ (in factored form), for $n = 0, 1, \ldots, k$.

Mathematica

```
cfcongruences[d_, k_] :=
  Module[{n, alpha, a, p, P, Q, g,},
    P[0] = 0;
    Q[0] = 1;
    p[-1] = 1;
    p[0] = Floor[Sqrt[d]];
    Do[
      P[n + 1] = (Floor[(P[n] + Sqrt[d])/Q[n]])Q[n] - P[n];
      Q[n + 1] = (d - (P[n + 1])^2)/Q[n];
      p[n + 1] = Floor[(P[n + 1] + Sqrt[d])/Q[n + 1]] p[n] +
        p[n - 1];
      Print[{n, Mod[p[n], d],
        FactorInteger[(-1)^(n - 1)Q[n + 1]]}],
      {n, 0, k}]]
```

Here are the corresponding *Maple* commands.

Maple

```
cfcongruences:=proc(d,k)
  local n, alpha, a, p, P, Q;
  P[0]:=0;
  Q[0]:=1;
  p[-1]:=1;
  p[0]:=floor(sqrt(d));
  for n from 0 to k do
    P[n+1]:=(floor((P[n]+sqrt(d))/Q[n]))*Q[n]-P[n];
    Q[n+1]:=(d-(P[n+1])]^2)/Q[n];
    p[n+1]:=floor((P[n+1]+sqrt(d))/Q[n+1])*p[n]+p[n-1];
    print([n,modp(p[n],d),ifactor((-1)^(n-1)*Q[n+1])]);
  end do;
end proc:
```

We now give the output when these procedures are evaluated on the pair $d = 320813$ and $k = 15$ (i.e., when we evaluate `cfcongruences[320813, 15]` in *Mathematica* or `cfcongruences(320813,15);` in *Maple*).

0	566	$-1 \cdot 457$
1	1133	$19 \cdot 23$
2	2832	$-1 \cdot 101$
3	29453	857
4	32285	$-1 \cdot 2^2 \cdot 53$
5	158593	449
6	28658	$-1 \cdot 2^2 \cdot 79$
7	244567	$11 \cdot 13$
8	136562	$-1 \cdot 659$
9	60316	$2^2 \cdot 109$
10	196878	$-1 \cdot 19 \cdot 31$
11	257194	353
12	69640	$-1 \cdot 521$
13	6021	$2^2 \cdot 11 \cdot 13$
14	75661	$-1 \cdot 251$
15	233004	839

With $n = 7$ we have

$$244567^2 \equiv 11 \cdot 13 \pmod{320813},$$

and with $n = 13$ we have

$$6021^2 \equiv 2^2 \cdot 11 \cdot 13 \pmod{320813}.$$

Putting these two together, we have

$$(244567 \cdot 6021)^2 \equiv (2 \cdot 11 \cdot 13)^2 \pmod{320813}.$$

We compute

$$\gcd(320813, 244567 \cdot 6021 + (2 \cdot 11 \cdot 13)) = 593.$$

Our procedure has led us to the proper divisor 593 of 320813. ∎

Some comments are in order. First, we cannot always expect to have success by merely considering the first 15 congruences. As the list of congruences grows, it becomes impractical to attempt to observe an appropriate combination of congruences. Second, if the period of the continued fraction expansion is small, we may not have enough congruences to produce a factorization. For instance, if we attempt to use the procedure above to factor the nth Fermat number F_n, we find

$$\sqrt{F_n} = [a_0; \overline{a_1}].$$

In such situations, we can instead consider the continued fraction expansion of $\sqrt{kd}$ for any positive integer k. Thus, we are using the procedure to find divisors of kd, which we hope will lead to divisors of d itself. For details on how to deal with both of these issues, we refer the reader to the paper by Morrison and Brillhart [25] in which they use this method to factor F_7.

Exercises

8.27 Given that the equation

$$x^2 - y^2 = 58853$$

has a solution $x = 243$ and $y = 14$, find a factorization of 58853.

$\diamond$**8.28** Given that the congruence

$$x^2 \equiv y^2 \pmod{62059}$$

has a solution $x = 20603$ and $y = 7$, find a factorization of 62059.

$\diamond$**8.29** Let $n = 15397$. Given the following set of congruences, find a factorization of n.

$$\begin{aligned}
124^2 &\equiv (-1) \cdot 3 \cdot 7 \pmod{n} \\
1365^2 &\equiv 2^2 \cdot 47 \pmod{n} \\
1489^2 &\equiv (-1) \cdot 47 \pmod{n} \\
7321^2 &\equiv 2^2 \cdot 3 \cdot 7 \pmod{n} \\
734^2 &\equiv (-1) \cdot 139 \pmod{n} \\
8055^2 &\equiv 67 \pmod{n} \\
9502^2 &\equiv (-1) \cdot 2^2 \pmod{n} \\
2591^2 &\equiv 3^3 \cdot 7 \pmod{n}
\end{aligned}$$

$\diamond$**8.30** Use `cfcongruences` (or your own procedure) to produce a list of congruences leading to a solution to the congruence

$$x^2 - y^2 \equiv 0 \pmod{4659829},$$

and a factorization of 4659829.

8.6 Notes

Continued fraction expansions of e

It is not surprising that e has a nice infinite simple continued fraction expansion, and it is perhaps also not surprising that Euler discovered it. The expansion is

$$e = [2; 1, 2, 1, 1, 4, 1, 1, 6, 1, 1, \ldots].$$

One can generalize the notion of continued fractions to allow the "numerators" to be numbers other than 1. Thus, we may consider expressions of the form

$$a_0 + \cfrac{b_1}{a_1 + \cfrac{b_2}{a_2 + \ddots + \cfrac{b_{n-1}}{a_{n-1} + \cfrac{b_n}{a_n}}}}.$$

Such an expression can be written in the form $\mathcal{P}_n/\mathcal{Q}_n$, where $\mathcal{P}_n$ and $\mathcal{Q}_n$ are defined recursively in a fashion similar to p_n and q_n (see Exercise 8.14). When a sequence of such $\mathcal{P}_n/\mathcal{Q}_n$ converges, we get an infinite generalized continued fraction expression of a number.

With a little bit of effort (see Exercise 8.31), one can show that e also has the following generalized continued fraction expansion:

$$e = 2 + \cfrac{2}{2 + \cfrac{3}{3 + \cfrac{4}{4 + \cfrac{5}{\ddots}}}}.$$

Continued fraction expansion of $\tan x$

In 1761 Johann Heinrich Lambert (1728–1777) (see [20]) obtained a generalized continued fraction of $\tan x$:

$$\tan x = \cfrac{x}{1 - \cfrac{x^2}{3 - \cfrac{x^2}{5 - \cfrac{x^2}{\ddots}}}}. \tag{8.72}$$

Lambert also showed that the if x is rational, then the right side of (8.72) is irrational. Since $\tan(\pi/4) = 1$, we may conclude that $\pi/4$ (and hence π) is irrational. This was, in fact, the first proof of the irrationality of π.

Srinivasa Ramanujan

Among his many contributions to the theory of numbers, Srinivasa Ramanujan is particularly noted for his wizardry with continued fraction expansions.

Ramanujan was born in India in 1887, and by the time he had reached his early twenties was publishing work in the *Journal of the Indian Mathematical Society*, despite having had little formal education at the university level. Ramanujan had a great gift for spotting patterns and making generalizations based on a few sample computations. However, his lack of a formal mathematical training made it difficult for him to provide (or see the need for) rigorous proofs for his results.

In 1912 he began sending letters to prominent British mathematicians of the day to get some feedback on his work. His attempts were, for the most part, rebuffed until he wrote to Hardy. The letter included a list of theorems with either no proofs or vague justifications. Some of these included continued fraction expansions such as

$$\left(\sqrt{\frac{5+\sqrt{5}}{2}}-\frac{\sqrt{5}+1}{2}\right)e^{2\pi/5}=\cfrac{1}{1+\cfrac{e^{-2\pi}}{1+\cfrac{e^{-4\pi}}{1+\cfrac{e^{-6\pi}}{\ddots}}}}.$$

At first Hardy did not know what to make of Ramanujan's claims. Eventually, he decided there was something to them, for as he said, "if they were not true, no one would have had the imagination to invent them." In 1914 Hardy arranged for Ramanujan to come to Cambridge, where the two began collaborative work.

Ramanujan suffered through poor health during his time at Cambridge and ultimately succumbed in 1920 shortly after returning to India. He left behind a collection of notebooks containing a vast quantity of unpublished theorems and other observations. A great amount of effort has been spent in the years since pouring over these manuscripts and filling in missing proofs.

Exercises

⋆8.31 Show that

$$e=2+\cfrac{2}{2+\cfrac{3}{3+\cfrac{4}{4+\cfrac{5}{\ddots}}}}.$$

Hint: Apply Exercise 8.15 to e^{-1} and then use Exercise 8.14(c).

⋆8.32 In a manner similar to the way we defined continued fraction expansions, we can define "continued square root expansions." For n a positive integer, let k^2 be the largest square that divides n. Then

$$n = k^2 n' = k^2 \sqrt{1 + (n')^2 - 1},$$

where $n' = n/k^2$. This process is continued with n replaced by $(n')^2 - 1$; for example,

$$\begin{aligned} 3 &= \sqrt{1+8} \\ &= \sqrt{1+4\sqrt{4}} \\ &= \sqrt{1+4\sqrt{1+3}}. \end{aligned}$$

At this point, we have completed a loop. Thus, we have the representation

$$3 = \sqrt{1 + 4\sqrt{1 + 1\sqrt{1 + 4\sqrt{1 + \cdots}}}}. \tag{8.73}$$

(a) Prove that the expression on the right side of (8.73) converges to 3.

(b) Find

$$\sqrt{1 + 2\sqrt{1 + 3\sqrt{1 + 4\sqrt{1 + \cdots}}}}.$$

This problem is due to Ramanujan.

Chapter 9

Diophantine Equations

> Cubum autem in duos cubos, aut quadratoquadratum in duos quadratoquadratos & generaliter nullam in infinitum ultra quadratum potestatem in duos eiusdem nominis fas est diuidere cuius rei demonstrationem mirabilem sane detexi. Hanc marginis exiguitas non caperet.
>
> [It is impossible to separate a cube into two cubes, or a fourth power into two fourth powers, or in general, any power higher than the second into two like powers. I have discovered a truly marvelous proof of this, which this margin is too narrow to contain.]
>
> PIERRE DE FERMAT (1823–1891)

A Diophantine equation is a polynomial equation with integer coefficients in which the variables are required to take integer values.

In this chapter, we investigate several types of Diophantine equations, starting with linear equations and progressing to other, more intricate types. In particular, we examine the Pythagorean formula, Pell's equation, and several equations of Fermat. Along the way, we discover beautiful uses of recurrence relations, mathematical induction (in the guise of the method of descent), and continued fractions. We also investigate Gaussian integers, which are complex numbers that behave much like ordinary integers.

9.1 Linear equations

We began our study of Diophantine equations in Chapter 2 at which point we classified solutions to linear Diophantine equations. As a result of Theorem 2.26, we know that the Diophantine equation

$$ax + by = c \tag{9.1}$$

has solutions if and only if $g = \gcd(a, b)$ divides c. We further showed (Theorem 2.30) that if (x_0, y_0) is a solution to (9.1), then any other solution is of the form $(x_0 + mb/g, y_0 - ma/g)$, for some integer m.

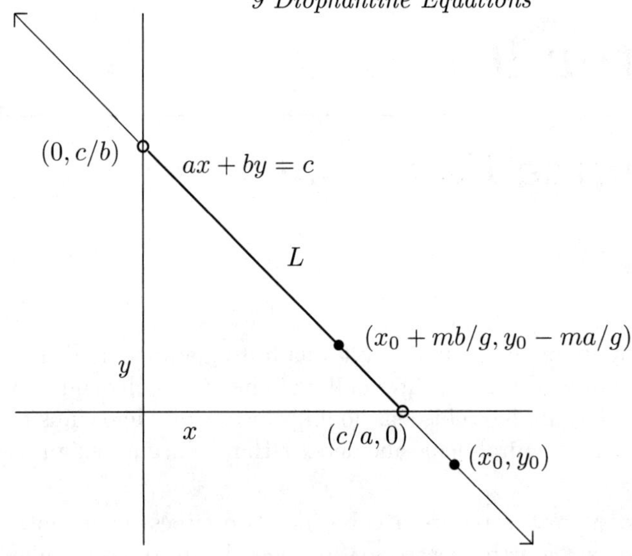

FIGURE 9.1: Solutions to $ax + by = c$.

In some situations, we are assured of a solution to (9.1) with x and y *positive* integers.

THEOREM 9.1

Suppose that a and b are positive integers. Let $g = \gcd(a, b)$. If $g \mid c$ and $c > ab/g$, then the equation $ax + by = c$ has a solution in positive integers x and y.

PROOF The equation $ax + by = c$ represents a line in the Euclidean plane passing through the points $(c/a, 0)$ and $(0, c/b)$. Let L be the part of the line lying in the first quadrant; that is, L is the (open) line segment joining $(c/a, 0)$ and $(0, c/b)$. See Figure 9.1. We wish to show that L contains a point (x, y) such that x and y are integers. The length l of L is given by

$$l = \sqrt{\left(\frac{c}{a}\right)^2 + \left(\frac{c}{b}\right)^2} = \frac{c\sqrt{a^2 + b^2}}{ab}.$$

Since $\gcd(a, b) \mid c$, the equation has an integer solution (x_0, y_0), and we know from Theorem 2.30 that all integer solutions may be written in the form $(x_0 + mb/g, y_0 - ma/g)$. The Euclidean distance between two solutions corresponding to consecutive values of m is $\sqrt{a^2/g^2 + b^2/g^2} = \sqrt{a^2 + b^2}/g$. By hypothesis, this quantity is less than l. It follows that there is some integer solution (x, y) on L. ∎

Example 9.2 Consider the equation

$$12x + 13y = 200.$$

Since $\gcd(12, 13) = 1$, and $200 > 12 \cdot 13$, there exists a solution in positive integers x and y. Indeed, $(x, y) = (8, 8)$ is such a solution. □

Exercises

9.1 Find a solution (x, y) to the equation

$$100x + 97y = 9701,$$

with x and y positive integers.

◇**9.2** Use a computer to find a solution (x, y) to the equation

$$123123123x + 1999y = 2093119078,$$

with x and y positive integers.

9.3 Sylvester sent this number theory puzzle to the *Education Times*: "I have a large number of stamps to the value of 5d and 17d only. What is the largest denomination which I cannot make up with a combination of these two different values?" Solve Sylvester's puzzle. (The 'd' is the old abbreviation for 'penny' in British currency.)

9.4 Show that it is always possible to solve the equation

$$\phi(pq) = rp + sq,$$

in nonnegative integers r, s, where p and q are distinct primes. (See the discussion at the end of Section 6.4.)

9.5 Suppose that $g = \gcd(a, b) \mid c$. Show that the Diophantine equation

$$ax + by = c$$

has a solution (x_0, y_0) with $0 \le x_0 < |b|/g$.

⋆**9.6** Suppose that $g = \gcd(a, b) \mid c$. Show that the Diophantine equation

$$ax + by = c$$

has a solution (x_0, y_0) satisfying

$$x_0^2 + y_0^2 \le \frac{c^2}{a^2 + b^2} + \frac{a^2 + b^2}{4g^2}.$$

†**9.7** Show that the Diophantine equation

$$a_1x_1 + a_2x_2 + \cdots + a_nx_n = c$$

has a solution if and only if $\gcd(a_1, a_2, \ldots, a_n) \mid c$.

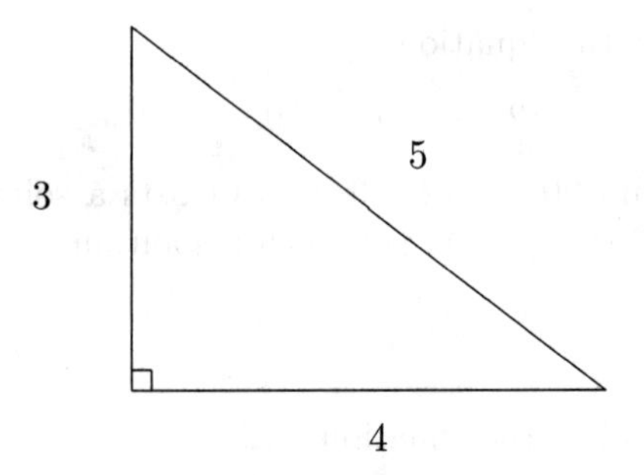

FIGURE 9.2: A familiar right triangle.

9.2 Pythagorean triples

Let's now turn our attention to degree two equations. We begin with perhaps the most famous of all degree two equations:

$$x^2 + y^2 = z^2, \quad x, y, z > 0. \tag{9.2}$$

As solutions to (9.2) provide right triangles with integer side lengths, we call such a solution (x, y, z) a *Pythagorean triple*. The Pythagorean triple $(3, 4, 5)$ is certainly a familiar one; see Figure 9.2. Note that for any positive integer k,

$$(3k)^2 + (4k)^2 = (5k)^2,$$

and so equation (9.2) has an infinite number of solutions. In fact, if (x, y, z) is any Pythagorean triple, so is (xk, yk, zk) for any positive integer k. Thus, any Pythagorean triple gives rise to an infinite family of Pythagorean triples through scalar multiplication.

How can we describe the set of Pythagorean triples? Let (x, y, z) be a Pythagorean triple, and define $g = \gcd(x, y, z)$. Since $x^2 + y^2 = z^2$, we see that $(x/g)^2 + (y/g)^2 = (z/g)^2$. Hence, any Pythagorean triple is a scalar multiple of a triple of numbers whose greatest common divisor is 1. Now suppose that $\gcd(x, y, z) = 1$ and p is a prime number that is a common divisor of x and y. Then the relation $x^2 + y^2 = z^2$ implies that p divides z^2. Since p is prime, we conclude that p divides z. This contradicts our assumption that $\gcd(x, y, z) = 1$. A similar argument shows that any prime that divides x and z divides y, and any prime that divides y and z divides x. Hence, the assumption $\gcd(x, y, z) = 1$ forces $\gcd(x, y) = \gcd(x, z) = \gcd(y, z) = 1$.

DEFINITION 9.3 *A Pythagorean triple* (x, y, z) *is* primitive *if* $\gcd(x, y, z) = 1$*, or equivalently if* $\gcd(x, y) = \gcd(x, z) = \gcd(y, z) = 1$.

Any Pythagorean triple can be expressed in the form (kx, ky, kz), where (x, y, z) is a primitive Pythagorean triple. We now focus on describing the set of primitive Pythagorean triples. Suppose that (x, y, z) is a primitive Pythagorean triple. First observe that the equation $x^2 + y^2 = z^2$ forces at least one of the three numbers to be even. Since at most one of the three numbers is even, in fact exactly one is even. We claim that z must be one of the odd values. If not, that is, if z is even and x and y are odd, we have $x^2 \equiv y^2 \equiv 1 \pmod 4$ while $z^2 \equiv 0 \pmod 4$. Thus, when we reduce $x^2 + y^2 = z^2$ modulo 4 we obtain $1 + 1 \equiv 0 \pmod 4$, and so we have a contradiction.

We now know that, given a primitive Pythagorean triple (x, y, z), either x or y is even. By swapping x and y if necessary, we may suppose, without loss of generality, that x is even. Now we rearrange our equation $x^2 + y^2 = z^2$ as

$$(z + y)(z - y) = x^2.$$

Since y and z are both odd, $z - y$ and $z + y$ are both even. So we can rewrite the above equation as

$$\left(\frac{z+y}{2}\right)\left(\frac{z-y}{2}\right) = \left(\frac{x}{2}\right)^2,$$

where both factors on the left side are integers. We claim that the gcd of these two integers is in fact 1. For if d is a positive divisor of these two numbers, it divides their sum and difference. Since

$$\left(\frac{z+y}{2}\right) + \left(\frac{z-y}{2}\right) = z$$
$$\left(\frac{z+y}{2}\right) - \left(\frac{z-y}{2}\right) = y,$$

we see that d is a common divisor of y and z, and so must be 1. We now know that $(z + y)/2$ and $(z - y)/2$ are two relatively prime numbers whose product is a square.

LEMMA 9.4
Let a and b be positive numbers such that $\gcd(a, b) = 1$ *and* $ab = c^2$. *Then a and b are perfect squares.*

PROOF Suppose that a has the canonical factorization

$$a = \prod_{i=1}^{k} p_i^{\alpha_i}.$$

We must show that α_i is even for all i. Since $\gcd(a, b) = 1$, we see that p_i does not divide b. Hence, the fundamental theorem of arithmetic implies that

$p_i^{\alpha_i}$ is the power of p_i appearing in the factorization of c^2. If $p_i^{\beta_i}$ is the power of p_i appearing in the factorization of c, then $\alpha_i = 2\beta_i$, and so α_i must be even for all i.

The same argument holds for b. □

We apply the lemma to our situation and find

$$\frac{z+y}{2} = s^2$$
$$\frac{z-y}{2} = t^2,$$

for some positive integers s and t. Since $(z+y)/2$ and $(z-y)/2$ are relatively prime, so are s and t. We see that $s^2t^2 = (x/2)^2$, and so

$$x = 2st.$$

Solving for y and z in terms of s and t we find

$$y = s^2 - t^2$$
$$z = s^2 + t^2.$$

Thus, we have a description of (x, y, z) in terms of the parameters s and t.

There are restrictions on what values of s and t can actually occur. We have already noted that s and t are relatively prime. Also, the fact that y is odd implies that exactly one of s and t is even. Finally, $s > t$ since y is a positive number. If we start with a pair of numbers s and t satisfying these conditions and define x, y, and z in terms of s and t as above, then we obtain a primitive Pythagorean triple. (A proof is called for in the exercises.) We summarize our work in the following theorem.

THEOREM 9.5

The triple (x, y, z) is a primitive Pythagorean triple where x is even if and only if

$$x = 2st$$
$$y = s^2 - t^2$$
$$z = s^2 + t^2,$$

for some relatively prime positive integers s and t satisfying $s > t > 0$ and s or t is even.

Example 9.6 Let's find all primitive Pythagorean triples in which one of the variables has the value 15. By Theorem 9.5, all such positive solutions have the form $(2st, s^2 - t^2, s^2 + t^2)$, for some $s > t > 0$. Since 15 is not the sum of two squares, we will need to solve $s^2 - t^2 = 15$. Factoring, we have

$$(s-t)(s+t) = 15.$$

Noting that $s+t > s-t$, we see that there are two possibilities for the above factorization. If we set $s-t=1$ and $s+t=15$, we find $s=8$ and $t=7$, which leads us to the Pythagorean triple $(112, 15, 113)$. We could also take $s-t=3$ and $s+t=5$. This gives $s=4$ and $t=1$ and produces the Pythagorean triple $(8, 15, 17)$. □

Exercises

9.8 Describe the set of Pythagorean triples (x, y, z) in which x is even and $z = x+1$.

9.9 Given an odd prime p, prove that there exists exactly one Pythagorean triangle with one leg of length p.

9.10 Find a primitive Pythagorean triple (x, y, z) with $x = 100$.

†9.11 Complete the proof of Theorem 9.5: Show that if s and t are positive relatively prime integers and either s or t is even, then $(2st, s^2-t^2, s^2+t^2)$ is a primitive Pythagorean triple.

†9.12 (a) Show that for any rational number m, the point

$$\left(\frac{1-m^2}{m^2+1}, \frac{2m}{m^2+1}\right)$$

is on the unit circle.

(b) Show that every rational point (i.e., point with rational coordinates) on the unit circle other than $(-1, 0)$ is of the form given in part (a) for some rational number m.

Hint: Given a rational point on the unit circle, construct the line passing through this point and the point $(-1, 0)$. The slope of this line is a rational number m.

(c) Use part (b) to provide an alternate proof of the classification of primitive Pythagorean triples (Theorem 9.5).

9.3 Gaussian integers

In this section we begin a study of the Diophantine equation

$$x^2 + y^2 = n. \tag{9.3}$$

Classically, the question "For which n does a solution to equation (9.3) exist?" would have been phrased "Which integers can be written as the sum of two squares?" Let's consider the question for the first twenty numbers.

Can be written as a sum of two squares	Cannot be written as a sum of two squares
1, 2, 4, 5, 8, 9, 10, 13, 16, 17, 18, 20	3, 6, 7, 11, 12, 14, 15, 19

No obvious pattern emerges to tell us how to separate the two cases. Upon closer inspection, we notice that the set of numbers that cannot be written as a sum of two squares contains an arithmetic progression: 3, 7, 11, 15, 19. In fact this pattern continues.

THEOREM 9.7
If $n \equiv 3 \pmod 4$, *then* n *is not a sum of two squares.*

PROOF Observe that for any integer x, we have $x^2 \equiv 0$ or $1 \pmod 4$. Similarly, $y^2 \equiv 0$ or $1 \pmod 4$. Thus, the only possibilities for $x^2 + y^2 \pmod 4$ are 0, 1, and 2, and so the congruence

$$x^2 + y^2 \equiv 3 \pmod 4$$

cannot hold. ∎

To complete the study of the Diophantine equation in (9.3), we are left to consider numbers that are not congruent to $3 \pmod 4$. This case turns out to be more involved. To simplify matters, we begin by restricting our attention to the case in which n is a prime number. We first observe that 2 is a sum of two squares ($2 = 1^2 + 1^2$). In light of Theorem 9.7, we can say that if an odd prime p is a sum of two squares, then $p \equiv 1 \pmod 4$. We will be able to prove that the converse of this statement is also true. This result will follow from some arithmetic properties of an enlargement of the set of integers known as the Gaussian integers.

Let $\mathbf{C}$ denote the set of complex numbers.

DEFINITION 9.8 *A* Gaussian integer *is an element* $a + bi$ *of* $\mathbf{C}$, *where* a *and* b *are integers.*

Example 9.9 The numbers $2 + i$, $7 - 4i$, 6, and 0 are Gaussian integers. ∎

We denote the set of Gaussian integers as $\mathbf{Z}[i] = \{a+bi \colon a, b \in \mathbf{Z}\}$. Our goal now is to show that the Gaussian integers share many of the properties of the ordinary (rational) integers. Accordingly, we will need to define such things as divisibility, primes, and factorization for the Gaussian integers. First, we observe that the sum and difference of any two Gaussian integers is a Gaussian integer, and the product of any two Gaussian integers is a Gaussian integer.

Addition and multiplication are commutative and associative operations, and the usual distributive law holds (i.e., $\alpha \cdot (\beta + \gamma) = \alpha \cdot \beta + \alpha \cdot \gamma$). There are identity elements for both operations, and every element has an additive inverse.

Note: A set endowed with addition and multiplication operations satisfying the properties listed above is called a *commutative ring.*

DEFINITION 9.10 *A nonzero Gaussian integer α* divides *a Gaussian integer β, and we write $\alpha \mid \beta$, if there exists a Gaussian integer γ such that $\beta = \alpha\gamma$.*

Example 9.11 The Gaussian integer $2 + i$ divides the Gaussian integer $1 + 3i$, since $(2 + i)(1 + i) = 1 + 3i$. ▯

In order to define Gaussian primes, we need to determine which Gaussian integers play the role that 1 and -1 do for the rational integers.

DEFINITION 9.12 *An element ϵ of $\mathbf{Z}[i]$ is a* unit *if $\epsilon\lambda = 1$, for some λ in $\mathbf{Z}[i]$.*

The Gaussian integers ± 1 and $\pm i$ are units in $\mathbf{Z}[i]$ as $1 \cdot 1 = 1$, $(-1)(-1) = 1$ and $(i)(-i) = 1$. In fact, these are the only units in this ring. We could check this directly. However, we can obtain this result more elegantly by introducing the norm function.

DEFINITION 9.13 *The* norm *of a complex number $x + yi$ is*

$$N(x + yi) = x^2 + y^2.$$

Note: The norm of a Gaussian integer is a nonnegative rational integer. Notice that $N(z) = z \cdot \overline{z}$, where $\overline{z} = x - yi$ (the complex conjugate of z). We also point out that the norm of a Gaussian integer is the sum of two squares of rational integers.

PROPOSITION 9.14

For any $z_1, z_2 \in \mathbf{C}$, we have $N(z_1 z_2) = N(z_1)N(z_2)$.

PROOF Let $z_1 = x_1 + y_1 i$ and $z_2 = x_2 + y_2 i$. Then

$$\begin{aligned} N(z_1 z_2) &= N((x_1 + y_1 i)(x_2 + y_2 i)) \\ &= N((x_1 x_2 - y_1 y_2) + (x_1 y_2 + y_1 x_2) i) \\ &= (x_1 x_2 - y_1 y_2)^2 + (x_1 y_2 + y_1 x_2)^2 \\ &= x_1^2 x_2^2 + y_1^2 y_2^2 + x_1^2 y_2^2 + y_1^2 x_2^2 \\ &= (x_1^2 + y_1^2)(x_2^2 + y_2^2) \\ &= N(z_1) N(z_2). \end{aligned}$$

□

Note: The proof of the identity $N(z_1 z_2) = N(z_1)N(z_2)$ gives us information about representing numbers as sums of two squares. We investigate this topic in the next section, but for now we note that if m and n are each a sum of two squares, $m = a^2 + b^2$ and $n = c^2 + d^2$, then mn is the also a sum of two squares:

$$mn = (a^2 + b^2)(c^2 + d^2) = (ac - bd)^2 + (ad + bc)^2.$$

This allows us to represent, for example, $85 = 5 \cdot 17$ as a sum of two squares:

$$85 = 5 \cdot 17 = (2^2 + 1^2)(4^2 + 1^2) = (8 - 1)^2 + (2 + 4)^2 = 7^2 + 6^2.$$

COROLLARY 9.15

Let α and β be Gaussian integers. If $\alpha \mid \beta$, then $N(\alpha) \mid N(\beta)$.

PROOF If $\alpha \mid \beta$, then $\beta = \alpha\gamma$, for some Gaussian integer γ, and applying Proposition 9.14, we have $N(\beta) = N(\alpha)N(\gamma)$. The result follows instantly. □

COROLLARY 9.16

A Gaussian integer ϵ is a unit if and only if $N(\epsilon) = 1$. The units in the ring of Gaussian integers are ± 1 and $\pm i$.

PROOF If $\epsilon = a + bi$ and $N(\epsilon) = 1$, then $a^2 + b^2 = 1$, so that $\epsilon = 1, -1, i$, or $-i$. As we mentioned earlier, all of these numbers are units. If $\epsilon = a + bi$ is a unit, then $\epsilon \mid 1$, so that by Corollary 9.15, $N(\epsilon) \mid 1$. This implies that $N(\epsilon) = 1$. □

DEFINITION 9.17 *Two Gaussian integers α and β are* associates *if $\alpha = \epsilon\beta$, for some unit ϵ.*

Note: As there are four units in $\mathbf{Z}[i]$, a nonzero Gaussian integer has four associates.

Example 9.18 The Gaussian integers $2 + i$ and $(2 + i) \cdot i = -1 + 2i$ are associates. □

Every nonzero Gaussian integer has an associate of the form $a + bi$, where $a > 0$ and $b \geq 0$.

Example 9.19 Let $\alpha = -2 + 3i$. Then $\alpha \cdot -i = 3 + 2i$ is an associate with positive real part and nonnegative imaginary part. □

DEFINITION 9.20 *A nonzero Gaussian integer is a* Gaussian composite *if it is the product of two Gaussian nonunits. A nonzero Gaussian integer that is neither composite nor a unit is a* Gaussian prime.

Note: Associates of Gaussian primes are Gaussian primes.

Example 9.21 The number 2 is a Gaussian composite:

$$2 = (1 + i)(1 - i).$$

Can we break this factorization down any further? Are $1 + i$ and $1 - i$ composite? Suppose that $1 + i = \alpha\beta$, for some Gaussian integers α and β. Taking norms, we have

$$2 = N(\alpha)N(\beta).$$

It follows that either $N(\alpha) = 1$, in which case α is a unit, or $N(\beta) = 1$, in which case β is a unit. Thus $1 + i$ is a Gaussian prime. Likewise, $1 - i$ is a Gaussian prime. □

We can generalize the argument from this example.

PROPOSITION 9.22
If α is a Gaussian integer and $N(\alpha) = p$ where p is a rational prime, then α is a Gaussian prime.

A formal proof of the proposition is called for in the exercises.

We would like to characterize all Gaussian primes. Which rational integers are Gaussian primes? As a rational composite number is clearly also a Gaussian composite, we need only investigate rational prime numbers.

Example 9.23 The number 3 is a Gaussian prime. For if $3 = \alpha\beta$, then $9 = N(3) = N(\alpha)N(\beta)$. If α and β are not units, then $N(\alpha) = N(\beta) = 3$. However, this would imply that $a^2 + b^2 = 3$, for $\alpha = a + bi$, and this equation clearly has no solutions. ▯

THEOREM 9.24
A rational prime is a Gaussian composite if and only if it is a sum of two squares.

PROOF If a prime p is expressible as the sum of two squares, say, $p = a^2 + b^2$, then it is a Gaussian composite, for $p = (a + bi)(a - bi)$.

Now suppose that p factors as $p = (a + bi)(c + di)$, with $a + bi$ and $c + di$ not units. Then

$$p^2 = N(a + bi)N(c + di).$$

Since $a + bi$ and $c + di$ are not units, we have $N(a + bi) = N(c + di) = p$. Therefore, $p = N(a + bi) = a^2 + b^2$. ▯

Note: The representation above is unique up to a choice of signs and order of terms. For if $p = a^2 + b^2$, then $N(a \pm bi) = p$, and so $a \pm bi$ are Gaussian primes. Hence, $p = (a + bi)(a - bi)$ is a prime factorization of p. Since a prime factorization in $\mathbf{Z}[i]$ is unique up to associates and order of factors (see Theorem 9.34), if $p = c^2 + d^2$, then $c + di$ must equal one of $a + bi$ or $a - bi$ or their associates. Therefore, (c, d) is one of the pairs (a, b), $(a, -b)$, $(-a, b)$, $(-a, -b)$, (b, a), $(-b, a)$, $(b, -a)$, or $(-b, -a)$.

In light of Theorems 9.7 and 9.24, we have the following corollary.

COROLLARY 9.25
If p is a rational prime and $p \equiv 3 \pmod 4$, then p is also a Gaussian prime.

Example 9.26 The primes 19, 23, and 31 are Gaussian primes. ▯

We proved in Chapter 2 that there are infinitely many primes of the form $4n + 3$. By the previous theorem, these primes are Gaussian primes.

THEOREM 9.27
There are infinitely many Gaussian primes.

Now we discuss the algebraic structure of $\mathbf{Z}[i]$. We find that we can define a version of the Euclidean algorithm for Gaussian integers.

THEOREM 9.28 (Division algorithm for Gaussian integers)
Let α and β be Gaussian integers, with $\beta \neq 0$. There exist Gaussian integers γ and ρ such that

$$\alpha = \beta\gamma + \rho,$$

and $N(\rho) < N(\beta)$.

Note: In general, γ and ρ are not determined uniquely.

PROOF Let $\alpha = a + bi$ and $\beta = c + di$. Define $z = \alpha/\beta$. Note that z is a complex number, but not necessarily a Gaussian integer. Then

$$z = \frac{a+bi}{c+di} = \frac{ac+bd}{c^2+d^2} + \frac{bc-ad}{c^2+d^2}i.$$

Define $x = (ac+bd)/(c^2+d^2)$ and $y = (bc-ad)/(c^2+d^2)$, let m and n be rational integers such that $|x-m| \leq 1/2$ and $|y-n| \leq 1/2$, and set $\gamma = m+ni$ and $\rho = \alpha - \beta\gamma$. We will show that $N(z-\gamma) < 1$, for then

$$N(\rho) = N(\alpha - \beta\gamma) = N(\beta)N(z-\gamma) < N(\beta).$$

We find that

$$N(z-\gamma) = N((x-m)+(y-n)i) = (x-m)^2 + (y-n)^2 \leq \frac{1}{4} + \frac{1}{4} < 1.$$

$\square$

Example 9.29 Let's perform the division algorithm on $\alpha = 12 + 8i$ and $\beta = 4 - i$. We compute

$$\frac{12+8i}{4-i} = \frac{(12+8i)(4+i)}{(4-i)(4+i)} = \frac{40}{17} + \frac{44}{17}i.$$

Thus, we take $\gamma = 2 + 3i$ and

$$\rho = 12 + 8i - (4-i)(2+3i) = 1 - 2i.$$

We check that $N(\rho) = 1^2 + (-2)^2 = 5$ is less than $N(\beta) = 4^2 + 1^2 = 17$. $\square$

Given a pair of Gaussian integers α and β, we would like to define a greatest common divisor of the pair as we did for rational integers. As in the division algorithm, we will use the norm to determine the sense in which one such Gaussian integer is greater than another. However, this does not lead to a unique element. Recall that the greatest common divisor of a pair of rational numbers is a linear combination of the two. Using this case for guidance, consider the set of Gaussian integers of the form $\alpha\xi + \beta\lambda$ where ξ and λ are Gaussian integers. Let δ be a number expressible in this form with smallest possible positive norm. Applying the division algorithm in a manner analogous to the rational case, we can show that δ divides both α and β. Furthermore, if δ' is any other common divisor of α and β, then it must also divide δ. If we write $\delta = \delta'\gamma$, then

$$N(\delta) = N(\delta')N(\gamma).$$

Thus, δ has the greatest norm among common divisors of α and β. Moreover, if $N(\delta') = N(\delta)$, then γ is a unit, and so δ and δ' are associates. We can now give a sensible definition of the greatest common divisor of two Gaussian integers.

DEFINITION 9.30 *The* greatest common divisor *(*gcd*) of two Gaussian integers α and β (not both 0) is the Gaussian integer of largest norm that divides both α and β. As this number is well defined only up to multiplication by a unit, we assume that the* gcd *has positive real and nonnegative imaginary parts.*

Example 9.31 The Gaussian integer $1 + i$ divides both $3 + 5i$ and $1 + 3i$. As $N(3+5i) = 34$ and $N(1+3i) = 10$, no Gaussian integer with norm greater than 2 can divide this pair of numbers. Hence, $\gcd(3 + 5i, 1 + 3i) = 1 + i$. □

The Euclidean algorithm for Gaussian integers is based on repeated use of the division relation. We implement this algorithm in *Mathematica* and *Maple*, using recursive programming in the former and procedural programming in the latter.

Mathematica

```
f[w_, z_] := f[z, w - z(Round[Re[w/z]] + I Round [Im[w/z]])]
f[w_, 0] := w
ourgcd[w_, z_] := First[Select[{f[w, z], -f[w, z],
  I f[w, z], -I f[w, z]}, (Re[#] > 0 && Im[#] >= 0) &]]
```

Maple

```
ourgcd:=proc(w,z)
  local a,x;
  a:=[w,z];
  while not a[2]=0 do
    a:=[a[2],a[1]-a[2]*
    (round(Re(a[1]/a[2]))+I*round(Im(a[1]/a[2])))];
  end do;
  op(select(x->evalb(Re(x)>0 and Im(x)>=0),
  {a[1],-a[1],I*a[1],-I*a[1]}));
end proc:
```

Testing our code, we find that $\gcd(3+5i, 1+3i) = 1+i$.

Based on our discussion preceding the definition of the greatest common divisor, we conclude that gcd is a linear combination.

THEOREM 9.32

For any Gaussian integers α and β (not both 0), there exist Gaussian integers ξ and λ such that

$$\alpha\xi + \beta\lambda = \gcd(\alpha, \beta).$$

Recall that we used the rational version of the theorem to prove Euclid's lemma. We can deduce the analogous result for Gaussian integers in a similar way.

COROLLARY 9.33

If $\pi \in \mathbf{Z}[i]$ is prime and $\pi \mid \alpha\beta$, then $\pi \mid \beta$ or $\pi \mid \beta$.

Euclid's lemma is the key step in proving the unique factorization of integers in $\mathbf{Z}$. Again, one can deduce a Gaussian integer version arguing in the same way.

THEOREM 9.34

Every nonzero, nonunit Gaussian integer can be written as a product of Gaussian primes. Furthermore, such a factorization is unique up to associates of the factors and order of the factors.

Corollary 9.33 is the result we need to complete our study of equation (9.3) in the case where n is prime.

THEOREM 9.35
If p is a rational prime and $p \equiv 1 \pmod 4$, then p is not a Gaussian prime.

PROOF Since $p \equiv 1 \pmod 4$, Theorem 3.37 implies that there exists a rational integer x such that

$$x^2 \equiv -1 \pmod p.$$

Thus, p divides x^2+1. If p were prime in $\mathbf{Z}[i]$, then p would necessarily divide $x+i$ or $x-i$ (by Corollary 9.33). Since $p(a+bi) \neq x \pm i$ for any choice of rational integers a and b, we see that p cannot be prime in $\mathbf{Z}[i]$. ∎

We summarize Theorems 9.7, 9.24, and 9.35.

THEOREM 9.36
Let p be a rational prime. The following are equivalent:

(i) p is a Gaussian composite;

(ii) p is a sum of two squares;

(iii) $p \equiv 1 \pmod 4$.

Since all primes congruent to 1 modulo 4 are sums of two squares, we know that there are an infinite number of primes that can be written in the form m^2+n^2. As mentioned in the Introduction, it is believed that there are actually an infinite number of primes of the form $1+n^2$. However, this appears to be a much more difficult assertion to prove.

We mention now a fast algorithm for writing a prime $p \equiv 1 \pmod 4$ as a sum of two squares. (Recall from the note following Theorem 9.24 that such a representation is unique up to a choice of signs and order of terms.)

(1) Find a quadratic nonresidue g modulo p. Perhaps the simplest way to do this is just to take various residues g modulo p in turn and compute $g^{(p-1)/2}$. If this quantity is -1, then g is a quadratic nonresidue. A random number chosen from the set $\{1, 2, \ldots, p-1\}$ has probability $1/2$ of being a quadratic residue modulo p. So if we choose k numbers, the probability that none of them are nonresidues is $1/2^k$, which we can make as small as we want.

(2) Let $x \equiv g^{(p-1)/4} \pmod p$. Then $x^2 \equiv -1 \pmod p$. (Recall that Wilson's theorem, too, gives us a way to find a solution to the congruence $x^2 \equiv -1 \pmod p$, but this method would be impractical for large p.)

(3) Compute $a+bi = \gcd(p, x+i)$. Then $(a+bi) \mid p$, so that $N(a+bi) = p$, and $p = a^2+b^2$.

Example 9.37 Let $p = 541$ (a prime congruent to 1 modulo 4). Using a computer, we find that 2 is a quadratic nonresidue modulo p. To do this, we check that

$$2^{(p-1)/2} \equiv -1 \pmod{p}.$$

Now we set $x = 2^{(p-1)/4}$ and find that $x \equiv 52 \pmod{p}$. Finally, we compute $\gcd(p, x+i) = 10 + 21i$, yielding the answer $541 = 10^2 + 21^2$. ▯

We now give the complete classification of Gaussian primes.

THEOREM 9.38

A Gaussian integer π is a Gaussian prime if and only if π is an associate of a rational prime number congruent to $3 \pmod 4$ *or $N(\pi)$ is a rational prime number.*

Note: In the latter case, since $N(\pi) = p = a^2+b^2$, we see that p is necessarily 2 or else of the form $4n + 1$.

PROOF If $N(\pi)$ is a prime integer, then π is prime by Proposition 9.22. We have already shown that associates of rational prime numbers of the form $4n + 3$ are Gaussian primes.

Suppose that π is a Gaussian prime. Then, since $\pi\overline{\pi} = N(\pi) \in \mathbf{Z}$, π is a divisor of a rational prime p. Hence, $N(\pi)$ divides $N(p) = p^2$, and $N(\pi) = p$ or $N(\pi) = p^2$. In the former case, we are finished. In the latter case, $\pi \cdot \overline{\pi} = p^2$. Since $\pi \mid p$, we may write $\pi\alpha = p$, for some α. Taking norms, we find that α is a unit and hence π is an associate of p. ▯

We conclude with an application of Gaussian integers to Pythagorean triangles, providing a second proof of the characterization of Pythagorean triples found in Section 9.2 (Theorem 9.5). We want to determine solutions to the equation

$$x^2 + y^2 = z^2,$$

with $\gcd(x, y) = 1$. The equation factors as $(x + yi)(x - yi) = z^2$. Since $\gcd(x, y) = 1$, we obtain $\gcd(x + yi, x - yi) = 1$. (A common divisor of $x + yi$ and $x - yi$ is a common divisor of $2x$ and $2y$, and such a divisor must also divide the odd number z^2 and hence be 1.) It follows (by unique factorization, as in Lemma 9.4) that $x + yi$ and $x - yi$ are each perfect squares, up to multiplication by units. There are two cases: $x + yi = \alpha^2$ and $x - yi = \beta^2$, or $x + yi = i\alpha^2$ and $x - yi = i\beta^2$. In the first case, let $\alpha = u + vi$; then $x + yi = (u^2 - v^2) + 2uvi$, and hence $x = u^2 - v^2$, $y = 2uv$, and $z = u^2 + v^2$. The second case is similar.

Exercises

9.13 Show that if $n \equiv 3$, 4, 5, or 6 (mod 9), then n is not the sum of two cubes.

9.14 Prove that $\sqrt{2}$ is not the ratio of two Gaussian integers.

9.15 Which of the following Gaussian integers are divisible by the Gaussian integer $1+i$?

(a) $-2+4i$

(b) $3+5i$

(c) $5+19i$

9.16 Which of the following Gaussian integers are associates?

$$3,\ 3i,\ -3,\ -3i,\ 4+i,\ -1+4i,\ 1+4i,\ -4+i$$

9.17 Determine which of the following are Gaussian primes.

(a) 31

(b) 37

(c) $-1+i$

(d) $1-6i$

(e) $5+3i$

9.18 Find prime factorizations in $\mathbf{Z}[i]$ of the following.

(a) 15

(b) $1-3i$

(c) $12+18i$

9.19 Let $p \equiv 1 \pmod 4$ be a rational prime. Show that p is the product of two Gaussian primes that are not associates.

†**9.20** Prove Proposition 9.22.

9.21 Why does the fact that $(3+i)(3-i) = 2\cdot 5$ not contradict Theorem 9.34 on the (almost) unique factorization of Gaussian integers?

9.22 Find Gaussian integers γ and ρ satisfying

$$58-19i = (4+7i)\gamma + \rho,$$

where $N(\rho) < N(4+7i)$.

9.23 Compute $\gcd(8+8i, 6-30i)$.

◇**9.24** Using a computer, write $p = 1234567891234567891\,49$ as a sum of two squares.

†⋆**9.25** The ring $\mathbf{Z}[\sqrt{5}i] = \{a + b\sqrt{5}i\colon a, b \in \mathbf{Z}\}$ is not a unique factorization domain because, for example, $2 \cdot 3 = (1+\sqrt{5}i)(1-\sqrt{5}i)$. What goes wrong?

9.4 Sums of squares

Now that we know which primes are the sum of two squares, we consider composite numbers. Suppose that two numbers m and n can be written as a sum of two squares, i.e., $m = a^2 + b^2$ and $n = c^2 + d^2$ for some integers a, b, c, and d. Taking the product and performing some clever factoring, we find

$$\begin{aligned} mn &= (a^2+b^2)(c^2+d^2) \\ &= (ad+bc)^2 + (ac-bd)^2. \end{aligned} \tag{9.4}$$

Note: We have already remarked in Section 9.3 that this equation is equivalent to the product rule for the norm of complex numbers.

We have proved the following result.

PROPOSITION 9.39
If m and n are sums of two squares, then so is mn.

What kinds of numbers do we now know can be written as a sum of two squares? If $p_1, \ldots, p_l$ are prime numbers all congruent to 1 mod 4, then $p_1^{\alpha_1} \ldots p_l^{\alpha_l}$ is a sum of squares for any nonnegative integers $\alpha_1, \ldots, \alpha_l$. Since $2 = 1^2 + 1^2$, we have that $2^{\alpha_0} p_1^{\alpha_1} \ldots p_l^{\alpha_l}$ is also a sum of two squares for any nonnegative integer α_0. Certainly, if we multiply a sum of squares by a perfect square, then the result is still a sum of squares. The next theorem tells us that these operations generate all numbers that can be written as a sum of two squares.

THEOREM 9.40
Let n be a positive number with canonical factorization

$$n = \prod_{i=1}^{k} p_i^{\alpha_i}.$$

Then n is a sum of two squares if and only if α_i is even whenever $p_i \equiv 3 \pmod 4$.

PROOF If α_i is even whenever $p_i \equiv 3 \pmod 4$, then n is a product of numbers that are sums of two squares and hence by Proposition 9.39 is itself a sum of two squares. Now suppose that α_i is odd for some $p_i \equiv 3 \pmod 4$, but $n = a^2 + b^2$, for some integers a and b. Let $g = \gcd(a, b)$. If we set $c = a/g$, $d = b/g$, and $m = n/g^2$, then c and d are relatively prime and

$$m = c^2 + d^2. \tag{9.5}$$

Note that the power of p_i dividing m is also odd and, in particular, is at least 1. We claim that c is not divisible by p_i. If it were, then it would also divide d since $d^2 = m - c^2$. As c and d are relatively prime, this cannot occur. According to Corollary 3.12, there exists an x such that

$$cx \equiv d \pmod{p_i}.$$

Reducing equation (9.5) mod p_i, we have

$$0 \equiv c^2 + d^2 \equiv c^2 + (cx)^2 \pmod{p_i}.$$

We conclude that p_i divides $c^2(1 + x^2)$. Since p_i does not divide c, it must divide $1 + x^2$. Theorem 3.35 tells us that the congruence $x^2 \equiv -1 \pmod{p_i}$ does not have a solution, and so we have reached a contradiction. ◻

Let's now consider the representation of an integer as a sum of more than two squares. Which positive numbers can be written as a sum of three squares? All of the first six positive numbers are sums of three squares, but 7 is not. In fact, if $a \equiv 7 \pmod 8$, then a cannot be written as a sum of three squares. One can give a classification of the set of numbers that can be written as a sum of three squares along the lines of Theorem 9.40 (see Chapter Notes). This result requires a bit more machinery than we have at hand right now. However, we are in a position to study the question, "Which positive numbers can be written as the sum of four squares?" The answer, provided by Lagrange in 1770, is all of them.

THEOREM 9.41 (Lagrange)
Every positive integer can be written as a sum of four squares.

PROOF As in the case of sums of two squares, the product of two sums of four squares can also be written as a sum of four squares. For if $m = x^2 + y^2 + z^2 + w^2$ and $n = a^2 + b^2 + c^2 + d^2$, then we obtain Euler's four-squares identity:

$$\begin{aligned} mn = {} & (xa + yb + zc + wd)^2 + (xb - ya + zd - wc)^2 \\ & + (xc - za + wb - yd)^2 + (xd - wa + yc - zb)^2. \end{aligned} \tag{9.6}$$

To complete the proof of the theorem, we only need to show that all primes can be written as a sum of four squares. Certainly, 2 is a sum of four squares, so we focus on odd primes. Before we proceed, we will need a lemma due to Euler. In fact, the rest of the proof of the theorem is essentially a proof published by Euler subsequent to the one given by Lagrange.

***LEMMA* 9.42**

Let p be an odd prime. Then there exist integers x, y, and k satisfying

$$x^2 + y^2 + 1 = kp$$

with $0 < k < p$.

PROOF We need to find values of x and y such that $x^2 + y^2 + 1 \equiv 0 \pmod{p}$ and $x^2 + y^2 + 1 < p^2$. We break the proof into two cases depending on the congruence class of $p \pmod 4$. If $p \equiv 1 \pmod 4$, then $\left(\frac{-1}{p}\right) = 1$, and the congruence

$$x^2 \equiv -1 \pmod{p}$$

has a solution. In fact, we can find a solution for which $0 < x < p/2$. For such an x, we have

$$x^2 + 1 < p^2/4 + 1 < p^2.$$

Since p divides $x^2 + 1$, we obtain a solution to our equation with $y = 0$.

Now suppose that $p \equiv 3 \pmod 4$. Let a be the smallest positive quadratic nonresidue modulo p. Since $\left(\frac{-1}{p}\right) = -1$, we must have $\left(\frac{-a}{p}\right) = 1$. Thus, there exists an x satisfying

$$x^2 \equiv -a \pmod{p},$$

with $0 < x < p/2$. By assumption, $a - 1$ is a quadratic residue and so there exists a y satisfying

$$y^2 \equiv a - 1 \pmod{p},$$

with $0 < y < p/2$. We have $x^2 + y^2 \equiv -1 \pmod{p}$ as desired. Furthermore,

$$x^2 + y^2 + 1 < (p/2)^2 + (p/2)^2 + 1 < p^2,$$

and so we obtain a solution in this case as well. ∎

Returning now to the proof of our theorem, the lemma tells us we can at least find integers x, y, z, and w such that

$$x^2 + y^2 + z^2 + w^2 = kp$$

for some positive $k < p$. Choose such x, y, z, and w so that the value of k is as small as possible. If $k = 1$, then we have our desired representation of p. Let's assume that $k > 1$ and work to find a contradiction. In fact, we will show that given a solution, we can find another solution with a smaller value of k. (This method of proof is known as an "infinite descent argument.") First, we note that k cannot be even. If it were, then an even number of x, y, z, and w would be odd. Thus, reordering if necessary, $x+y$, $x-y$, $z-w$, and $z+w$ would be even, and

$$\left(\frac{x+y}{2}\right)^2 + \left(\frac{x-y}{2}\right)^2 + \left(\frac{z+w}{2}\right)^2 + \left(\frac{z-w}{2}\right)^2 = (k/2)p.$$

This would contradict the minimality of k.

Now let a, b, c, and d be the integers satisfying

$$\begin{aligned} a &\equiv x \pmod{k} \\ b &\equiv y \pmod{k} \\ c &\equiv z \pmod{k} \\ d &\equiv w \pmod{k}, \end{aligned}$$

and $0 \le |a|, |b|, |c|, |d| < k/2$. (We may take a strict inequality since we know that k is odd.) We see that $a^2+b^2+c^2+d^2 \equiv x^2+y^2+z^2+w^2 \equiv 0 \pmod{k}$. Consequently, we can write $a^2+b^2+c^2+d^2 = lk$, for some nonnegative integer l. Now

$$a^2+b^2+c^2+d^2 < (k/2)^2+(k/2)^2+(k/2)^2+(k/2)^2 = k^2.$$

Thus, we see that $l < k$. We can also see that $l \neq 0$. Otherwise, we would have $a=b=c=d=0$, which would imply that k^2 divides $x^2+y^2+z^2+w^2$. This cannot be the case as $x^2+y^2+z^2+w^2 = kp$, with $0 < k < p$.

We recall from earlier in our proof that the product of $(a^2+b^2+c^2+d^2)$ and $(x^2+y^2+z^2+w^2)$ is also a sum of four squares. In fact, if we take

$$\begin{aligned} X &= xa+yb+zc+wd \\ Y &= xb-ya+zd-wc \\ Z &= xc-za+wb-yd \\ W &= xd-wa+yc-zb, \end{aligned}$$

then we obtain

$$X^2+Y^2+Z^2+W^2 = (lk)(kp).$$

To complete the descent, we just need to show that we can divide through by k^2. Now

$$\begin{aligned} X &\equiv x^2+y^2+z^2+w^2 \equiv 0 \pmod{k} \\ Y &\equiv xy-yx+zw-wz \equiv 0 \pmod{k} \\ Z &\equiv xz-zx+wy-yw \equiv 0 \pmod{k} \\ W &\equiv xw-wx+yz-zy \equiv 0 \pmod{k}. \end{aligned}$$

We conclude that k divides X, Y, Z, and W, so lp is, in fact, the sum of squares of four integers. That is,

$$(X/k)^2+(Y/k)^2+(Z/k)^2+(W/k)^2 = lp.$$

Since $0 < l < k$, we have a contradiction. □

In 1770 Edward Waring (1736–1798) suggested the study of a generalization of this question. He conjectured that every positive number can be written as a sum of nine cubes. Furthermore, any positive number can be written as a sum of 19 fourth powers. Waring goes on to claim that for any positive k, there exists a number $g(k)$ such that any positive number can be written as a sum of $g(k)$ kth powers. The question of whether $g(k)$ exists is known as Waring's problem and was answered in the affirmative by David Hilbert (1862–1943) in 1906. As for the computation of values of $g(k)$, Arthur Wieferich (1884–1954) confirmed in 1909 that $g(3) = 9$, but it was not until 1986 that Waring's assertion $g(4) = 19$ was verified. This computation was a result of work by Ramachandran Balasubramanian, François Dress, and Jean-Marc Deshouillers. Not surprisingly, Euler made contributions to the study of Waring's problem. He proved the following theorem, which gives a lower bound for $g(k)$. See Exercise 9.31.

THEOREM 9.43
For $k \geq 2$, we have $g(k) \geq 2^k + \lfloor (3/2)^k \rfloor - 2$.

Notice that for $k = 2$, 3, and 4, the bounds from the theorem are the actual values of $g(k)$. In fact, this trend continues. The lower bound for $g(k)$ provided by the theorem is known to be the actual value of $g(k)$ for all $k \leq 471600000$. (See [19].)

Exercises

9.26 Show that every number greater than 169 can be written as the sum of five nonzero squares.

Hint: Use the fact that 169 can be written as a sum of one, two, three, or four nonzero squares.

9.27 Show that if $n \equiv 7 \pmod 8$, then n is *not* the sum of three squares. That is, show that the Diophantine equation

$$x^2 + y^2 + z^2 = n$$

has no solution.

9.28 Find a positive number that can be written as a sum of nine nonnegative cubes, but no fewer than nine.

9.29 Show that there are no integer solutions to the equation

$$x_1^8 + x_2^8 + \cdots + x_{30}^8 = 3199.$$

⋆9.30 (Gauss's Circle Problem) Let $r(n)$ be the number of ordered pairs of integers (x, y) for which $x^2 + y^2 = n$. Show that

$$\lim_{N\to\infty} \frac{1}{N^2} \sum_{n=0}^{N^2} r(n) = \pi.$$

⋆9.31 Prove Theorem 9.43.

Hint: Show that $\lfloor (3/2)^k \rfloor 2^k - 1$ cannot be written as a sum of fewer than $2^k + \lfloor (3/2)^k \rfloor - 2$ kth powers.

†⋆9.32 A *quaternion* is a number of the form

$$q = a + bi + cj + dk,$$

where a, b, c, $d \in \mathbf{R}$, and i, j, k are subject to the rules $ij = k$, $jk = i$, $ki = j$, $ji = -k$, $ik = -j$, $kj = -i$, and $i^2 = j^2 = k^2 = -1$.

(a) Prove that the set Q of quaternions is a division ring, i.e., a field without commutative multiplication.

(b) The *norm* $N(q)$ of a quaternion $q = a + bi + cj + dk$ is defined as

$$N(q) = a^2 + b^2 + c^2 + d^2.$$

Show that for any q_1, $q_2 \in Q$,

$$N(q_1 q_2) = N(q_1)N(q_2).$$

(c) Show that this relation is equivalent to Euler's four-squares identity (9.6).

9.5 The case $n = 4$ in Fermat's Last Theorem

Recall that Fermat's Last Theorem asserts that for any integer $n > 2$ the equation

$$x^n + y^n = z^n \tag{9.7}$$

has no nontrivial solutions. That is, the equation has no solutions in which x, y, and z are all positive. The complete proof of this theorem draws on many different areas of mathematics and in particular deep results from the theories of elliptic curves and modular forms. We will have more to say about this in Chapter 11. In this section, we will provide an elementary proof of the theorem in the special case $n = 4$. (The case $n = 3$ is more difficult and requires the use of algebraic number theory.) This was the case for which a proof was uncovered among the papers of Fermat. We will use Fermat's method of descent to prove the theorem in this case. Assuming that a solution to the equation exists, we make a series of calculations to obtain a "smaller" solution. Since we are working with positive numbers, this eventually leads to a contradiction.

We will actually prove a slightly stronger statement than that of the $n = 4$ case of Fermat's Last Theorem.

THEOREM 9.44

The Diophantine equation

$$x^4 + y^4 = z^2 \tag{9.8}$$

has no solutions for which x, y, and z are all nonzero.

PROOF Following the method of descent, we suppose that we have a solution $x^4 + y^4 = z^2$ for which x, y and z are all nonzero. The goal will be to construct a new solution (X, Y, Z), where $0 < Z < z$. This procedure can be repeated as many times as we like, which will force a contradiction since there are only a finite number of integers between 0 and z.

Let $g = \gcd(x, y)$. Then g^4 divides $x^4 + y^4$ and hence z^2. We conclude (see Exercise 9.33) that g^2 divides z. We now have a new integer solution to our equation

$$(x/g)^4 + (y/g)^4 = (z/g^2)^2.$$

Thus, given any solution (x, y, z), one may obtain a new solution in which the greatest common divisor of the first two variables is 1. Thus, we can and will suppose that $\gcd(x, y) = 1$ for our solution.

Our method for constructing new solutions will consist of two applications of Theorem 9.5. (Thus, we will be using our description of the solutions to the equation $x^2 + y^2 = z^2$ to prove that the equation $x^4 + y^4 = z^4$ has no solutions!) We have

$$(x^2)^2 + (y^2)^2 = z^2.$$

Since $\gcd(x, y) = 1$, we also have $\gcd(x^2, y^2) = 1$, and so (x^2, y^2, z) is a primitive Pythagorean triple. By interchanging x and y, if necessary, we may suppose that x^2 is even and y^2 is odd. By Theorem 9.5, there exist relatively prime integers s and t such that

$$\begin{aligned} x^2 &= 2st \\ y^2 &= s^2 - t^2 \\ z &= s^2 + t^2. \end{aligned}$$

Observe that $t^2 + y^2 = s^2$. Since $\gcd(s, t) = 1$, we have another primitive Pythagorean triple. Since y is odd, t must be even, and s must be odd. It follows that $\gcd(s, 2t) = 1$. Since $x^2 = s(2t)$, we may apply Lemma 9.4 to find integers u and v such that

$$s = u^2 \tag{9.9}$$
$$2t = v^2 \tag{9.10}$$

Now we return to our second Pythagorean triple and again apply Theorem 9.5 to obtain relatively prime integers S and T such that

$$t = 2ST \tag{9.11}$$
$$y = S^2 - T^2 \tag{9.12}$$
$$s = S^2 + T^2. \tag{9.13}$$

From (9.11) we have $2t = 4ST$. Combining this with (9.10) yields

$$(v/2)^2 = ST.$$

Since $\gcd(S, T) = 1$, we may again make use of Lemma 9.4 to obtain X and Y satisfying

$$S = X^2$$
$$T = Y^2.$$

We compute

$$X^4 + Y^4 = S^2 + T^2 = s = u^2.$$

Thus, if we take $Z = u$, then (X, Y, Z) is a solution to (9.8). We check that this is a smaller solution:

$$Z \leq u^2 = s < s^2 + t^2 = z.$$

Thus, $Z < z$ and we have successfully made a descent.

∎

COROLLARY 9.45
The equation

$$x^4 + y^4 = z^4 \tag{9.14}$$

has no solutions with x, y, and z all nonzero.

PROOF A nonzero solution (x, y, z) to equation (9.14) leads to a nonzero solution (x, y, z^2) to (9.8), which we now know does not exist. ∎

In fact, the theorem tells us that no solution exists to Fermat's equation whenever n is divisible by 4.

COROLLARY 9.46
If n is divisible by 4, then the equation

$$x^n + y^n = z^n \tag{9.15}$$

has no solutions with x, y, and z all nonzero.

PROOF Let $n = 4m$. A solution (x, y, z) to equation (9.15) leads to a nonzero solution (x^m, y^m, z^{2m}) to equation (9.8), which cannot exist. □

Exercises

9.33 Show that if a^2 divides b^2 then a divides b.

9.34 Show that the equation $x^2 + 2xy^2 = z^4$ has no solution in nonzero integers.

Hint: Add y^4 to both sides.

9.35 Show that the Diophantine equation

$$x^4 - y^4 = z^2$$

has no nonzero solutions.

Hint: Use an argument similar to the one given in the proof of Theorem 9.44.

9.36 Prove that there is no right triangle with integer sides and perfect square area.

Hint: Use Exercise 9.35

9.37 Suppose that (x, y, z) is a Pythagorean triple. Show that at most one of x, y, and z is a perfect square.

9.6 Pell's equation

The *Pell equation* is a Diophantine equation of the form

$$x^2 - d\,y^2 = 1. \tag{9.16}$$

Given d, we would like to find all integer pairs (x, y) that satisfy the equation. Since any solution (x, y) yields multiple solutions $(\pm x, \pm y)$, we may restrict our attention to solutions where x and y are nonnegative integers.

We usually take d in (9.16) to be a positive nonsquare integer. Otherwise, there are only uninteresting solutions: if $d < 0$, then $(x, y) = (\pm 1, 0)$ in the case $d < -1$, and $(x, y) = (0, \pm 1)$ or $(\pm 1, 0)$ in the case $d = -1$; if $d = 0$, then $x = \pm 1$ (y arbitrary); and if d is a nonzero square, then $d\,y^2$ and x^2 are consecutive squares, implying that $(x, y) = (\pm 1, 0)$.

Notice that the Pell equation always has the trivial solution $(x, y) = (1, 0)$.

We now investigate an illustrative case of Pell's equation and its solution involving recurrence relations.

Example 9.47 Let's consider the Pell equation

$$x^2 - 3y^2 = 1. \tag{9.17}$$

We denote the trivial solution $(1, 0)$ as (x_0, y_0), and the solution $(2, 1)$ as (x_1, y_1).

We define $\{(x_n, y_n)\}$, for $n \geq 0$, so that

$$x_n + \sqrt{3}\,y_n = (x_1 + \sqrt{3}\,y_1)^n = (2 + \sqrt{3})^n. \tag{9.18}$$

(This is consistent with our previous definition for $n = 0, 1$.) It is easy to see that

$$x_n - \sqrt{3}\,y_n = (2 - \sqrt{3})^n, \tag{9.19}$$

and hence,

$$x_n^2 - 3y_n^2 = (x_n + \sqrt{3}\,y_n)(x_n - \sqrt{3}\,y_n) = (2 + \sqrt{3})^n (2 - \sqrt{3})^n = 1.$$

Therefore, (x_n, y_n) is a solution to (9.17), for each $n \geq 0$. (We will later prove that these are *all* the positive solutions.)

From (9.18) and (9.19), we obtain explicit formulas for x_n and y_n:

$$\begin{aligned} x_n &= \frac{1}{2}(2 + \sqrt{3})^n + \frac{1}{2}(2 - \sqrt{3})^n \\ y_n &= \frac{1}{2\sqrt{3}}(2 + \sqrt{3})^n - \frac{1}{2\sqrt{3}}(2 - \sqrt{3})^n. \end{aligned} \tag{9.20}$$

We can effectively calculate the (x_n, y_n). From (9.18), we obtain the recurrence relation

$$x_n + \sqrt{3}\,y_n = (x_{n-1} + \sqrt{3}\,y_{n-1})(2 + \sqrt{3}), \quad n \geq 1, \tag{9.21}$$

which in turn yields the recurrence relations

$$\begin{aligned} x_n &= 2x_{n-1} + 3y_{n-1} \\ y_n &= x_{n-1} + 2y_{n-1}, \quad n \geq 1, \\ x_0 &= 1, \; y_0 = 0. \end{aligned} \tag{9.22}$$

In these recurrence relations, x_n and y_n each depend on previous values of both x_n and y_n.

The recurrence relations (9.22) enable us to obtain further solutions to the Pell equation. Using a computer, we evaluate $\{x_n, y_n\}$, for $0 \leq n \leq 10$.

Mathematica

```
x[0] = 1; y[0] = 0;
x[n_] := x[n] = 2x[n - 1] + 3y[n - 1]
y[n_] := y[n] = x[n - 1] + 2y[n - 1]
```

Maple

```
pellxy:=proc(n)
  option remember:
  if n=0 then
    return [1,0]
    else
    return [2*pellxy(n-1)[1]+3*pellxy(n-1)[2],
    pellxy(n-1)[1]+2*pellxy(n-1)[2]]
  end if;
end proc:
```

We record the solutions in the table below.

n	x_n	y_n
0	1	0
1	2	1
2	7	4
3	26	15
4	97	56
5	362	209
6	1351	780
7	5042	2911
8	18817	10864
9	70226	40545
10	262087	151316

It is possible (see Exercises) to turn the recurrences (9.22) into "pure recurrences," where each sequence is defined in terms of previous values of the same sequence:

$$\begin{aligned} x_n &= 4x_{n-1} - x_{n-2} \\ y_n &= 4y_{n-1} - y_{n-2}, \quad n \geq 2. \end{aligned} \tag{9.23}$$

We see that $\{x_n\}$ and $\{y_n\}$ are defined by the same recurrence relation, but with different initial values.

□

The set of solutions to the general Pell equation (9.16) can be obtained in the same manner as in our example. Suppose that we have a solution (x_1, y_1) where x_1 and y_1 are both positive and x_1 (and hence y_1) is as small as possible. We define $\{(x_n, y_n)\}$, for $n \geq 0$, so that

$$x_n + \sqrt{d}\, y_n = (x_1 + \sqrt{d}\, y_1)^n. \tag{9.24}$$

Since

$$x_n - \sqrt{d}\, y_n = (x_1 - \sqrt{d}\, y_1)^n, \tag{9.25}$$

we see that

$$x_n^2 - d\, y_n^2 = (x_1 + \sqrt{d}\, y_1)^n (x_1 - \sqrt{d}\, y_1)^n = (x_1^2 - d\, y_1^2)^n = 1,$$

and therefore (x_n, y_n) is a solution of (9.16) for all $n \geq 0$.

From (9.24) and (9.25), we find explicit formulas for the solutions:

$$\begin{aligned} x_n &= \frac{1}{2}(x_1 + \sqrt{d}\, y_1)^n + \frac{1}{2}(x_1 - \sqrt{d}\, y_1)^n \\ y_n &= \frac{1}{2\sqrt{d}}(x_1 + \sqrt{d}\, y_1)^n - \frac{1}{2\sqrt{d}}(x_1 - \sqrt{d}\, y_1)^n. \end{aligned} \tag{9.26}$$

As $0 < x_1 - \sqrt{d}\, y_1 < 1$, we see that the growth rate of x_n and y_n are given by

$$\begin{aligned} x_n &\sim \frac{1}{2}(x_1 + \sqrt{d}\, y_1)^n \\ y_n &\sim \frac{1}{2\sqrt{d}}(x_1 + \sqrt{d}\, y_1)^n. \end{aligned} \tag{9.27}$$

From (9.24), we obtain the recurrence relation

$$x_n + \sqrt{d}\, y_n = (x_{n-1} + \sqrt{d}\, y_{n-1})(x_1 + \sqrt{d}\, y_1), \quad n \geq 1. \tag{9.28}$$

THEOREM 9.48

The solutions x_n and y_n of the Pell equation (9.16) satisfy the recurrence relations

$$\begin{aligned} x_n &= x_1 x_{n-1} + d\, y_1 y_{n-1} \\ y_n &= y_1 x_{n-1} + x_1 y_{n-1}, \quad n \geq 2, \\ x_0 &= 1,\ y_0 = 0. \end{aligned} \tag{9.29}$$

The recurrence relations (9.29) yield the "pure recurrences"

$$\begin{aligned} x_n &= 2x_1x_{n-1} - x_{n-2} \\ y_n &= 2x_1y_{n-1} - y_{n-2}, \quad n \geq 2 \end{aligned} \tag{9.30}$$

(see Exercises).

We have not yet described how to find x_1 and y_1. As in the special case above, $\{x_n\}$ and $\{y_n\}$ are defined by the same recurrence relation, but with different initial values. Also note that d does not occur in the formula. The form of the common recurrence (9.30) makes sense because we know from (9.26) that the roots of the characteristic equation for $\{x_n\}$ and $\{y_n\}$, say, r_1 and r_2, satisfy $r_1 + r_2 = 2x_1$ and $r_1r_2 = 1$.

The sequence $\{(x_n, y_n)\}$ can also be produced via matrix multiplication:

$$\begin{pmatrix} x_n \\ y_n \end{pmatrix} = \begin{pmatrix} x_1 & d\,y_1 \\ y_1 & x_1 \end{pmatrix} \begin{pmatrix} x_{n-1} \\ y_{n-1} \end{pmatrix} = \begin{pmatrix} x_1 & d\,y_1 \\ y_1 & x_1 \end{pmatrix}^n \begin{pmatrix} 1 \\ 0 \end{pmatrix}. \tag{9.31}$$

Similarly, the sequence satisfies the matrix recurrence relation

$$\begin{aligned} \begin{pmatrix} x_n & d\,y_n \\ y_n & x_n \end{pmatrix} &= \begin{pmatrix} x_1 & d\,y_1 \\ y_1 & x_1 \end{pmatrix} \begin{pmatrix} x_{n-1} & d\,y_{n-1} \\ y_{n-1} & x_{n-1} \end{pmatrix} \\ &= \begin{pmatrix} x_1 & d\,y_1 \\ y_1 & x_1 \end{pmatrix}^n \begin{pmatrix} 1 & 0 \\ 0 & 1 \end{pmatrix} \\ &= \begin{pmatrix} x_1 & d\,y_1 \\ y_1 & x_1 \end{pmatrix}^n. \end{aligned} \tag{9.32}$$

These matrix representations provide a convenient way to derive some of the properties of Pell equation solutions.

COROLLARY 9.49

The sequence $\{(x_n, y_n)\}$ satisfies the "addition/subtraction identities"

$$\begin{aligned} x_{m\pm n} &= x_mx_n \pm d\,y_my_n \\ y_{m\pm n} &= x_ny_m \pm x_my_n, \quad m \geq n. \end{aligned} \tag{9.33}$$

A proof is called for in the exercises; there are many methods, including using the relations (9.24) and (9.25), or the matrix relation (9.32).

COROLLARY 9.50

The sequence $\{(x_n, y_n)\}$ satisfies the "double angle identities"

$$\begin{aligned} x_{2n} &= 2x_n^2 - 1 \\ y_{2n} &= 2x_ny_n, \quad n \geq 0. \end{aligned} \tag{9.34}$$

PROOF Let $m = n$ in Corollary 9.49. □

Two basic questions arise as a result of our investigations thus far:

(1) Do all solutions (with $x,\ y > 0$) of a Pell equation arise via the recurrence relations (9.29)?

(2) How do we smallest the first nontrivial solution, (x_1, y_1)?

There is an example of Pell's equation, called the "special Pell equation" (first studied extensively by Fermat), in which we can find the smallest nontrivial solution easily.

Example 9.51 The "special Pell equation"

$$x^2 - (a^2 - 1)y^2 = 1 \tag{9.35}$$

has the solution $(x_1, y_1) = (a, 1)$. This is clearly the smallest nontrivial solution (since $y_1 = 1$). Example 9.47 is the case $a = 2$. Knowing the initial values, we can write the recurrence relations (9.30) in a complete form:

$$\begin{aligned} x_n &= 2ax_{n-1} - x_{n-2} \\ y_n &= 2ay_{n-1} - y_{n-2}, \quad n \geq 2, \\ x_0 &= 1,\ y_0 = 0 \\ x_1 &= a,\ y_1 = 1. \end{aligned} \tag{9.36}$$

Note: A variant of the special Pell equation is employed in the solution to Hilbert's Tenth Problem, discussed in Chapter 12.

□

In order to answer the above questions (1) and (2), we invoke a geometric setting. In this context, we will see some of our previous analysis of Pell's equation duplicated. The solutions to Pell's equation lie on the hyperbola

$$x^2 - d\,y^2 = 1. \tag{9.37}$$

See Figure 9.3. There is an obvious symmetry in the two branches of the hyperbola and in the reflection of the hyperbola in the x-axis. Indeed, these symmetries are apparent in the solutions to the equation $x^2 - d\,y^2 = 1$; that is, if (x, y) is a solution then $(\pm x, \pm y)$ are solutions. Given these symmetries, we can restrict our attention to solutions for which x is positive. Under this restriction, solutions come in pairs $(x, \pm y)$.

There is a way to "add" any two points on the hyperbola to obtain another point on the hyperbola. This process will allow us to generate new solutions to Pell's equation from given solutions.

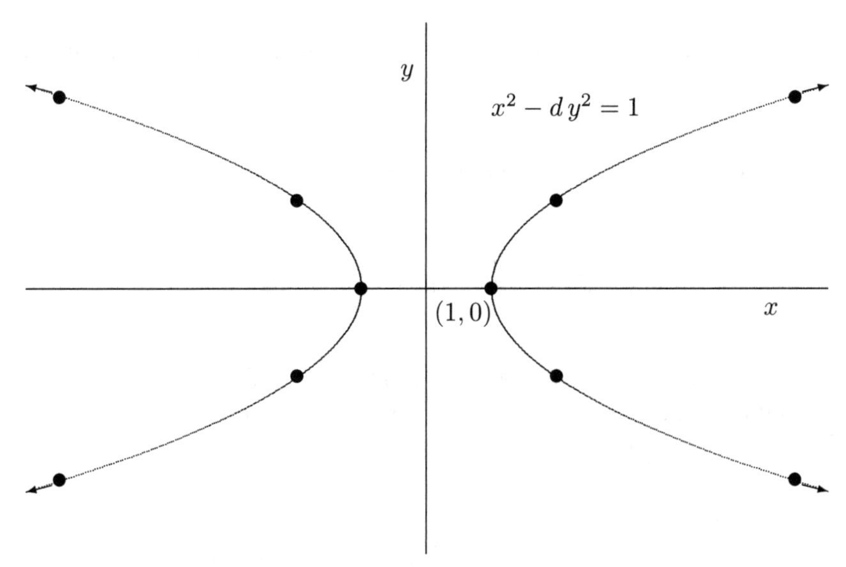

FIGURE 9.3: Solutions to the Pell equation $x^2 - d\,y^2 = 1$.

Earlier in the section, we used one solution of Pell's equation to generate new ones. In a similar way we can take two distinct solutions and put them together to obtain a new solution (and hence infinitely many solutions).

Suppose that (x, y) and (x', y') are two points on the hyperbola. We define another point (x'', y'') by the formula

$$x'' + \sqrt{d}\,y'' = (x + \sqrt{d}\,y)(x' + \sqrt{d}\,y'), \tag{9.38}$$

or equivalently,

$$\begin{aligned} x'' &= xx' + d\,yy' \\ y'' &= yx' + xy'. \end{aligned} \tag{9.39}$$

We leave it as an exercise to check that (x'', y'') is a solution to the equation $x^2 - d\,y^2 = 1$. Notice that if $x = x' = x_1$ and $y = y' = y_1$, then $x'' = x_2$ and $y'' = y_2$.

This definition is turned into an "addition" operation $\oplus$ by the rule

$$(x, y) \oplus (x', y') = (x'', y''). \tag{9.40}$$

We now proceed to show that $\oplus$ is a group operation for the set of points on the right branch of the hyperbola. It is useful to parameterize these points:

$$x = \cosh\theta, \quad y = \frac{1}{\sqrt{d}}\sinh\theta, \quad \theta \in \mathbf{R}. \tag{9.41}$$

The positive values of θ correspond to the top half of the curve, the negative values of θ to the bottom half.

With this parameterization, $\oplus$ has a natural meaning in terms of angle addition. That is, if (x, y) and (x', y') correspond to angles θ and θ', respectively, then $(x, y) \oplus (x', y')$ corresponds to angle $\theta + \theta'$. This is easy to check:

$$\cosh(\theta + \theta') = \cosh\theta \cosh\theta' + \sinh\theta \sinh\theta' = xx' + d\,yy',$$

and

$$\frac{1}{\sqrt{d}} \sinh(\theta + \theta') = \frac{1}{\sqrt{d}} \sinh\theta \cosh\theta' + \frac{1}{\sqrt{d}} \cosh\theta \sinh\theta' = yx' + xy'.$$

Note: The parameterization (9.41) explains the familiar form of the "addition/subtraction identity" (9.33), "double angle identity" (9.34), and the addition formula (9.38).

THEOREM 9.52

The set of points on the right branch of the hyperbola $x^2 - d\,y^2 = 1$ is a commutative group under $\oplus$, with identity element $(1, 0)$. Furthermore, the points with integer coordinates form a subgroup.

PROOF Closure comes from the fact that $(x, y) \oplus (x', y')$ corresponds to a point with coordinates $(\cosh\phi, \frac{1}{\sqrt{d}} \sinh\phi)$, therefore lying on the hyperbola.

Associativity of $\oplus$ follows from the associativity of angle addition.

The element $(1, 0)$, corresponding to the angle $\theta = 0$, serves as an identity.

We need to check that each element has an inverse. The inverse of (x, y) is $(x, -y)$, which we check as follows: $(x, y) \oplus (x, -y) = (x^2 - d\,y^2, xy - xy) = (1, 0)$. In terms of angles, the inverse of the element corresponding to θ is the element corresponding to $-\theta$.

Commutativity of $\oplus$ follows from commutativity of angle addition.

The points with integer coordinates form a subgroup, because if (w, x) and (y, z) are ordered pairs of integers, so are $(w, x) \oplus (y, z)$ and $(w, -x)$. ∎

Let the group of points on the right branch of the hyperbola be denoted G, and the subgroup of points with integer coordinates be denoted H. In fact, H is a cyclic group. That is, every element of H can be obtained by adding a fixed element of H to itself an appropriate number of times. We now state this result as a theorem and give the proof.

THEOREM 9.53

The integer points on the right branch of the hyperbola $x^2 - d\,y^2 = 1$ (i.e., the solutions to Pell's equation with x positive) form a cyclic group with identity $(1, 0)$. Furthermore, the group is generated by the solution (x_1, y_1), where $y_1 > 0$ and x_1 is the smallest x-coordinate greater than 1 among all integer solutions.

PROOF Let θ be the smallest positive angle corresponding to an integer point on the hyperbola. We claim that H is the cyclic subgroup generated by θ. Suppose that there is an integer point on the hyperbola that corresponds to an angle θ' that is not a multiple of θ. Then $m\theta < \theta' < (m+1)\theta$, for some integer m. Hence, the angle $\theta' - m\theta$ corresponds to an integer point (because of closure) and is positive (due to the first inequality) and smaller than θ (due to the second inequality). This is a contradiction. Therefore, θ generates all integer points. ∎

Example 9.54 Let's return to the equation $x^2 - 3y^2 = 1$ of Example 9.47. We have already noted that $(2, 1)$ is a solution to the equation. In fact, this is the smallest positive solution because if $x = 1$ we have only the trivial solution $(1, 0)$. Therefore, by Theorem 9.53, the point $(2, 1)$ generates all solutions to (9.47). The complete set of solutions (with $x, y > 0$) is given by the recurrence relations (9.29). ∎

Exercises

9.38 Find three positive solutions to the Pell equation $x^2 - 5y^2 = 1$.

9.39 Use the fact that $(80, 9)$ is a solution to

$$x^2 - 79y^2 = 1$$

to obtain an additional positive solution to this Diophantine equation.

9.40 Prove the validity of the recurrence relations (9.23) and (9.30).

9.41 Suppose that (x, y) and (x', y') are two solutions to the equation $x^2 - dy^2 = 1$. Prove that (x'', y'') is another solution, where

$$\begin{aligned} x'' &= xx' + dyy' \\ y'' &= yx' + xy'. \end{aligned}$$

9.42 Prove the formula (9.33).

9.43 Show that the Diophantine equation

$$x^2 - dy^2 = -1$$

has no integral solutions when $d \equiv 3 \pmod 4$.

9.44 Suppose that the Diophantine equation

$$x^2 - dy^2 = -1$$

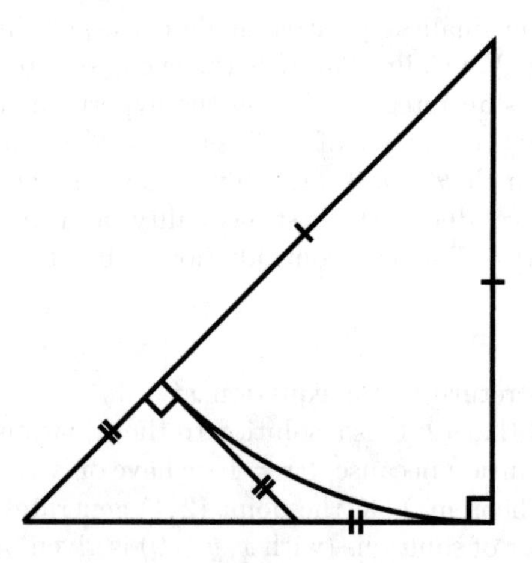

FIGURE 9.4: The number $\sqrt{2}$ is irrational.

does have a solution, and let (x_1, y_1) be the smallest such positive solution. Show that (x_n, y_n) is a positive solution to the equation if and only if

$$x_n + y_n\sqrt{d} = (x_1 + y_1\sqrt{d})^n,$$

for some odd positive integer n.

9.45 As in $\mathbf{Z}[i]$, an element α in $\mathbf{Z}[\sqrt{m}]$ is said to be a unit if there exists an element β in $\mathbf{Z}[\sqrt{m}]$ such that $\alpha\beta = 1$. Discuss the connection between units in the ring $\mathbf{Z}[\sqrt{m}]$ and solutions to Pell's equation.

9.7 Continued fraction solution of Pell's equation

Now we show how to find a solution to Pell's equation, $x^2 - dy^2 = 1$, via the continued fraction expansion of $\sqrt{d}$.

Example 9.55 We showed a classic proof of the irrationality of $\sqrt{2}$ in Chapter 2. Now we look at another proof, one that is visual and does not require a parity (even/odd) argument.

Suppose that $\sqrt{2}$ is a rational number. Then there exists an isosceles right triangle with integer sides (Figure 9.4). We see from the figure that the smaller isosceles right triangle also has integer sides. But this descent cannot continue forever. Therefore, $\sqrt{2}$ is irrational.

Labeling this diagram, we see that the proof can be given in a purely algebraic way: if $a/b = \sqrt{2}$, then $(2b - a)/(a - b) = \sqrt{2}$. But the numerator and denominator are smaller numbers. This descent cannot occur forever.

A feature of the proof is the transformation $(a, b) \to (2b - a, a - b)$. The inverse of this transformation is $(a, b) \to (a+2b, a+b)$. This latter transformation yields closer and closer approximations to $\sqrt{2}$ (starting with a pair such as $(a, b) = (1, 1)$). We see that if a/b tends to l, then $l = (l + 2)/(l + 1)$, and hence $l = \sqrt{2}$. We also observe that if $a/b > \sqrt{2}$, then $(2b - a)/(a - b) < \sqrt{2}$, while if $a/b < \sqrt{2}$, then $(2b - a)/(a - b) > \sqrt{2}$. Indeed, starting with the pair $(1, 1)$, the successive pairs (a, b) are solutions to the "alternating Pell equation" $x^2 - 2y^2 = \pm 1$. The fractions produced by this process are

$$\frac{1}{1}, \frac{3}{2}, \frac{7}{5}, \frac{17}{12}, \frac{41}{29}, \frac{99}{70}, \ldots.$$

These fractions are the convergents of the continued fraction $[1; \overline{2}]$, the continued fraction expansion of $\sqrt{2}$. □

Now we turn to the second question; that is, how do we find a generator of the group of integer solutions to Pell's equation? In light of the preceding discussion, it is perhaps not surprising that we can use continued fractions. We illustrate with an example.

Example 9.56 Let's find a solution to the Pell equation $x^2 - 17y^2 = 1$. The basic idea is that if $x^2 - 17y^2 = 1$, then $x^2 \doteq 17y^2$, and $x/y \doteq \sqrt{17}$. We find the continued fraction expansion of $\sqrt{17}$:

$$\begin{aligned}
\sqrt{17} &= 4 + (\sqrt{17} - 4) \\
&= 4 + \frac{1}{\frac{1}{\sqrt{17}-4}} \\
&= 4 + \frac{1}{\sqrt{17} + 4} \\
&= 4 + \frac{1}{8 + (\sqrt{17} - 4)} \\
&= 4 + \cfrac{1}{8 + \cfrac{1}{8 + \cdots}} \\
&= [4; \overline{8}].
\end{aligned}$$

The first approximation is

$$\sqrt{17} \doteq 4 + \frac{1}{8} = \frac{33}{8}.$$

We obtain the solution $(x, y) = (33, 8)$:

$$33^2 - 17 \cdot 8^2 = 1089 - 1088 = 1.$$

While it is not necessarily the case that every approximation gives rise to a solution to the equation, it is true that every positive solution will be given by a further approximation. ∎

THEOREM 9.57

Let (x, y) be a pair of positive integers that give a solution to the Pell equation $x^2 - d\,y^2 = 1$. Then x/y is a convergent of the continued fraction expansion of $\sqrt{d}$.

PROOF Factoring the Pell equation, we have

$$(x - \sqrt{d}\,y)(x + \sqrt{d}\,y) = 1. \tag{9.42}$$

From this, we may deduce that $x - y\sqrt{d}$ is positive and consequently

$$\frac{x}{y} > \sqrt{d}. \tag{9.43}$$

We may rewrite (9.42) as

$$x - y\sqrt{d} = \frac{1}{x + y\sqrt{d}}.$$

Thus, we have

$$\begin{aligned}
\frac{x}{y} - \sqrt{d} &= \frac{1}{y(x + \sqrt{d}\,y)} \\
&= \frac{1}{y^2(x/y + \sqrt{d})} \\
&< \frac{1}{\sqrt{d}\,2y^2} \qquad \text{(by inequality (9.43))} \\
&< \frac{1}{2y^2}.
\end{aligned}$$

Since $x/y - \sqrt{d}$ is positive, we can write

$$\left|\frac{x}{y} - \sqrt{d}\right| < \frac{1}{2y^2}.$$

It now follows from Theorem 8.26 that x/y is a convergent of the continued fraction expansion of $\sqrt{d}$. ∎

We now make use of Theorem 8.43 to describe which convergents give rise to solutions to Pell's equation.

THEOREM 9.58

Let d be a positive integer that is not a perfect square. Let s be the period of the continued fraction expansion of $\sqrt{d}$, and let p_n/q_n be the nth convergent for this continued fraction. Every positive solution to the Pell equation $x^2 - d\,y^2 = 1$ is of the form (p_n, q_n), for some n. Furthermore,

- *If s is even, the complete set of positive solutions is*

$$\{(p_{ts-1}, q_{ts-1}) : t \text{ a positive integer}\}.$$

- *If s is odd, the complete set of positive solutions is*

$$\{(p_{2ts-1}, q_{2ts-1}) : t \text{ a positive integer}\}.$$

PROOF By Theorem 9.57, all positive solutions to the Pell equation must be of the form (p_n, q_n), where p_n/q_n is a convergent of the continued fraction expansion of $\sqrt{d}$. Let P_n and Q_n be the usual quantities associated with the continued fraction expansion of $\sqrt{d}$ (see Lemma 8.34). By Theorem 8.43, (p_n, q_n) is a solution to the Pell equation $x^2 - d\,y^2 = 1$ if and only if $(-1)^{n-1}Q_{n+1} = 1$. The following lemma will allow us to determine when the latter condition holds.

LEMMA 9.59

Let d be a positive integer that is not a perfect square. Let s be the period of the continued fraction expansion of $\sqrt{d}$, and define Q_n as in Lemma 8.34 for $\alpha = \sqrt{d}$. Then for $n \geq 0$, we have $Q_n > 0$ and $Q_n = 1$ if and only if $s \mid n$.

PROOF We begin by arguing that we may work with the continued fraction expansion of $\lfloor\sqrt{d}\rfloor + \sqrt{d}$ instead of $\sqrt{d}$. Using the definitions given in Lemma 8.34, we have the table below.

α	$\sqrt{d}$	$\lfloor\sqrt{d}\rfloor + \sqrt{d}$
P_0	0	$\lfloor\sqrt{d}\rfloor$
Q_0	1	1
a_0	$\lfloor\sqrt{d}\rfloor$	$2\lfloor\sqrt{d}\rfloor$
P_1	$\lfloor\sqrt{d}\rfloor - 0 = \lfloor\sqrt{d}\rfloor$	$2\lfloor\sqrt{d}\rfloor - \lfloor\sqrt{d}\rfloor = \lfloor\sqrt{d}\rfloor$
Q_1	$d - \lfloor\sqrt{d}\rfloor^2$	$d - \lfloor\sqrt{d}\rfloor^2$

Because of the recursive definitions of a_n, α_n, P_n, and Q_n, they will be the same for $\sqrt{d}$ and $\lfloor\sqrt{d}\rfloor + \sqrt{d}$ for $n \geq 1$. For the rest of the proof we may, therefore, work with the quantities a_n, α_n, P_n, and Q_n determined by the continued fraction expansion of $\lfloor\sqrt{d}\rfloor + \sqrt{d}$ (including the values for $n = 0$).

By Lemma 8.34,

$$\lfloor\sqrt{d}\rfloor + \sqrt{d} = [a_0; a_1, \ldots, a_{s-1}, \alpha_s].$$

Recall that $\lfloor\sqrt{d}\rfloor + \sqrt{d}$ is a reduced quadratic irrational, and so by Theorem 8.40, its continued fraction expansion is purely periodic. Furthermore, its period is s (i.e., it has the same period as the expansion for $\sqrt{d}$). We conclude that $\alpha_s = \lfloor\sqrt{d}\rfloor + \sqrt{d}$. Since $\alpha_s = (P_s + \sqrt{d})/Q_s$, Exercise 8.23 implies $Q_s = 1$ and $P_s = \lfloor\sqrt{d}\rfloor$. The hypotheses of Lemma 8.35 hold with $N = 0$. Thus, Equations (8.56) and (8.57) hold and, by induction, we have

$$P_{r+ts} = P_r \tag{9.44}$$

$$Q_{r+ts} = Q_r \tag{9.45}$$

for any nonnegative integers r and t. In particular, $Q_n = Q_0 = 1$ when $s \mid n$. We also note, from the proof of Theorem 8.38, that $Q_n > 0$ for all sufficiently large n. Thus, (9.45) implies that $Q_n > 0$ for all n.

Now suppose, by way of contradiction, that $Q_n = 1$ for some n that is not divisible by s. There exist nonnegative integers r and t with $0 < r < s$ such that $n = r + ts$. It follows from (9.45) that $Q_r = 1$. Now

$$\lfloor\sqrt{d}\rfloor + \sqrt{d} = [a_0; a_1, \ldots, a_{r-1}, \alpha_r].$$

Since $\lfloor\sqrt{d}\rfloor + \sqrt{d}$ is purely periodic, α_r also has a periodic continued fraction expansion. By Theorem 8.40, α_r must be reduced. Thus $-1 < \alpha_r' < 0$, or

$$-1 < P_r - \sqrt{d} < 0.$$

We conclude that $P_r = \lfloor\sqrt{d}\rfloor$, and so

$$P_r = P_0,$$

$$Q_r = Q_0.$$

From Lemma 8.35, we now have $a_{k+r} = a_k$, for all $k \geq 0$. Since $0 < r < s$, we have contradicted the fact that s is the period of $\lfloor\sqrt{d}\rfloor + \sqrt{d}$. ∎

We now complete the proof of the theorem. By the lemma, (p_n, q_n) is a solution if and only if

(i) n is odd, and

(ii) the period s of the continued fraction expansion of $\sqrt{d}$ divides $n + 1$.

Condition (ii) holds if and only if we can write

$$n = rs - 1,$$

for some positive integer r. If s is even, then such an n will satisfy (i) for all positive integers r. If s is odd, condition (i) holds exactly when r is even. ∎

Example 9.60 Let's use the theorem to find the smallest positive solution to the Pell equation

$$x^2 - 13y^2 = 1.$$

We will make use of the procedure from Example 8.39 to compute the continued fraction expansion of $\sqrt{13}$.

Mathematica

```
quadirrcontinuedfraction[0, 1, 13]

{3, {1, 1, 1, 1, 6}}
```

Maple

```
quadirrcontinuedfraction(0, 1, 13);

[3, [1, 1, 1, 1, 6]]
```

We see that the period is 5, and so Theorem 9.58 tells us that the smallest positive solution to the Pell equation is (p_9, q_9). From our computer computation, we have

$$\frac{p_9}{q_9} = [3; 1, 1, 1, 1, 6, 1, 1, 1, 1].$$

Using `ourfromcontinuedfraction`, we find that the convergent is 649/180. Thus

$$649^2 - 13 \cdot 180^2 = 1,$$

and this is the smallest positive solution to our Pell equation. ∎

Exercises

9.46 Determine the continued fraction expansion of $\sqrt{7}$ and use your answer to find a solution to the Diophantine equation

$$x^2 - 7y^2 = 1.$$

9.47 Determine the continued fraction expansion of $\sqrt{10}$ and use your answer to find a solution to the Diophantine equation

$$x^2 - 10y^2 = 1.$$

◇**9.48** Determine the continued fraction expansion of $\sqrt{19}$ and use your answer to find a solution to the Diophantine equation

$$x^2 - 19y^2 = 1.$$

◇**9.49** Determine the continued fraction expansion of $\sqrt{29}$ and use your answer to find a solution to the Diophantine equation

$$x^2 - 29y^2 = 1.$$

9.50 Let d be an integer that is not a perfect square. Show that the infinite simple continued fraction expansion of $\sqrt{d}$ has the form

$$\sqrt{d} = [a_0; \overline{a_1, \ldots, a_s}],$$

where $a_s = 2a_0$.

9.8 The *abc* conjecture

In this section, we examine one of the most famous unsolved problems in number theory, the *abc* conjecture. Suppose that a, b, and c are integers satisfying

$$a + b = c. \tag{9.46}$$

That is, (a, b, c) is a solution to the linear Diophantine equation $x + y = z$. Any common divisor of two of the values necessarily divides the third, and so by canceling out any such common factors, we may assume that a, b, and c are pairwise relatively prime. We exhibit several solutions in the table below along with the prime divisors of a, b, and c.

Solution	Prime divisors of a, b, and c
$2 + 7 = 9$	2, 3, 7
$4 - 9 = -5$	2, 3, 5
$8 + 25 = 33$	2, 3, 5, 11

Notice that in each of these cases, the product of the prime divisors of a, b, and c is greater than the values of any of the three variables. We might wonder if this holds in general. That is, given pairwise relatively prime a, b, and c satisfying $a + b = c$, is it true that

$$\max\{a, b, c\} < \prod_{p \mid abc} p? \tag{9.47}$$

Let's perform a computer search to investigate. Notice that by rearranging and renaming variables, if necessary, we may assume that $1 \le a \le b < c$, in which case $\max\{a, b, c\} = c$. We search for counterexamples to (9.47) among triples (a, b, c) with $1 \le a \le b \le 50$.

Mathematica

```
Do[If[GCD[a, b] == 1,
  producttemp =
    Abs[Product[FactorInteger[a*b*(a + b)][[i]][[1]],
      {i, 1, Length[FactorInteger[a*b*(a + b)]]}]];
  If[a + b > producttemp,
    Print[{a, b, a + b, {producttemp}}]]],
  {b, 1, 50}, {a, 1, b}]

{1, 8, 9, {6}}
{5, 27, 32, {30}}
{1, 48, 49, {42}}
{32, 49, 81, {42}}
```

Maple

```
for b from 1 to 50 do
  for a from 1 to b do
    if gcd(a,b)=1 then
      producttemp:=product(ifactors(a*b*(a+b))[2][i][1],
        i=1..nops(ifactors(a*b*(a+b))[2]));
        if a+b>producttemp then print(a,b,a+b,[producttemp]);
        end if
    end if;
  end do;
end do;

1, 8, 9, [6]
5, 27, 32, [30]
1, 48, 49, [42]
32, 49, 81, [42]
```

Thus, we see that inequality in (9.47) does not hold in general, but counterexamples appear to be somewhat sparse. In 1985 Joseph Oesterle and David Masser conjectured that the inequality will hold in general if we increase the right side just slightly. To simplify the notation, we introduce a term.

DEFINITION 9.61 *The radical of a positive integer n is the quantity*

$$\operatorname{rad}(n) = \prod_{p|n} p.$$

So for example, $\text{rad}(40) = 2 \cdot 5 = 10$.

CONJECTURE (*abc* **conjecture**)
For every $\varepsilon > 0$, there exist a constant K_ε such that

$$c < K_\varepsilon(\text{rad}(abc))^{1+\varepsilon} \tag{9.48}$$

for all pairwise relatively prime positive numbers a, b, c satisfying $a + b = c$.

We may take ε as small as we like, but consider the case $\varepsilon = 1$. The conjecture predicts the existence of a number K_1 such that

$$c < K_1(\text{rad}(abc))^2 \tag{9.49}$$

for all pairwise relatively prime positive numbers a, b, and c satisfying $a+b = c$. In fact, there are no known counterexamples to the inequality in (9.49) with $K_1 = 1$.

On the surface, the conjecture is a statement about a restricted set of solutions to a linear Diophantine equation. Diving a little deeper, we find connections to famous Diophantine problems, both solved and unsolved. Perhaps there is no more famous Diophantine problem than that of Fermat's Last Theorem. Suppose, in contradiction to the theorem, that there existed an integer $n \geq 3$ and positive numbers x, y, and z (which we may assume to be pairwise relatively prime) satisfying

$$x^n + y^n = z^n.$$

Taking $a = x^n$, $b = y^n$, and $c = z^n$, the *abc* conjecture predicts that

$$z^n < K_1\left(\text{rad}((xyz)^n)\right)^2 < K_1(xyz)^2 < K_1 z^6.$$

We conclude that

$$z^{n-6} < K_1.$$

Since $z \geq 2$, this excludes all but finitely many values n. If, as suggested above, we may take $K_1 = 1$, then the condition becomes $z^n < z^6$, and so $n < 6$. That is, if the *abc* conjecture holds with $K_1 = 1$, then we immediately obtain a proof of Fermat's Last Theorem in the case $n \geq 6$. Note: Elementary arguments exist for the cases $n = 3$, 4, and 5.

Let's consider what might be a worst-case scenario for the *abc* conjecture. Suppose that $a+b = c$ where a, b, and c are all powers of small numbers. More specifically, consider the case $a = 1$, $b = x^p$, and $c = y^q$ for some numbers p and q. Then $\text{rad}(abc) = \text{rad}(xy)$, and we would expect the value of $\text{rad}(xy)$ to be small relative to y^q. Thus, numerous solutions to the equation

$$1 + x^p = y^q \tag{9.50}$$

could pose a problem for the *abc* conjecture. Notice that solutions to the equation provide us with a pair of consecutive numbers that are perfect powers. For instance, with $x = 2$, $p = 3$, $y = 3$, and $q = 2$, we obtain a solution to (9.50) corresponding to the consecutive perfect powers 8 and 9. In 1844 Eugène Catalan (1814–1894) conjectured that these are the only pair of consecutive perfect powers. While this conjecture did not remain an open problem quite as long as Fermat's, it did defy the efforts of mathematicians for over 150 years. In 2002 Preda Mihailescu succeeded in furnishing a proof of the conjecture.

THEOREM (Catalan's Conjecture)
The only pair of consecutive numbers of the form x^p and y^q with $p, q \geq 2$ is the pair 8 and 9.

Given what the *abc* conjecture has to say about Fermat's Last Theorem and Catalan's conjecture, it is natural to wonder what the conjecture implies more generally about perfect powers that can be written as a sum of two other perfect powers. For instance we have

$$2^5 + 7^2 = 3^4.$$

Thus, we wish to look for solutions to the equation

$$x^p + y^q = z^r \tag{9.51}$$

where x, y, and z are relatively prime. If $p = q = r = 2$, then this amounts to looking for primitive Pythagorean triples. As we have seen, there are an infinite number of such triples. We wish to impose a restriction on the exponents to require some or all of them to be sufficiently large. Let's consider exponents p, q, and r such that

$$\frac{1}{p} + \frac{1}{q} + \frac{1}{r} < 1.$$

In fact, the minimum value of a sum satisfying the inequality is 41/42 (Exercise 9.52). The *abc* conjecture predicts that a solution to (9.51) must satisfy

$$z^r < K_\epsilon (xyz)^{1+\epsilon}.$$

Now $x < z^{r/p}$ and $y < z^{r/q}$, and so the inequality above implies that

$$z^r < K_\epsilon \left((z^r)^{(1/p+1/q+1/r)} \right)^{1+\epsilon}.$$

If we take $\epsilon = 1/42$ and apply the assumption $1/p + 1/r + 1/q < 41/42$, then we have

$$z^r < K_{1/42} (z^r)^{1763/1764}.$$

Since $K_{1/42}$ is a constant, we see that there are finitely many possible values for z and hence finitely many possible solutions to (9.51). Thus, the *abc* conjecture implies the following conjecture.

CONJECTURE (Fermat–Catalan Conjecture)
There are finitely many triples of relatively prime numbers x^p, y^q, z^r satisfying

$$x^p + y^q = z^r$$

with $1/p + 1/q + 1/r < 1$.

Unlike Fermat's Last Theorem and the original Catalan Conjecture, this remains an open problem. There are a total of ten known triples satisfying the conditions of the conjecture, including the two that we have already seen (i.e., $1^p + 2^3 = 3^2$ and $2^5 + 7^2 = 3^4$).

Exercises

9.51 Show that for any integer $k \geq 0$, the number $3^{2^k} - 1$ is divisible by 2^{k+1}. Use this to show that there are infinitely many triples of relatively prime numbers a, b, c such that $a + b = c$ and

$$c > \prod_{p|abc} p.$$

9.52 (a) Show that if p, q, and r are positive integers satisfying

$$\frac{1}{p} + \frac{1}{q} + \frac{1}{r} > 1,$$

then (p, q, r) must equal $(2, 3, 3)$, $(2, 3, 4)$, $(2, 3, 5)$, $(2, 2, k)$, or some permutation of one of these cases.

(b) Show that if p, q, and r are positive integers satisfying

$$\frac{1}{p} + \frac{1}{q} + \frac{1}{r} = 1,$$

then (p, q, r) must equal $(2, 4, 4)$, $(2, 3, 6)$, $(3, 3, 3)$, or some permutation of one of these cases.

(c) Show that if p, q, and r are positive integers satisfying

$$\frac{1}{p} + \frac{1}{q} + \frac{1}{r} < 1,$$

then

$$\frac{1}{p} + \frac{1}{q} + \frac{1}{r} \leq \frac{41}{42}.$$

◇**9.53** Find as many of the remaining eight triples satisfying the conditions of the Fermat–Catalan conjecture as you can.

9.54 Let d, e, and f be fixed positive integers. Show that the *abc* conjecture implies that there are finitely many relatively prime integers x, y, z satisfying

$$dx^p + ey^p = fz^p,$$

with $p \geq 4$.

9.9 Notes

Diophantus

Diophantus of Alexandria (c. 200–c. 284 B.C.E.) studied equations with integer or rational solutions. One of his triumphs was the complete parameterization of solutions to the equation for a right triangle: $x^2 + y^2 = z^2$. We have called these solutions Pythagorean triples.

Diophantus wrote what are generally considered to be the first algebra books. Many of his contributions were posed as problems with solutions. His most well-known book, *Arithmetica*, is the one in which Fermat inscribed his famous marginal note (see the next Note).

Diophantus expressed his mathematical analysis in words rather than symbols. This was certainly a limitation, and great successes were to follow centuries later when place notation as well as other basics of number theory and algebra had been worked out. Nonetheless, Diophantus' early contributions are remarkable for their ingenuity and served as a catalyst for further mathematical thought.

Pierre de Fermat

One of the most important figures in the history of number theory is a man for whom mathematics was a hobby rather than a profession. Pierre de Fermat (1601–1665) made his living practicing law, but he is more famous for his contributions to mathematics and, in particular, to number theory. Fermat had no interest in publishing results during his lifetime but did correspond frequently with other mathematicians of the day. Much of the work of Fermat that has been published is the result of efforts made by Fermat's son Samuel after Fermat died.

Fermat was particularly inspired by Diophantus and made many notes in his copy of the *Arithmetica*. One such note is the famous quote at the beginning of this chapter. Thus, Fermat states his famous "Last Theorem," while providing a tantalizing reference to a proof that was never located among any of his collected notes. This was actually fairly typical of the notes from Fermat that we have. He makes many statements but provides few proofs.

Virtually all of the Diophantine equations presented in this chapter were considered by Fermat. We saw in Section 9.5 a demonstration of the method of descent pioneered by Fermat to show that a Diophantine equation has no solution. Fermat challenged other mathematicians to show that Pell's equation has an infinite number of solutions. Here, again, Fermat indicates that he is in possession of a proof but no record of one exists. Fermat does provide some examples of rather large solutions, which indicates that he likely knew how to generate solutions in general. The fact that every integer can be written as a sum of four squares was also known to Fermat. In this case, he at least provided an outline for the proof. Several of the statements in the outline were subsequently verified by Euler, but the proof was not completed (at least a published proof) until Lagrange arrived on the scene.

While he is best known for his work in number theory, Fermat made contributions to other areas of mathematics as well. Fermat was the first to use tangent lines to determine the maximum and minimum values of polynomials. His method is essentially the same one taught in calculus courses today. In fact, his contributions compelled Lagrange to credit Fermat as the inventor of calculus.

Three squares and triangular numbers

In Section 9.5 we bypassed the issue of which numbers can be represented as a sum of three squares. In fact, a number n can be written as the sum of three squares if and only if $n \neq 4^m(8k+7)$, for integers m and k. Given this result, one can deduce Theorem 9.41. One only needs to show that any prime is a sum of four squares. Certainly 2 is, and any prime congruent to 1 mod 4 is a sum of two squares by Theorem 9.5. If $p \equiv 3 \pmod 4$, then either p or $p-4$ is congruent to 3 mod 8. The result quoted above implies that either p or $p-4$ is a sum of three squares from which it follows p is a sum of four squares.

Gauss also used the sum of three squares theorem above to obtain another interesting result on the geometry of numbers. Loosely speaking, n is a triangular number if one can arrange n dots into the shape of a triangle. We arrange the dots in rows so that row 1 has one dot, row 2 has two dots, and so on. See Figure 9.5. The total number of dots in such an arrangement with k rows is the sum $1+2+\cdots+k$. This leads us to a more formal definition of a triangular number as any number of the form $k(k+1)/2$. One can ask the analogous question about representation of integers as sums of triangular numbers. What is the smallest number t such that any positive integer can be written as the sum of t triangular numbers? An entry in Gauss' diary (next to which he wrote "Eureka!") provides the answer:

$$n = \triangle + \triangle + \triangle$$

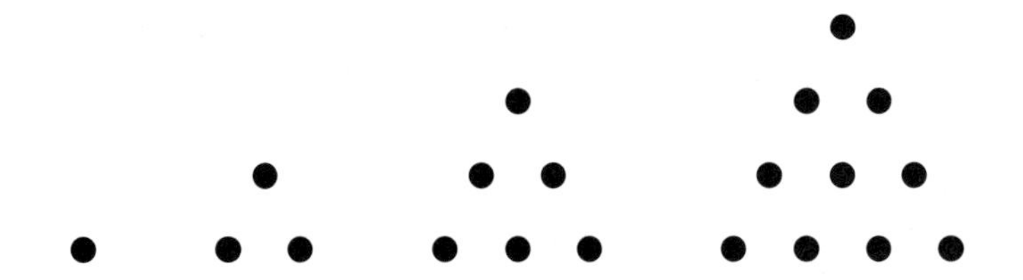

FIGURE 9.5: The first four triangular numbers: 1, 3, 6, 10.

(i.e., $t = 3$). From the perspective of Diophantine equations, one can state the result as follows.

THEOREM

Given any positive integer n, there exist nonnegative integers x, y, and z such that

$$n = x(x+1)/2 + y(y+1)/2 + z(z+1)/2.$$

The proof follows easily from the three squares theorem, which implies that there exist nonnegative integers a, b, and c such that

$$8n + 3 = a^2 + b^2 + c^2.$$

Reducing the above equation mod 8, we see that a, b, and c must all be odd. Thus, we can write

$$\begin{aligned} a &= 2x + 1 \\ b &= 2y + 1 \\ c &= 2z + 1 \end{aligned}$$

for some nonnegative integers x, y, and z. With a little algebra, we have

$$\frac{(2x+1)^2 + (2y+1)^2 + (2z+1)^2 - 3}{8} = \frac{x(x+1)}{2} + \frac{y(y+1)}{2} + \frac{z(z+1)}{2},$$

as desired.

Exercises

9.55 The number 36 is both a square number and a triangular number as

$$36 = 6^2 = \frac{8(8+1)}{2}.$$

Show that there are an infinite number of positive integers that are simultaneously square numbers and triangular numbers.

Hint: Complete the square on $8m^2 = 4n^2 + 4n$ and make a substitution to arrive at a Pell equation.

9.56 Show that there are an infinite number of triangular numbers t such that $2t$ is also a triangular number. The number 3 is one such example. What is the next such number?

◊**9.57** Let $s(n) = 1^2 + 2^2 + \cdots + n^2$. Use a computer to find a value of $n > 1$ for which $s(n)$ is a perfect square. It can be proved that there is only one such value.

History of Pell's equation

Pell's equation $x^2 - d\,y^2 = 1$ was probably first studied in the case $x^2 - 2y^2 = 1$. Early mathematicians, upon discovering that $\sqrt{2}$ is irrational, realized that although one cannot solve the equation $x^2 - 2y^2 = 0$ in integers, one can at least solve the "next best thing." Two of the early investigators of Pell equations were the Indian mathematicians Brahmagupta (c. 665) and Bhaskara (c. 12th century). In particular, Bhaskara studied Pell's equation for the values $d = 8$, 11, 32, 61, and 67. Some of these equations lead to basic solutions that are quite large. For example, Bhaskara found the solution $x = 1776319049$, $y = 22615390$, for $d = 61$.

Fermat generated interest in the Pell equation and worked out some of the basic theory. Although it was Lagrange who discovered the complete theory of the equation $x^2 - dy^2 = 1$, Euler named the equation after John Pell (1611–1685). He did so apparently because Pell was instrumental in writing a book containing the equation.

Brahmagupta has left us with this intriguing challenge: "A person who can, within a year, solve $x^2 - 92y^2 = 1$ is a mathematician."

Exercises

◊**9.58** Solve Brahmagupta's challenge.

The p-adic numbers

When we say that two rational numbers are "close," we mean that the difference of the two numbers is small. If we have a sequence of rational numbers

$$a_1, a_2, a_3, \ldots$$

that converges to a rational number, then the terms of the sequence are "getting closer." That is, the difference between a_m and a_n is as small as we like, provided that m and n are sufficiently large. The converse of this statement is not true. For instance, the terms in the sequence of convergents in the continued fraction expansion of $\sqrt{3}$ are "getting closer," but the sequence does not converge to a rational number since $\sqrt{3}$ is irrational.

Recall that a sequence is *Cauchy* if the difference between a_m and a_n is as small as we like, provided that m and n are sufficiently large. All convergent sequences are Cauchy, but, as our example shows, not all Cauchy sequences in $\mathbf{Q}$ converge.

However, if we complete the set of rational numbers by adding in the numbers that are limits of Cauchy sequences, we obtain a set in which all Cauchy sequences do converge. In fact, this enlarged set is the set of real numbers. If we change our perspective on what it means for a number to be small, we obtain new completions of the set of rational numbers.

Let p be a prime number. Given a nonzero rational number x, we can write $x = (a/b)p^n$, where a and b are integers not divisible by p and n is some integer. We define the p-adic valuation on $\mathbf{Q}$ by the formula

$$|x|_p = |(a/b)p^n|_p = p^{-n}.$$

We also set $|0|_p = 0$. So, for example, we have $|12/35|_2 = 1/4$ while $|12/35|_7 = 7$. We now think of a rational number as small if its p-adic valuation is small. Thus, an integer that is divisible by a high power of p is small while a rational number whose denominator is a multiple of a high power of p is big. Two numbers are close if the p-adic valuation of their difference is small. With $p = 5$, the numbers 3 and 28 are close since $|3-28|_5 = 1/25$, but 3 and 16/5 are not so close since $|3 - 16/5|_5 = 5$.

Let $a_1 = 3$ and notice that

$$|a_1^2 - 2|_7 = \frac{1}{7}.$$

Thus, with this new notion of closeness, we can say that 3 is a decent estimate of $\sqrt{2}$. Now if we take $a_2 = 10$, then

$$|a_2^2 - 2|_7 = \frac{1}{7^2},$$

and so 10 is an even better approximation of $\sqrt{2}$. Suppose that a_n is an integer satisfying

$$|a_n^2 - 2|_7 \leq \frac{1}{7^n}.$$

If we set $a_{n+1} = a_n + 7^n k$, then for an appropriate choice of k, we will have

$$|a_{n+1}^2 - 2|_7 \leq \frac{1}{7^{n+1}}$$

(Exercise 9.60). Thus, we have a Cauchy sequence with our new metric. This sequence will not converge since $\sqrt{2}$ is irrational, but we can again enlarge the set of rational numbers to obtain a set in which all Cauchy sequences do converge. We call this the set of *p-adic* numbers and denote it $\mathbf{Q}_p$. We note that $\mathbf{Q}_p$ is different from $\mathbf{R}$. For instance, one can show that $\mathbf{Q}_7$ will not contain $\sqrt{3}$.

As we have seen in this chapter, it can be difficult to decide whether a given Diophantine equation has integer solutions, or even rational solutions.

However, it is often much easier to determine solvability of equations over $\mathbf{Q}_p$. What makes this particularly useful is that in many general situations the existence of nontrivial solutions to an equation in $\mathbf{R}$ and $\mathbf{Q}_p$ is enough to guarantee the existence of nontrivial solutions in $\mathbf{Q}$. This idea is referred to as the "Hasse principle," named after Helmut Hasse (1898–1979).

The Hasse principle is known to hold for equations involving quadratic forms. That is, if $f(x_1, x_2, \ldots, x_n)$ is a polynomial in which all of the terms have degree exactly 2, then

$$f(x_1, x_2, \ldots, x_n) = 0 \tag{9.52}$$

has a nontrivial rational solution if and only if it has a nontrivial solution in $\mathbf{R}$ and $\mathbf{Q}_p$ for all p.

By clearing out denominators, we can see that when f is a quadratic form, (9.52) will have a nontrivial rational solution if and only if it has a nontrivial integer solution.

If f is a quadratic form in at least five variables, then (9.52) will necessarily have nontrivial solutions in $\mathbf{Q}_p$ for all p. Thus, by the Hasse principle, it will have nontrivial solutions in $\mathbf{Q}$ if and only if it has nontrivial solutions in $\mathbf{R}$. As an example, if a_1, a_2, a_3, and a_4 are positive numbers, then the equation

$$a_1x_1^2 + a_2x_2^2 + a_3x_3^2 + a_4x_4^2 - x_5^2 = 0$$

has real solution $(1, 1, 1, 1, \sqrt{a_1 + a_2 + a_3 + a_4})$, and so it will necessarily have a nontrivial rational (and hence integer) solution as well.

Exercises

9.59 Show that for any integer a, we have

$$|a^2 - 3|_7 = 1.$$

9.60 Show that there exists a sequence of integers a_1, a_2, ... satisfying

$$|a_n^2 - 2|_7 \leq 1/7^n.$$

Hint: Use induction. Given a_n, set $a_{n+1} = a_n + k7^n$ and show that for some choice of k, the following congruence holds:

$$a_{n+1}^2 \equiv 2 \pmod{7^{n+1}}.$$

†**9.61** Show that the following properties of the p-adic valuation hold.

(a) $|xy|_p = |x|_p|y|_p$.

(b) $|x + y|_p \leq \max\{|x|_p, |y|_p\}$ with equality when $|x|_p \neq |y|_p$.

9.62 Let $f(x)$ be a polynomial with integer coefficients and suppose that

$$| f(a) |_p < |(f'(a))^2|_p$$

for some integer a and prime p. (Here $f'(x)$ is the usual derivative of $f(x)$.) Then one can show that $f(x)$ has a root in $\mathbf{Q}_p$. This result is known as Newton's lemma as the proof is based on the idea of Newton's method, named after Isaac Newton (1643–1727), for approximating real roots of functions. Use Newton's lemma to show that the polynomial

$$f(x) = (x^2 - 2)(x^2 - 17)(x^2 - 34)$$

has a root in $\mathbf{Q}_p$ for all p.

Note that $f(x)$ also has real roots but no rational roots, and so the Hasse principle fails in this case.

Part III

Advanced Topics

Chapter 10

Analytic Number Theory

> [T]he sole end of science is the honor of the human mind, and ... under this title a question about numbers is worth as much as a question about the system of the world.
>
> CARL JACOBI (1804–1851)

One of the central questions in analytic number theory is, how are the primes distributed among the set of natural numbers? At first glance, obtaining a satisfactory answer to this question might seem hopeless. For, on the one hand, there exist runs of one million (or more) consecutive composite numbers (Proposition 2.37), while on the other hand, it is believed that there are an infinite number of pairs of prime numbers separated by only one composite number (the twin primes conjecture). In fact, there are rather deep results that provide elegant answers to this question. In this chapter, we will investigate some of these results, and along the way we will become acquainted with some of the basic tools of analytic number theory.

10.1 Sum of reciprocals of primes

In Section 2.6 we saw, using an argument dating back to Euclid, that there are an infinite number of primes. One way to begin to investigate how the primes are distributed is to compare them to another infinite subset of the integers, for instance, the set of perfect squares. One can say that squares form a rather "small" subset of the natural numbers, since if we sum the reciprocals of squares we get the convergent p-series

$$\sum_{n=1}^{\infty} \frac{1}{n^2}.$$

By comparison, the sum of the reciprocals of all positive integers is the p-series

$$\sum_{n=1}^{\infty} \frac{1}{n},$$

which diverges. The following theorem shows that the set of prime numbers is a much denser subset of the natural numbers than the set of squares.

THEOREM 10.1

Let p_1, p_2, ... be the sequence of prime numbers. Then $\sum_{m=1}^{\infty} \frac{1}{p_m}$ *diverges.*

PROOF Fix a positive number N and let p_M be the largest prime dividing any number less than or equal to N. Then every number less than or equal to N can be written as a product of powers of prime numbers less than or equal to p_M. Thus, we have

$$\sum_{n=1}^{N} \frac{1}{n} \leq \prod_{m=1}^{M} \left(1 + \frac{1}{p_m} + \frac{1}{p_m^2} + \frac{1}{p_m^3} + \cdots \right). \tag{10.1}$$

Each term in the product on the right side is a geometric series, and so we can rewrite this inequality as

$$\sum_{n=1}^{N} \frac{1}{n} \leq \prod_{m=1}^{M} \left(\frac{1}{1 - \frac{1}{p_m}} \right).$$

We know that the sum on the left will diverge when we let N go to infinity. We will manipulate the right side to extract the series of reciprocals of primes. We begin by taking logarithms of both sides to obtain

$$\log \left(\sum_{n=1}^{N} \frac{1}{n} \right) \leq \sum_{m=1}^{M} \log \left(\frac{1}{1 - \frac{1}{p_m}} \right). \tag{10.2}$$

We observe that the left side will still tend to infinity as $N \to \infty$. The function $\log(1 - x)$ has Taylor series expansion

$$\log(1 - x) = -\sum_{k=1}^{\infty} \frac{x^k}{k},$$

which is valid for $|x| < 1$. Using this formula, we can expand each term on the right side of (10.2), which leads to the inequality

$$\log \left(\sum_{n=1}^{N} \frac{1}{n} \right) \leq \sum_{m=1}^{M} \sum_{k=1}^{\infty} \frac{1}{k p_m^k}.$$

With some reordering, we may rewrite the right side as

$$\sum_{m=1}^{M} \frac{1}{p_m} + \sum_{m=1}^{M} \sum_{k=2}^{\infty} \frac{1}{k p_m^k}.$$

Thus, we have our series of reciprocals of primes. To show that it diverges as $M \to \infty$, it is enough to show that

$$\sum_{m=1}^{\infty} \sum_{k=2}^{\infty} \frac{1}{kp_m^k} \tag{10.3}$$

converges. We have

$$\begin{aligned} \sum_{m=1}^{\infty} \sum_{k=2}^{\infty} \frac{1}{kp_m^k} &< \sum_{m=1}^{\infty} \sum_{k=2}^{\infty} \frac{1}{p_m^k} \\ &= \sum_{m=1}^{\infty} \frac{1}{p_m(p_m - 1)} \\ &< \sum_{m=1}^{\infty} \frac{1}{(p_m - 1)^2}. \end{aligned}$$

This last sum is a subseries of the series of reciprocals of squares of integers, and so must converge. Hence, the series in (10.3) converges as well.

□

Euler gave the first proof of this theorem and noted that it implies Euclid's theorem on the infinitude of the set of primes. While this new proof is quite a bit more involved than Euclid's original one, it can be generalized to prove the existence of certain infinite arithmetic progressions of primes. We will explore this generalization in greater detail in Section 10.7.

Exercises

◇**10.1** Use a computer to find the first value of x for which

$$\sum_{p \leq x} \frac{1}{p} > 2$$

and a value of x for which

$$\sum_{p \leq x} \frac{1}{p} > 3.$$

10.2 Show that the sum

$$\sum_{p} \frac{1}{p^2}$$

converges while the sum

$$\sum_{p,q} \frac{1}{pq}$$

diverges. (The first sum is taken over all primes p and the second sum is taken over all pairs of distinct primes p and q.)

10.3 Show that there are an infinite number of primes whose base-10 representation contains the digit "1" at least once. Show that there are an infinite numbers of primes whose base-10 representation contains the block of digits "1776" at least once.

$\star$**10.4** Let α be the real number whose decimal expansion is the list of prime numbers:

$$\alpha = 0.23571113\ldots.$$

Show that α is irrational.

10.2 Orders of growth of functions

Before we continue work on proving facts about the set of prime numbers, we introduce some new terminology that will be useful for comparing functions.

DEFINITION 10.2 *For functions $f(x)$ and $\phi(x)$ we write*

$$f(x) = O(\phi(x)),$$

for $x \geq a$ and say $f(x)$ is "big oh" of $\phi(x)$ if

$$|f(x)| < C\phi(x)$$

for some positive constant C for $x \geq a$.

For example, $x^2 + 3x + 2 = O(x^2)$ for $x \geq 1$, since

$$|x^2 + 3x + 2| \leq x^2 + 3x^2 + 2x^2 < 7x^2,$$

for $x \geq 1$. We also write

$$f(x) = g(x) + O(\phi(x))$$

if $f(x) - g(x) = O(\phi(x))$. One may think of $g(x)$ as an approximation of $f(x)$ with the error for this approximation bounded by a constant multiple of $\phi(x)$. In many cases we will simply be interested in the long-term behavior of the functions under consideration, and so the particular value of a will not be important. In such situations, we will simply write $f(x) = g(x) + O(\phi(x))$ with the understanding that $|f(x) - g(x)| < C\phi(x)$ for some constant C for sufficiently large values of x.

Example 10.3 For $x \geq 1$, the function $\log x$ can be approximated by the sum $\sum_{n=1}^{\lfloor x \rfloor} 1/n$, and in fact,

$$\log x = \sum_{n=1}^{\lfloor x \rfloor} \frac{1}{n} + O(1) \tag{10.4}$$

for $x > 1$. To see this, recall from calculus that

$$\log x = \int_1^x \frac{1}{t}\, dt.$$

On the one hand, we have

$$\log x \leq \int_1^{\lfloor x \rfloor + 1} \frac{1}{t}\, dt \leq \sum_{n=1}^{\lfloor x \rfloor} \frac{1}{n},$$

since left Riemann sums provide overestimates for integrals of decreasing functions. On the other hand, right Riemann sums are underestimates, and so

$$\log x \geq \int_1^{\lfloor x \rfloor} \frac{1}{t}\, dt \geq \sum_{n=2}^{\lfloor x \rfloor} \frac{1}{n}.$$

Thus, the sum $\sum_{n=1}^{\lfloor x \rfloor} 1/n$ provides an estimate for $\log x$ with an error of no more than 1, and so (10.4) holds.

□

If $f(x) = O(g(x))$ and $g(x) = O(f(x))$, then there exist positive constants C_1 and C_2 such that

$$C_1|g(x)| < |f(x)| < C_2|g(x)|. \tag{10.5}$$

In this case we say that $f(x)$ and $g(x)$ have the same "order of growth." For example, $x^2 + x + 2$ and x^2 have the same order of growth.

Exercises

10.5 Prove that $x^3 + 2x^2 + x + 1 = O(x^3)$. Prove that $x^3 + 2x^2 + x + 1$ and x^3 have the same order of growth.

10.6 Show that for $x > 1$,

$$\sum_{p \leq x} \frac{1}{p} > \log(\log x) - 2.$$

10.7 Show that if $f(x) = O(g(x))$ and $g(x) = O(h(x))$, then $f(x) = O(h(x))$.

10.8 For a fixed function $\phi(x)$, write $f(x) \sim_\phi g(x)$ if $f(x) = g(x) + O(\phi(x))$. Show that $\sim_\phi$ is an equivalence relation on the set of real functions.

10.3 Chebyshev's theorem

Another way to measure the density of the set of prime numbers is to attempt to compute the fraction of prime numbers less than a number x. Recall that the function $\pi(x)$ gives the number of primes less than or equal to x, and so our question is what is

$$\lim_{x\to\infty} \frac{\pi(x)}{x}? \tag{10.6}$$

Recall that in Section 2.6 we gave a statement of the Prime Number Theorem, which asserts that

$$\lim_{x\to\infty} \frac{\pi(x)}{x/\log x} = 1, \tag{10.7}$$

and so the limit in (10.6) must be 0. We will have more to say about the Prime Number Theorem in Section 10.5. For now we will prove a weaker version of the Prime Number Theorem due to Pafnuty Chebyshev (1821–1894), which also implies that the limit in (10.6) is 0.

THEOREM 10.4 (Chebyshev's theorem)
There exist positive constants C_1 and C_2 such that

$$C_1\frac{x}{\log x} \le \pi(x) \le C_2\frac{x}{\log x}$$

for sufficiently large x.

Chebyshev's theorem can be restated by saying that $\pi(x)$ and $x/\log x$ have the same order of growth.

To estimate $\pi(x)$ we will analyze a pair of related functions known as Chebyshev functions.

DEFINITION 10.5 *For a positive real number x,*

$$\vartheta(x) = \sum_{p\le x} \log p, \tag{10.8}$$

where the sum is over primes p less than or equal to x, and

$$\psi(x) = \sum_{p^m\le x} \log p, \tag{10.9}$$

where the sum is over prime powers p^m less than or equal to x.

Example 10.6 We have $\vartheta(10) = \log 2 + \log 3 + \log 5 + \log 7$ and $\psi(10) = 3\log 2 + 2\log 3 + \log 5 + \log 7$. ▯

We now record an equivalent formulation for each of these functions. We first observe that

$$\vartheta(x) = \log\left(\prod_{p\le x} p\right). \tag{10.10}$$

Second, a prime power p^m contributes a $\log p$ term in the sum in (10.9) exactly when $m \le \log x/\log p$. It follows that

$$\psi(x) = \sum_{p\le x}\left\lfloor\frac{\log x}{\log p}\right\rfloor \log p. \tag{10.11}$$

The following theorem gives the connection between $\vartheta(x)$ and $\pi(x)$ that we will need to prove Chebyshev's theorem.

THEOREM 10.7
For $x > 1$,

$$\frac{\vartheta(x)}{\log x} \le \pi(x) \le \frac{2\vartheta(x)}{\log x} + \sqrt{x}.$$

PROOF There are $\pi(x)$ terms in the sum in (10.8), and each term is less than or equal to $\log x$. Hence,

$$\vartheta(x) \le \pi(x)\log x,$$

and the first inequality now follows.

In the other direction, let δ be a number satisfying $0 < \delta < 1$. (In this proof, we will ultimately need only the case $\delta = 1/2$, but later on, in the proof of Theorem 10.18, we shall need the following more general computation.) We have

$$\vartheta(x) \ge \sum_{x^{1-\delta}<p\le x} \log p \ge (\pi(x) - \pi(x^{1-\delta}))\log x^{1-\delta}.$$

Since $\pi(x^{1-\delta}) \le x^{1-\delta}$, it follows that

$$\vartheta(x) \ge (\pi(x) - x^{1-\delta})(1-\delta)\log x.$$

We may rearrange this as

$$\pi(x) \le \frac{\vartheta(x)}{(1-\delta)\log x} + x^{1-\delta}. \tag{10.12}$$

Taking $\delta = 1/2$, we obtain the second inequality. ▯

To prove Chebyshev's theorem, it will be sufficient to show that $\vartheta(x)$ is of the same order of growth as x. To achieve this result, we will also need to estimate $\psi(x)$, which will also turn out to be of the same order of growth as x. We begin by obtaining an overestimate for $\vartheta(x)$.

THEOREM 10.8
If n is a positive integer, then $\vartheta(n) < 2n\log 2$.

PROOF We first observe that the inequality holds for $n = 1$ and $n = 2$ as

$$\vartheta(1) = 0 < 2\log 2,$$
$$\vartheta(2) = \log 2 < 4\log 2.$$

We now proceed by induction and suppose that the result holds for all $n < N$, for some $N > 2$. If N is even, then

$$\vartheta(N) = \vartheta(N-1) < 2(N-1)\log 2 < 2N\log 2.$$

For the case in which N is odd we need a lemma.

LEMMA 10.9
If k is a positive integer then

$$\vartheta(2k+1) - \vartheta(k+1) < 2k\log 2.$$

PROOF If we set $K = \binom{2k+1}{k+1} = \binom{2k+1}{k}$, then

$$2^{2k+1} = (1+1)^{2k+1} = \sum_{i=0}^{2k+1} \binom{2k+1}{i} > 2K. \tag{10.13}$$

Now suppose that p is a prime satisfying $k+1 < p \leq 2k+1$. Then

$$p \mid (2k+1)(2k)(2k-1)\ldots(k+2),$$

but p does not divide $(k+1)!$. Hence, p divides K, and moreover the product of all such p divides K. In particular, this product is smaller than K. In light of (10.10), we may now conclude that

$$\vartheta(2k+1) - \vartheta(k+1) \leq \log K.$$

Applying (10.13), it follows that

$$\vartheta(2k+1) - \vartheta(k+1) < \log((1/2)2^{2k+1}) = 2k\log 2.$$

∎

Returning to the proof of the theorem, when N is odd, we can write $N = 2k+1$, for some $k > 0$. By the lemma, we have

$$\vartheta(N) < 2k \log 2 + \vartheta(k+1),$$

and so, by our inductive hypothesis,

$$\vartheta(N) < 2k \log 2 + 2(k+1) \log 2 = 2N \log 2.$$

$\square$

COROLLARY 10.10
For $x \geq 1$, $\vartheta(x) \leq 2x \log 2$, and so $\vartheta(x) = O(x)$.

PROOF By the theorem,

$$\vartheta(x) = \vartheta(\lfloor x \rfloor) < 2\lfloor x \rfloor \log 2 \leq 2x \log 2.$$

$\square$

Our next step is to deduce an underestimate for $\psi(x)$.

LEMMA 10.11 (De Polignac's formula)
Let m be a positive number, and for any prime p, let

$$j(p, m) = \sum_{k=1}^{\infty} \left\lfloor \frac{m}{p^k} \right\rfloor. \tag{10.14}$$

Then

$$m! = \prod_{p} p^{j(p,m)}. \tag{10.15}$$

Thus, the exact power of p that divides $m!$ is $j(p, m)$.

PROOF We first observe that the sum in (10.14) is finite since $\lfloor m/p^k \rfloor = 0$ when $p^k > m$. As $j(p, m) = 0$ for all primes $p > m$, we see that the product in (10.15) is finite as well.

For any positive integer m and prime p, there are exactly $\lfloor m/p \rfloor$ multiples of p less than or equal to m. Similarly, there are $\lfloor m/p^2 \rfloor$ multiples of p^2 less than or equal to m, and in general, there are $\lfloor m/p^k \rfloor$ multiples of p^k less than or equal to m. Consequently, the highest power of p dividing $m!$ is exactly $j(p, m)$. $\square$

THEOREM 10.12
For all positive integers n, $\psi(2n) \geq n \log 2$.

PROOF As a result of the lemma,

$$\frac{(2n)!}{(n!)^2} = \prod_p p^{j(p,2n)-2j(p,n)}. \tag{10.16}$$

Let $e_p = j(p,2n) - 2j(p,n)$, and so

$$e_p = \sum_{k=1}^{\infty} \left\lfloor \frac{2n}{p^k} \right\rfloor - 2\left\lfloor \frac{n}{p^k} \right\rfloor. \tag{10.17}$$

By definition of the floor function,

$$\frac{2n}{p^k} - 1 < \left\lfloor \frac{2n}{p^k} \right\rfloor \leq \frac{2n}{p^k},$$

$$2\left(\frac{n}{p^k} - 1\right) < 2\left\lfloor \frac{n}{p^k} \right\rfloor \leq \frac{2n}{p^k},$$

and hence,

$$-1 < \left\lfloor \frac{2n}{p^k} \right\rfloor - 2\left\lfloor \frac{n}{p^k} \right\rfloor < 2.$$

Consequently, each term in the sum in (10.17) must be 0 or 1. Furthermore, both $\lfloor 2n/p^k \rfloor$ and $\lfloor n/p^k \rfloor$ are 0 when $p^k > 2n$. Thus, we have

$$0 \leq e_p \leq \left\lfloor \frac{\log(2n)}{\log p} \right\rfloor. \tag{10.18}$$

Putting this together with (10.16), we may conclude that

$$\log\left(\frac{(2n)!}{(n!)^2}\right) \leq \sum_{p \leq 2n} \left\lfloor \frac{\log(2n)}{\log p} \right\rfloor \log p.$$

From (10.11), the quantity on the right above is exactly $\psi(2n)$, and so we have shown that

$$\psi(2n) \geq \log\left(\frac{(2n)!}{(n!)^2}\right). \tag{10.19}$$

We observe that

$$\frac{(2n)!}{(n!)^2} = \frac{2n(2n-1)\ldots(n+1)}{n(n-1)(n-2)\ldots 1} \geq 2^n,$$

and so by (10.19),

$$\psi(2n) \geq \log(2^n) = n \log 2.$$

□

Putting these two theorems together, we can now show that both $\vartheta(x)$ and $\psi(x)$ are of the same order of growth as x.

THEOREM 10.13
There exist positive constants c_1, c_2, c_3, and c_4 such that

$$c_1 x < \vartheta(x) < c_2 x \tag{10.20}$$
$$c_3 x < \psi(x) < c_4 x \tag{10.21}$$

for sufficiently large x.

We have already proved the existence of c_2 (Corollary 10.10). Also, the existence of c_3 can be deduced from Theorem 10.12 (see Exercise 10.11). The existence of the remaining two constants will follow from the next lemma.

LEMMA 10.14
For $x \geq 1$,

$$0 \leq \psi(x) - \vartheta(x) \leq 2\sqrt{x}\log x,$$

and consequently $\psi(x) = \vartheta(x) + O(\sqrt{x}\log x)$.

PROOF The condition $p^m \leq x$ is equivalent to $p \leq x^{1/m}$, and so we have

$$\psi(x) = \vartheta(x) + \vartheta(x^{1/2}) + \cdots + \vartheta(x^{1/k}) + \vartheta(x^{1/(k+1)}),$$

where $x^{1/(k+1)} \leq 2$. This leads to the crude estimate

$$\psi(x) \leq \vartheta(x) + k\vartheta(x^{1/2}).$$

Now $x^{1/(k+1)} \leq 2$ holds exactly when

$$k \geq \frac{\log x}{\log 2} - 1$$

and, in particular, is true if we take $k = \lfloor \log x / \log 2 \rfloor$. Hence,

$$\psi(x) \leq \vartheta(x) + \left(\frac{\log x}{\log 2}\right)\vartheta(x^{1/2}).$$

Using Corollary 10.10 and the fact that $\vartheta(x) \leq \psi(x)$, we see that

$$0 \leq \psi(x) - \vartheta(x) \leq 2\sqrt{x}\log x$$

for $x \geq 1$.

∎

PROOF (of Theorem 10.13) We have already noted that the existence of c_2 and c_3 follows from previous work. As a result of the lemma, we have

$$\psi(x) - \vartheta(x) \leq 2\sqrt{x}\log x, \tag{10.22}$$

for $x \geq 1$. Thus, for sufficiently large x,

$$\begin{aligned}\psi(x) &\leq \vartheta(x) + 2\sqrt{x}\log x \\ &< c_2 x + 2\sqrt{x}\log x. \\ &< c_2 x + 2x.\end{aligned}$$

Consequently, we may take $c_4 = c_2 + 2$.

From (10.22) we also know

$$-2\sqrt{x}\log x + \psi(x) \leq \vartheta(x),$$

and so

$$-2\sqrt{x}\log x + c_3 x < \vartheta(x)$$

for sufficiently large x. We have $(c_3/2)x > 2\sqrt{x}\log x$ when x is sufficiently large, and hence,

$$(c_3/2)x < \vartheta(x).$$

Thus, we may take $c_1 = c_3/2$.

□

We now are in position to complete the proof of Chebyshev's theorem.

PROOF (of Theorem 10.4) Putting together Theorems 10.13 and 10.7, we see that there exist constants c_1 and c_2 such that

$$\frac{c_1 x}{\log x} < \pi(x) < \frac{2c_2 x}{\log x} + \sqrt{x}$$

for sufficiently large x. For $x > 1$,

$$\sqrt{x} < \frac{x}{\log x},$$

and so $\pi(x) < (2c_2 + 1)x/(\log x)$ for sufficiently large x. Thus, taking $C_1 = c_1$ and $C_2 = 2c_2 + 1$, we have the desired constants. □

Exercises

10.9 Compute $\vartheta(15)$ and $\psi(15)$.

10.10 Our statement of Chebyshev's theorem is that there exist constants C_1 and C_2 such that

$$C_1 \frac{x}{\log x} \leq \pi(x) \leq C_2 \frac{x}{\log x}$$

for sufficiently large x. Show that there exist C_1 and C_2 such that the inequalities hold for *all* $x \geq 2$.

10.11 Use Theorem 10.12 to show that

$$\psi(x) > \left(\frac{\log 2}{4}\right) x$$

for all $x > 2$.

10.12 (a) Prove that the exact power of 2 that divides $n!$ is $n - d(n)$, where $d(n)$ is the number of 1's in the binary representation of n.

(b) Prove that the exact power of 2 that divides the binomial coefficient $\binom{n}{k}$ is equal to the number of 'carries' when k and $n-k$ are added in binary.

(c) Prove that the binomial coefficient $\binom{n}{k}$ is odd if and only if the binary representation of n contains a 1 in every position where the binary representation of k contains a 1.

10.4 Bertrand's Postulate

With c_1 and c_2 as in Theorem 10.13 and x sufficiently large, we have

$$\vartheta(x) < c_2 x = c_1 \left(\frac{c_2 x}{c_1}\right) < \vartheta\left(\frac{c_2 x}{c_1}\right).$$

In particular, $\vartheta(x) \neq \vartheta\left(\frac{c_2 x}{c_1}\right)$, and so there must be a prime between x and $(c_2/c_1)x$. Thus, we have proved the following result.

PROPOSITION 10.15
There exists a positive constant C such that for any positive integer n, there exists a prime p satisfying

$$n < p \leq Cn.$$

In 1845 Joseph Bertrand (1822–1900) correctly conjectured that the statement above holds with $C = 2$. The result, commonly referred to as Bertrand's Postulate, was proved by Chebyshev in 1850.

THEOREM 10.16 (Bertrand's Postulate)
For any number $n > 1$, there exists a prime p satisfying

$$n < p < 2n.$$

PROOF Suppose, by way of contradiction, that there exists a number $n \geq 3$ for which there is no prime p satisfying $n < p < 2n$. Let

$$N = \frac{(2n)!}{(n!)^2}. \tag{10.23}$$

We will derive a pair of estimates of N (actually $\log N$) and show that they contradict each other for sufficiently large n.

If we write

$$N = \prod_p p^{e_p}, \tag{10.24}$$

as in the proof of Theorem 10.12, then we again have

$$e_p = \sum_p \left\lfloor \frac{2n}{p^k} \right\rfloor - 2\left\lfloor \frac{n}{p^k} \right\rfloor.$$

Suppose that p is a prime dividing N. Then p must divide the numerator on the right side of (10.23) and so is at most $2n$. Since n is a counterexample to the theorem, p must actually be less than or equal to n. We can place an even smaller upper bound on p. Suppose that $2n/3 < p \leq n$. Then

$$1 \leq \frac{n}{p} < \frac{3}{2}, \tag{10.25}$$

and

$$2 \leq \frac{2n}{p} < 3, \tag{10.26}$$

which means that

$$\left\lfloor \frac{2n}{p} \right\rfloor - 2\left\lfloor \frac{n}{p} \right\rfloor = 2 - 2 \cdot 1 = 0.$$

Furthermore, from (10.26), we see that $2n/p^2 < 3/p$, and since $p > 2n/3 \geq 2$, we have

$$\frac{2n}{p^2} < 1.$$

We conclude that $\lfloor 2n/p^k \rfloor = \lfloor n/p^k \rfloor = 0$, for $k \geq 2$, and so $e_p = 0$ for such p. Hence, the primes p dividing N satisfy $p \leq 2n/3$. As a consequence,

$$\sum_{p|N} \log p \leq \sum_{p \leq 2n/3} \log p.$$

The quantity on the right is exactly $\vartheta(2n/3)$, and so by Corollary 10.10,

$$\sum_{p|N} \log p \leq (4n/3)\log 2. \tag{10.27}$$

Next we provide an upper bound on the primes p that divide N to a power greater than 1 (i.e., primes for which $e_p \geq 2$). For such a prime p, (10.18) implies that

$$2 \leq \frac{\log(2n)}{\log p},$$

from which we may conclude that

$$p \leq \sqrt{2n}. \tag{10.28}$$

We now obtain our first estimate for $\log N$. From (10.24), we have

$$\log N = \sum_{e_p=1} \log p + \sum_{e_p \geq 2} e_p \log p.$$

Again from (10.18), we have $e_p \log p \leq \log(2n)$, and so

$$\log N \leq \left(\sum_{e_p=1} \log p \right) + \sqrt{2n} \log(2n).$$

Now, applying (10.27), we conclude that

$$\log N \leq (4n/3) \log 2 + \sqrt{2n} \log 2n. \tag{10.29}$$

For our second estimate, we begin by observing that

$$2^{2n} = 2 + \sum_{m=1}^{2n-1} \binom{2n}{m}.$$

Now $\binom{2n}{m} \leq \binom{2n}{n} = N$, for $m = 1, 2, \ldots, 2n-1$ (by Exercise 10.13), and so

$$2^{2n} \leq 2 + (2n-1)N \leq 2nN.$$

From this we obtain the estimate

$$2n \log 2 - \log(2n) \leq \log N. \tag{10.30}$$

Putting (10.29) and (10.30) together, we have

$$2n \log 2 - \log(2n) \leq (4n/3) \log 2 + \sqrt{2n} \log 2n,$$

which is equivalent to

$$2n \log 2 \leq 3(1 + \sqrt{2n}) \log(2n).$$

Since $n \geq 3$, we have $1 + \sqrt{2n} < 2\sqrt{n}$, and hence,

$$n \log 2 \leq 3\sqrt{n} \log(2n). \tag{10.31}$$

LEMMA 10.17
If $x \geq 10000$, *then* $\log(2x) \leq x^{1/4}$.

PROOF Let $f(x) = x^{1/4} - \log(2x)$. Then

$$f'(x) = \frac{1}{4x}(x^{1/4} - 4),$$

and so $f'(x) > 0$ when $x > 256$. Consequently, $f(x)$ is increasing when $x \geq 10000$, and since $f(10000) > 0$, it follows that $f(x)$ is positive for all $x \geq 10000$. ∎

Returning to the theorem, suppose that $n \geq 10000$. Using the lemma and (10.31), we have

$$n \log 2 < 3n^{3/4}.$$

Rearranging this inequality, we obtain

$$\frac{\log 2}{3} < n^{-1/4} \leq 10000^{-1/4} = \frac{1}{10}.$$

Thus, we have a contradiction (which was our goal).

It remains to check the theorem in the case $2 \leq n < 10000$. Observe that if p and p' are primes satisfying $p < p' < 2p$, then

$$n < p' < 2n,$$

for all integers satisfying $p \leq n < p'$. Thus, to check the remaining cases, it suffices to exhibit a list of primes

$$p_1, p_2, \ldots, p_l$$

such that $p_1 = 2$, $p_l > 10000$, and $p_i < p_{i+1} < 2p_i$ for $i = 1, \ldots, l-1$. The following list satisfies these criteria:

$$2, 3, 5, 7, 13, 23, 43, 83, 163, 317, 631, 1259, 2503, 5003, 9973, 19937.$$

Thus, the theorem holds for $2 \leq n < 10000$ as well. ∎

Exercises

10.13 Let k be a positive integer. Show that

$$\binom{k}{i} \leq \binom{k}{\lfloor k/2 \rfloor},$$

for $i = 0, 1, \ldots, k$.

10.14 Show that for any positive integer n there exist at least three primes with n digits.

10.15 Let $n \geq 2$ be an integer. Use Bertrand's Postulate to show that there exists a prime $p \leq n$ such that p does not divide the integer

$$n! + \frac{n!}{2} + \frac{n!}{3} + \cdots + \frac{n!}{n}.$$

Use this result to obtain a second proof (see Exercise 2.32) that the rational number

$$1 + \frac{1}{2} + \frac{1}{3} + \cdots + \frac{1}{n}$$

is not an integer.

10.5 The Prime Number Theorem

The Prime Number Theorem was first conjectured by Gauss as early as 1791, though he did not publish an account until late in his life. Credit is due to Legendre for first stirring up interest in the problem. In 1798 he suggested that $\pi(x)$ is asymptotic to the function

$$\frac{x}{\log x - 1.08366}.$$

By the latter half of the nineteenth century, the conjecture had become a central problem in number theory. Chebyshev proved that if the function

$$\frac{\pi(x)}{x/\log x}$$

tends to a limit as x goes to infinity, then this limit is 1. Bernhard Riemann (1826–1866) introduced complex analysis to the study of the problem by connecting it to the zeta function (which we shall discuss in Section 10.6). Finally, in 1896 Hadamard and de la Vallée Poussin each obtained proofs independently by making use of complex analysis.

Perhaps a testament to the importance of this theorem is that it continued to be studied even after a proof was discovered. The statement of the theorem involves elementary concepts (i.e., at the level of calculus). The proof provided by Hadamard and de la Vallée Poussin draws on mathematics at a much deeper level. In 1949 Selberg and Erdős succeeded in finding a proof that avoids the use of complex analysis. While this proof is "elementary," it is far from routine. We will not give the complete argument, but instead will be content to prove Chebyshev's partial result. We refer the reader to [23] for a complete, self-contained account of the Selberg–Erdős proof.

We begin by strengthening the link between the Prime Number Theorem and the Chebyshev functions.

THEOREM 10.18

$$\pi(x) \sim \frac{\vartheta(x)}{\log x} \sim \frac{\psi(x)}{\log x}$$

PROOF The first inequality in Theorem 10.7 tells us that

$$\frac{\pi(x)}{\vartheta(x)/\log x} \geq 1 \tag{10.32}$$

for $x > 1$. On the other hand, (10.12) from the proof of Theorem 10.7 gives us

$$\frac{\pi(x)}{\vartheta(x)/\log x} \leq \frac{1}{1-\delta} + \frac{x^{1-\delta}\log x}{\vartheta(x)},$$

for $0 < \delta < 1$ and $x > 1$. Now $\vartheta(x) \geq Cx$ for some positive constant C and x sufficiently large, and so

$$\frac{\pi(x)}{\vartheta(x)/\log x} \leq \frac{1}{1-\delta} + \frac{x^{1-\delta}\log x}{Cx} = \frac{1}{1-\delta} + \frac{\log x}{Cx^{\delta}}.$$

Hence,

$$1 \leq \frac{\pi(x)}{\vartheta(x)/\log x} \leq 1 + \epsilon + \frac{\log x}{Cx^{\delta}},$$

where $\epsilon = \delta/(1-\delta)$. We observe that

$$\lim_{x\to\infty} \frac{\log x}{Cx^{\delta}} = 0,$$

and so $\log x/(Cx^{\delta}) < \epsilon$ for sufficiently large x. We now have

$$1 \leq \frac{\pi(x)}{\vartheta(x)/\log x} \leq 1 + 2\epsilon,$$

for sufficiently large x. Since δ can be made arbitrarily small, so can ϵ, and so

$$\lim_{x\to\infty} \frac{\pi(x)}{\vartheta(x)/\log x} = 1.$$

Thus, we have shown that

$$\pi(x) \sim \frac{\vartheta(x)}{\log x}.$$

For the second asymptotic equivalence, we apply Lemma 10.14 to conclude

$$\left|\frac{\psi(x)}{\vartheta(x)} - 1\right| \leq \frac{2\sqrt{x}\log x}{\vartheta(x)}$$

for $x > 2$. Now by Theorem 10.13, we see that the quantity on the right side goes to 0 as x goes to infinity. Hence, $\vartheta(x) \sim \psi(x)$. ☐

We observe now that proving the Prime Number Theorem is equivalent to proving

$$\lim_{x\to\infty} \frac{\psi(x)}{x} = 1. \tag{10.33}$$

Again, we will only show that if the limit exists then it must be 1. So let's suppose that

$$\lim_{x\to\infty} \frac{\psi(x)}{x} = L, \tag{10.34}$$

for some L. We will show that L cannot be greater than 1. (A similar argument can be used to show that L cannot be less than 1, but we leave this as an exercise.) Suppose, by way of contradiction, that $L > 1 + \delta$ for some $\delta > 0$. Under this assumption, there exists a number N such that

$$\frac{\psi(x)}{x} > 1 + \delta$$

for $x \geq N$. We note that such an N must be greater than 2, and it follows that

$$\begin{aligned} \int_2^x \frac{\psi(t)}{t^2}\,dt &\geq \int_N^x \frac{\psi(t)}{t^2}\,dt \\ &\geq \int_N^x \frac{1+\delta}{t}\,dt \\ &= (1+\delta)(\log x - \log N). \end{aligned}$$

Hence,

$$\int_2^x \frac{\psi(t)}{t^2}\,dt - \log x \geq \delta \log x - A,$$

where A is some constant. This statement is contradicted by the following theorem.

THEOREM 10.19
For sufficiently large x,

$$\int_2^x \frac{\psi(t)}{t^2}\,dt = \log x + O(1). \tag{10.35}$$

We will break the proof of this theorem up into several steps. The first will be to relate the integral to a sum. We actually will prove a more general form of a transformation, which we will then apply to our special case.

THEOREM 10.20
Let $f(t)$ be a function whose derivative is continuous for $t \geq 1$, and let c_1, c_2, ... be a sequence of real numbers. If we set

$$C(x) = \sum_{n \leq x} c_n,$$

then

$$\sum_{n \leq x} c_n f(n) = f(x)C(x) - \int_1^x f'(t)C(t)\, dt. \tag{10.36}$$

PROOF We observe that $c_n = C(n) - C(n-1)$ (with the understanding that $C(0) = 0$). Thus, we can write the sum on the left side of (10.36) as

$$\begin{aligned}\sum_{n \leq x} c_n f(n) &= \sum_{n \leq x} (C(n) - C(n-1))f(n) \\ &= \left(\sum_{n \leq x-1} C(n)(f(n) - f(n+1)) \right) + C(\lfloor x \rfloor) f(\lfloor x \rfloor). \end{aligned} \tag{10.37}$$

Observe that $C(\lfloor x \rfloor) = C(x)$. Also, for $n \leq t < n+1$, we have $C(t) = C(n)$, and so

$$C(n)(f(n) - f(n+1)) = -\int_n^{n+1} C(t)f'(t)\, dt.$$

Equation (10.37) now becomes

$$\sum_{n \leq x} c_n f(n) = -\int_1^{\lfloor x \rfloor} C(t)f'(t)\, dt + C(x)f(\lfloor x \rfloor). \tag{10.38}$$

We also note that $C(t) = C(\lfloor x \rfloor)$ for $\lfloor x \rfloor \leq t \leq x$, and so

$$\int_{\lfloor x \rfloor}^x C(t)f'(t)\, dt = C(x)f(x) - C(x)f(\lfloor x \rfloor).$$

Hence,

$$-\int_1^{\lfloor x \rfloor} C(t)f'(t)\, dt = -\int_1^x C(t)f'(t)\, dt + C(x)f(x) - C(x)f(\lfloor x \rfloor).$$

Combining this with (10.38), we obtain (10.36). □

To apply this result to prove Theorem 10.19, we will need to be able to write

$$\psi(x) = \sum_{n \leq x} c_n,$$

for some sequence of numbers $c_1, c_2, \ldots$. The von Mangoldt function, $\Lambda(n)$, introduced in Exercise 6.34, will do the trick. We recall that

$$\Lambda(n) = \begin{cases} \log p & \text{if } n = p^k, \text{ where } p \text{ is prime and } k \geq 1 \\ 0 & \text{otherwise.} \end{cases}$$

We immediately see from the definition that

$$\psi(x) = \sum_{n \leq x} \Lambda(n). \tag{10.39}$$

Applying Theorem 10.20 with $c_n = \Lambda(n)$ and $f(t) = 1/t$, we obtain

$$\sum_{n \leq x} \frac{\Lambda(n)}{n} = \frac{\psi(x)}{x} + \int_1^x \frac{\psi(t)}{t^2}\, dt. \tag{10.40}$$

Note that $\psi(t) = 0$ for $t < 2$, and consequently the integral in (10.40) is the same as the integral in (10.35). Now by Theorem 10.13, $\psi(x)/x = O(1)$, and so we may rewrite (10.40) as

$$\int_2^x \frac{\psi(t)}{t^2}\, dt = \sum_{n \leq x} \frac{\Lambda(n)}{n} + O(1).$$

The estimate given in the next theorem will now complete the proof of Theorem 10.19.

THEOREM 10.21

For sufficiently large x,

$$\sum_{n \leq x} \frac{\Lambda(n)}{n} = \log x + O(1).$$

PROOF We will estimate $\sum_{n \leq x} \log n$ in two ways, and by piecing the estimates together we will obtain the result. First, by Lemma 10.11,

$$\prod_{n \leq x} n = \lfloor x \rfloor ! = \prod_p p^{j(p, \lfloor x \rfloor)}.$$

Consequently,

$$\begin{aligned}\sum_{n\le x}\log n &= \sum_{p}\log p\left(\sum_{k}\left\lfloor\frac{\lfloor x\rfloor}{p^k}\right\rfloor\right)\\ &= \sum_{p^k}\log p\left\lfloor\frac{x}{p^k}\right\rfloor\\ &= \sum_{n\le x}\Lambda(n)\left\lfloor\frac{x}{n}\right\rfloor\\ &= \sum_{n\le x}\Lambda(n)\left(\frac{x}{n}\right)+O\left(\sum_{n\le x}\Lambda(n)\right).\end{aligned}$$

Since $\sum_{n\le x}\Lambda(n)=\psi(x)$ and $\psi(x)=O(x)$ (by Theorem 10.13), we have

$$\sum_{n\le x}\log n = x\sum_{n\le x}\frac{\Lambda(n)}{n}+O(x). \tag{10.41}$$

For our second estimate, we make use of Theorem 10.20 with $c_n=1$ and $f(x)=\log x$. By the theorem, we have

$$\begin{aligned}\sum_{n\le x}\log n &= (\log x)\lfloor x\rfloor - \int_1^x (1/t)\lfloor t\rfloor\,dt\\ &= x\log x + O\left(\log x+\int_1^x 1\,dt\right).\end{aligned}$$

Now,

$$\log x+\int_1^x 1\,dt = O(x),$$

and so we have

$$\sum_{n\le x}\log n = x\log x+O(x). \tag{10.42}$$

Putting (10.41) and (10.42) together, we have

$$x\sum_{n\le x}\frac{\Lambda(n)}{n} = x\log x+O(x),$$

and so

$$\sum_{n\le x}\frac{\Lambda(n)}{n} = \log x+O(1).$$

□

Exercises

10.16 Show that $\sum_{p \leq x} \frac{\log p}{p} = \log x + O(1)$.

10.17 Use Theorem 10.20 to show that

$$\vartheta(x) = \pi(x) \log x - \int_1^x \frac{\pi(t)}{t}\, dt.$$

10.18 Use Theorem 10.20 to show that

$$\pi(x) = \frac{\vartheta(x)}{\log x} + \int_1^x \frac{\vartheta(t)\, dt}{t(\log t)^2}.$$

†10.19 (a) Show that the Prime Number Theorem is equivalent to the statement

$$\pi(x) \sim \frac{x}{\log \pi(x)}.$$

(b) Let p_n be the nth prime number. Use part (a) to show that the Prime Number Theorem is equivalent to the statement

$$p_n \sim n \log n.$$

10.6 The zeta function and the Riemann hypothesis

Chebyshev's work on the proof of the Prime Number Theorem left little doubt that the statement was true, but there was still work to be done to complete the proof. Riemann took up the problem where Chebyshev left off, and though he was also unable to finish the proof, he made a significant breakthrough that would eventually lead to proofs given by Hadamard and de la Vallée Poussin. Riemann's insights concerned a connection he made between the distribution of primes and the zeros of a function known as the zeta function. If these zeros behaved in a certain way, the Prime Number Theorem would follow. Riemann's conjecture about the zeros of the zeta function became known as the Riemann hypothesis. While de la Vallée Poussin and Hadamard were able to exploit the connection Riemann made to prove the Prime Number Theorem, they were unable to prove the full generality of his conjecture. Today, the Riemann hypothesis remains without proof and is regarded by many as the most important open problem in mathematics. In this section we will give a brief introduction to the zeta function and the connection between the Prime Number Theorem and the Riemann hypothesis.

Riemann's zeta function is a function of a complex variable traditionally denoted s. We will follow this convention and also that of denoting the real and imaginary parts of s as σ and t so that

$$s = \sigma + ti.$$

DEFINITION 10.22 *For* Re $s = \sigma > 1$, *let*

$$\zeta(s) = \sum_{n=1}^{\infty} \frac{1}{n^s}. \tag{10.43}$$

To see that the series converges, we observe

$$\left| \sum_{n=1}^{\infty} \frac{1}{n^s} \right| \leq \sum_{n=1}^{\infty} \frac{1}{|n^s|} = \sum_{n=1}^{\infty} \frac{1}{n^\sigma}.$$

This last series is just a p-series, and since $\sigma > 1$, it converges. The convergence is uniform on the half-plane Re $s \geq \delta$, for any $\delta > 1$, and so $\zeta(s)$ is analytic for Re $s > 1$.

Euler actually studied the zeta function before Riemann and, in fact, computed several values of the function. For instance, he found

$$\zeta(2) = \sum_{n=1}^{\infty} \frac{1}{n^2} = \frac{\pi^2}{6}. \tag{10.44}$$

Euler also connected the zeta function with the distribution of primes by rewriting the function as an infinite product. This expansion, which we hinted at in the proof of Theorem 10.1, is given below.

THEOREM 10.23
For Re $s = \sigma > 1$,

$$\zeta(s) = \prod_p \frac{1}{1 - p^{-s}}.$$

PROOF We again let $p_1, p_2, \ldots, p_M$ be the first M primes. The inequality in (10.1) can be made into an equality if we sum over positive integers whose divisors come from the list $p_1, p_2, \ldots, p_M$. Replacing n with n^s this becomes

$$\sum \frac{1}{(p_1^{e_1} \ldots p_M^{e_M})^s} = \prod_{m=1}^{M} \left(1 + \frac{1}{p_m^s} + \frac{1}{p_m^{2s}} + \frac{1}{p_m^{3s}} + \cdots \right). \tag{10.45}$$

We should note that the series given by the product is a rearrangement of the series on the left side of the equation, but since the series on the left converges absolutely, any rearrangement of it will converge to the same sum. As in the proof of Theorem 10.1, we observe that each factor on the right is a geometric series, and so we can rewrite (10.45) as

$$\sum \frac{1}{(p_1^{e_1} \dots p_M^{e_M})^s} = \prod_{m=1}^{M} \left(\frac{1}{1 - p_m^{-s}} \right).$$

All that remains is to show that the sum on the left converges to $\zeta(s)$ as $M \to \infty$. We compute

$$\begin{aligned} \left| \zeta(s) - \sum \frac{1}{(p_1^{e_1} \dots p_M^{e_M})^s} \right| &< \left| \sum_{n=p_M}^{\infty} \frac{1}{n^s} \right| \\ &\leq \sum_{n=p_M}^{\infty} \frac{1}{|n^s|} \\ &= \sum_{n=p_M}^{\infty} \frac{1}{n^\sigma}. \end{aligned}$$

Being the tail of a convergent p-series, the last sum tends to 0 as $M \to \infty$. ▯

Euler also considered an alternating version of the sum in (10.43):

$$\eta(s) = \sum_{n=1}^{\infty} \frac{(-1)^{n-1}}{n^s}. \tag{10.46}$$

This function can be related to the zeta function by observing that

$$\begin{aligned} \zeta(s) - \eta(s) &= \sum_{n=1}^{\infty} \frac{1}{n^s} + \sum_{n=1}^{\infty} \frac{(-1)^n}{n^s} \\ &= \sum_{m=1}^{\infty} \frac{2}{(2m)^s} \\ &= 2^{1-s}\zeta(s). \end{aligned}$$

Hence,

$$\zeta(s) = \frac{\eta(s)}{1 - 2^{1-s}} = \frac{1}{1 - 2^{1-s}} \sum_{n=1}^{\infty} \frac{(-1)^{n-1}}{n^s}. \tag{10.47}$$

One can show that the sum above converges for all complex numbers s with Re $s > 0$. Again, the convergence is uniform on the half-plane Re $s \geq \delta$, for any $\delta > 0$, and so the sum converges to an analytic function for Re $s > 0$. Thus, the function on the right side of (10.47) is analytic for all $s \neq 1$ in the half-plane Re $s > 0$. We may consequently regard (10.47) as an extension of the definition of $\zeta(s)$ to a larger domain. The theory of analytic continuation says that this is, in fact, the only way to extend the definition of $\zeta(s)$ to Re $s > 0$. Riemann took this a step further and showed that the definition of $\zeta(s)$ could be extended (in a unique way) to an analytic function defined everywhere except $s = 1$. He also provided a "functional equation" for $\zeta(s)$. If we let $\Gamma(s) = \int_0^\infty x^{s-1}e^{-x}\,dx$ (the "gamma function"), then for Re $s < 0$,

$$\zeta(s) = 2^s\pi^{s-1}\sin\left(\frac{\pi s}{2}\right)\Gamma(1-s)\zeta(1-s). \tag{10.48}$$

Since Re $(1-s) > 1$ when Re $s < 0$, this functional equation, together with our original definition of $\zeta(s)$, provides an explicit definition for $\zeta(s)$ on the half-plane Re $s < 0$. One can show (Exercise 10.20) that $\Gamma(n) = (n-1)!$ when n is a positive integer. Thus, the functional equation allows us to compute, for example,

$$\zeta(-1) = \frac{-\zeta(2)}{2\pi^2} = -\frac{1}{12}.$$

Notice that (10.48) implies that $\zeta(s) = 0$ when s is a negative even integer. One can use the Euler product representation of $\zeta(s)$ to show that $\zeta(s) \neq 0$ when Re $s > 1$. It is also possible to show that $\Gamma(s) \neq 0$ for Re $s > 0$, and so the functional equation implies that $\zeta(s) \neq 0$ for Re $s < 0$, except when s is a negative even integer. (These values of s are sometimes called the "trivial zeros" of $\zeta(s)$.) We have said nothing about zeros of $\zeta(s)$ in the strip $0 \leq$ Re $s \leq 1$. There do exist zeros in this region known as the "critical strip," but they are much more difficult to locate. The functional equation implies (with some additional consideration of $\Gamma(s)$) that for complex numbers s in the interior of the critical strip, $\zeta(s) = 0$ if and only if $\zeta(1-s) = 0$. Thus, the set of such zeros is symmetric about the "critical line" $s = 1/2$. Riemann's breakthrough was to relate these zeros to the Chebyshev function $\psi(x)$. He obtained the "explicit formula"

$$\psi(x) = x - \sum_\rho \frac{x^\rho}{\rho} - \frac{\zeta'(0)}{\zeta(0)} - \frac{1}{2}\log(1-x^{-2}),$$

where the sum is over all nontrivial zeros ρ of $\zeta(s)$. Recall that to prove the Prime Number Theorem, it is enough to show that

$$\lim_{x\to\infty} \psi(x)/x = 1.$$

In light of Riemann's formula, it would be enough to show that

$$\lim_{x\to\infty} \frac{1}{x}\left(\sum_\rho \frac{x^\rho}{\rho}\right) = 0. \tag{10.49}$$

In working toward a proof of this, Riemann made the following conjecture.

CONJECTURE (Riemann hypothesis)
If $s = \sigma + ti$ is a nontrivial zero of $\zeta(s)$, then $\sigma = 1/2$.

The conjecture predicts that not only are the nontrivial zeros in the critical strip symmetric about the line $\sigma = 1/2$, they actually all lie on the line. Hadamard and de la Vallée Poussin were able to prove that if $s = \sigma + ti$ is a nontrivial zero, then $\sigma < 1$. This small step toward the proof of the Riemann hypothesis is sufficient to prove (10.49) and hence the Prime Number Theorem. This however is not the end of the story in terms of the connections between the Riemann hypothesis and the Prime Number Theorem. In his calculations, Gauss found that the function

$$\mathrm{Li}(x) = \int_2^x \frac{dt}{\log t} \tag{10.50}$$

provides an even better estimate of $\pi(x)$ than that of $x/\log x$. One can show (see Exercise 10.22) that

$$\mathrm{Li}(x) \sim \frac{x}{\log x},$$

and so by the original version of the Prime Number Theorem,

$$\mathrm{Li}(x) \sim \pi(x).$$

Assuming the Riemann hypothesis, one can prove that the error in this estimate is $O(\sqrt{x}\log x)$. Amazingly, the converse of this statement also holds. That is, the Riemann hypothesis is true if and only if

$$\pi(x) = \mathrm{Li}(x) + O(\sqrt{x}\,\log x)$$

for sufficiently large x.

In the century since the work of Hadamard and de la Vallée Poussin, evidence for the conjecture has continued to mount. Hardy proved that there are an infinite number of zeros on the critical line. It is also known that at least a certain percentage of all nontrivial zeros are on the critical line. Currently, the best result in this direction is that of Brian Conrey who proved that at least 40% of all zeros in the critical strip satisfy $\sigma = 1/2$. A complete proof that all nontrivial zeros lie on the critical line continues to elude mathematicians.

Exercises

10.20 Show that if n is a positive integer, then $\Gamma(n) = (n-1)!$.

†**10.21** The sequence of "Bernoulli numbers" is defined recursively by letting $B_0 = 1$ and

$$B_n = -\frac{1}{n+1}\sum_{k=0}^{n-1}\binom{n+1}{k}B_k, \quad n \geq 1.$$

One can show that for any positive integer n,

$$\zeta(2n) = (-1)^{n+1}\frac{(2\pi)^{2n}}{2(2n)!}B_{2n}.$$

(a) Use a computer to find the first ten Bernoulli numbers.

(b) Apply the functional equation (10.48) to deduce a formula for $\zeta(-m)$ for odd positive integers m in terms of Bernoulli numbers. Use these formulas to compute $\zeta(4)$ and $\zeta(-3)$.

10.22 Show that $\mathrm{Li}(x) \sim x/\log x$.

Hint: Use integration by parts and estimate the resulting integral by breaking it into two integrals:

$$\int_2^x \frac{dt}{(\log t)^2} = \int_2^{\sqrt{x}} \frac{dt}{(\log t)^2} + \int_{\sqrt{x}}^x \frac{dt}{(\log t)^2}.$$

10.7 Dirichlet's theorem

The zeta and eta functions introduced in the last section are special examples from a class of functions known as Dirichlet series. In general, a Dirichlet series is one of the form

$$\sum_{n=1}^{\infty} \frac{a_n}{n^s}.$$

In general, the variable s is complex, but for the purposes of this section, we will need to consider only the case in which s is a real variable. In 1839 Dirichlet's study of these kinds of series enabled him to prove the following theorem.

THEOREM 10.24 (Dirichlet's theorem)
If $\gcd(a, m) = 1$, *then there are an infinite number of primes of the form* $a + km$.

We have already given a proof of this theorem in the case $m = 4$ (Propositions 2.38 and 3.38). In this section we will reprove the theorem in this case but this time making use of some of the ideas that Dirichlet developed in proving the general result.

We begin by revisiting Theorem 10.1 in which we proved that the series

$$\sum_p \frac{1}{p}$$

diverges. As we pointed out in Section 10.1, a consequence of this divergence is that there must be an infinite number of primes. We now reformulate the argument in the theorem in terms of the zeta function to obtain a slightly different proof of the infinitude of the set of primes. We begin by taking logarithms of the Euler product expansion of $\zeta(s)$ (Theorem 10.23). We have

$$\log \zeta(s) = \sum_p \log\left(\frac{1}{1-p^{-s}}\right) \tag{10.51}$$

for $s > 1$. (Compare with equation (10.2).) Again making use of the Taylor expansion of $\log(1-x)$, we obtain

$$\log\left(\frac{1}{1-p^{-s}}\right) = \sum_{k=1}^{\infty} \frac{1}{k(p^s)^k}.$$

With some reordering, we can rewrite (10.51) as

$$\log \zeta(s) = \sum_p \frac{1}{p^s} + \sum_p \sum_{k=2}^{\infty} \frac{1}{k(p^s)^k}.$$

Since $s > 1$, the latter sum is bounded above by

$$\sum_p \sum_{k=2}^{\infty} \frac{1}{kp^k}, \tag{10.52}$$

which we know to converge as a result of our work in the proof of Theorem 10.1. Thus, we may write

$$\log \zeta(s) = \sum_p \frac{1}{p^s} + O(1) \tag{10.53}$$

for $s > 1$. If there were only finitely many primes, then the sum on the right side would have finitely many terms and, in particular, would be bounded as s approaches 1 from the right. However,

$$\zeta(s) > \int_1^{\infty} \frac{1}{x^s}\, dx = \frac{1}{s-1},$$

and so $\log \zeta(s)$ approaches infinity as $s \to 1^+$.

To prove Dirichlet's theorem, it is sufficient to show that the quantity

$$\sum_{p \equiv a \pmod{m}} \frac{1}{p^s}$$

diverges as $s \to 1^+$. This is accomplished by relating the series to an appropriate combination of Dirichlet series.

Before we describe the additional Dirichlet series that we will need in the proof of the $m = 4$ case, we introduce an arithmetic function. Let

$$\chi(n) = \begin{cases} 1 \text{ if } & n \equiv 1 \pmod 4 \\ -1 \text{ if } & n \equiv 3 \pmod 4 \\ 0 \text{ if } & n \equiv 0 \text{ or } 2 \pmod 4. \end{cases} \tag{10.54}$$

PROPOSITION 10.25

The function χ is completely multiplicative. That is,

$$\chi(ab) = \chi(a)\chi(b) \tag{10.55}$$

for all integers a and b.

PROOF If either a or b is even, then both sides of (10.55) are 0. Suppose now that both a and b are odd. If $a \equiv b \pmod 4$, then $\chi(a) = \chi(b)$, and so $\chi(a)\chi(b) = 1$. Furthermore, $ab \equiv 1 \pmod 4$, and so $\chi(ab) = 1$. Finally, if $a \not\equiv b \pmod 4$, then $\chi(a) = -\chi(b)$ and so $\chi(a)\chi(b) = -1$. Also, $ab \equiv 3 \pmod 4$, which means that $\chi(ab) = -1$. ∎

We now introduce our new Dirichlet series. For $s > 0$, let

$$L(s, \chi) = \sum_{n=1}^{\infty} \frac{\chi(n)}{n^s}.$$

Observe that this series does converge for $s > 0$ since it is alternating. Furthermore, since

$$L(s, \chi) = 1 - \frac{1}{3^s} + \cdots,$$

we may conclude

$$1 - \frac{1}{3^s} < L(s, \chi) < 1. \tag{10.56}$$

The multiplicativity of $\chi(s)$ allows us to expand $L(s, \chi)$ into a product.

THEOREM 10.26

For $s > 0$,

$$L(s, \chi) = \prod_p \frac{1}{1 - \chi(p)p^{-s}}.$$

The proof is similar to that of Theorem 10.23. A proof of this, and in fact a more general product formula, is called for in the exercises.

Manipulating this product formula as we did in the beginning of the section with $\zeta(s)$, we find for $s > 1$,

$$\log L(s,\chi) = -\sum_p \log(1-\chi(p)p^{-s}) \tag{10.57}$$

$$= \sum_p \sum_{k=1}^{\infty} \frac{\chi(p)^k}{k(p^s)^k} \tag{10.58}$$

$$= \sum_p \frac{\chi(p)}{p^s} + \sum_p \sum_{k=2}^{\infty} \frac{\chi(p)^k}{k(p^s)^k}. \tag{10.59}$$

The rearrangement of the sum in the last step is justified as the double sum in (10.58) converges absolutely for $s > 1$. The absolute value of the second sum in (10.59) is bounded by the convergent series in (10.52). Thus, we have shown that

$$\log L(s,\chi) = \sum_p \frac{\chi(p)}{p^s} + O(1), \tag{10.60}$$

for $s > 1$.

PROPOSITION 10.27

There are an infinite number of primes of the form $1+4k$ and an infinite number of the form $3+4k$.

PROOF Our goal is to prove that as $s \to 1^+$, the quantity

$$\sum_{p \equiv a \pmod 4} \frac{1}{p^s}$$

diverges when $a = 1$ or 3. For such a, we observe that $a \equiv p \pmod 4$ if and only if $\chi(a)\chi(p) = 1$. In fact, for odd primes p,

$$1 + \chi(a)\chi(p) = \begin{cases} 2 \text{ if } & p \equiv a \pmod 4 \\ 0 \text{ if } & p \not\equiv a \pmod 4. \end{cases}$$

Thus, for odd primes we have

$$\frac{1}{p^s} + \chi(a)\frac{\chi(p)}{p^s} = \frac{1+\chi(a)\chi(p)}{p^s} = \begin{cases} \dfrac{2}{p^s} \text{ if } & p \equiv a \pmod 4 \\ 0 \text{ if } & p \not\equiv a \pmod 4. \end{cases}$$

Summing over all primes we obtain

$$\frac{1}{2}\left(\sum_p \frac{1}{p^s} + \chi(a)\sum \frac{\chi(p)}{p^s}\right) - \frac{1}{2^{s+1}} = \sum_{p \equiv a \pmod 4} \frac{1}{p^s} \tag{10.61}$$

for $a = 1$ or $a = 3$. Applying (10.53) and (10.60), we have

$$\sum_{p \equiv a \pmod{4}} \frac{1}{p^s} = \frac{\log \zeta(s)}{2} + \frac{\chi(a) \log L(s, \chi)}{2} + O(1). \tag{10.62}$$

We have already seen that $\log \zeta(s) \to \infty$ as $s \to 1^+$. On the other hand (10.56) implies that for $s > 1$,

$$2/3 < L(s, \chi) < 1,$$

and so $\log L(s, \chi)$ is bounded as $s \to 1^+$. Thus, the sum on the left side of (10.62) must go to infinity as $s \to 1^+$, which implies that there are an infinite number of primes of the form $a + 4k$.

□

A complete proof of Dirichlet's theorem makes use of a generalization of the relation given in (10.62). For any integer m, a *Dirichlet character modulo* m is a complex valued arithmetic function χ that satisfies

1. $\chi(a) = \chi(b)$ if $a \equiv b \pmod{m}$
2. $\chi(a) = 0$ if and only if $\gcd(a, m) \neq 1$
3. $\chi(ab) = \chi(a)\chi(b)$.

We call a Dirichlet character a "character" for short. For any m we always have the "trivial" character

$$\chi_0(a) = \begin{cases} 1 \text{ if } \gcd(a, m) = 1 \\ 0 \text{ otherwise.} \end{cases}$$

The function we worked with in the $m = 4$ case (defined in (10.54)) is in fact a character modulo 4. This and the trivial character turn out to be the only characters modulo 4. In general, there are $\phi(m)$ characters modulo m.

Associated to any such χ, we define a Dirichlet series, as before,

$$L(s, \chi) = \sum_{n=1}^{\infty} \frac{\chi(n)}{n^s},$$

which will converge for $s > 1$. (The proof of this convergence is called for in the exercises.) Working as in the $m = 4$ case, one can obtain a product formula for $L(s, \chi)$, which leads to the same estimate we obtained in the $m = 4$ case,

$$\log L(s, \chi) = \sum_{p} \frac{\chi(p)}{p^s} + O(1) \tag{10.63}$$

for $s > 1$. (The proofs of these facts are called for in the exercises as well.)

Observe that a Dirichlet character χ modulo m induces a map $\hat{\chi} : \mathbf{Z}_m \to \mathbf{C}$ defined by

$$\hat{\chi}(a) = \chi(a).$$

Moreover, if we restrict χ to $\mathbf{Z}_m^*$, then we obtain a group homomorphism from $\mathbf{Z}_m^*$ to $\mathbf{C}^*$ (i.e., a map that preserves the group operation). Conversely, if $\hat{\chi}$ is any homomorphism from $\mathbf{Z}_m^*$ to $\mathbf{C}^*$, then the arithmetic function

$$\chi(a) = \begin{cases} \hat{\chi}(a) & \text{if } \gcd(a,m) = 1 \\ 0 & \text{otherwise} \end{cases}$$

is a Dirichlet character modulo m. Thus, Dirichlet characters are in one-to-one correspondence with group homomorphisms from $\mathbf{Z}_m^*$ to $\mathbf{C}^*$. In general, if G is any finite commutative group, then one can show that

$$\sum_f \overline{f(a)} f(b) = \begin{cases} |G| & \text{if } a = b \\ 0 & \text{if } a \neq b \end{cases}$$

for all a and b in G, where $\overline{f(a)}$ is the complex conjugate of $f(a)$ and the sum is over all group homomorphisms f from G to $\mathbf{C}^*$. (See [2] for details.) In our case, this formula implies that

$$\sum_\chi \overline{\chi(a)} \chi(b) = \begin{cases} \phi(m) & \text{if } a \equiv b \pmod{m} \\ 0 & \text{if } a \not\equiv b \pmod{m} \end{cases} \tag{10.64}$$

when $\gcd(a, m) = 1$ and the sum is over all Dirichlet characters modulo m. It follows that

$$\begin{aligned} \sum_\chi \overline{\chi(a)} \sum_p \frac{\chi(p)}{p^s} &= \sum_p \frac{1}{p^s} \sum_\chi \overline{\chi(a)} \chi(p) \\ &= \phi(m) \sum_{p \equiv a \pmod{m}} \frac{1}{p^s}. \end{aligned}$$

Combining this result with (10.63), we obtain

$$\sum_{p \equiv a \pmod{m}} \frac{1}{p^s} = \frac{1}{\phi(m)} \sum_\chi \overline{\chi(a)} \log L(s, \chi) + O(1), \tag{10.65}$$

for $s > 1$. Thus, to prove Dirichlet's theorem it is enough to show that the sum on the right goes to infinity as $s \to 1^+$. The product formula for $L(s, \chi_0)$ is nearly the same as the Euler product expansion of $\zeta(s)$. Specifically, we have

$$\zeta(s) = L(s, \chi_0) \prod_{p \mid m} \frac{1}{1 - p^{-s}}.$$

Hence, $\log L(s, \chi_0) \to \infty$ as $s \to 1^+$. To complete the proof, one only needs to show that $L(s, \chi)$ is bounded and does not approach zero for nontrivial characters χ. This turns out to be the difficult part of the proof. Notice that this is similar to the difficulty in completing the proof of the prime number theorem. Recall that in that case the final issue to be resolved was the proof that the Riemann zeta function does not have zeros on the line Re $s = 1$.

Exercises

10.23 Dirichlet's theorem asserts that for any pair of relatively prime numbers a and m, there are an infinite number of primes of the form $a + km$. Show that to prove Dirichlet's theorem, it suffices to show that for any pair of relatively primes numbers a and m, there is at least one prime of the form $a + km$ with $k > 0$.

10.24 Use Dirichlet's theorem to show that for any positive number n there exists a prime number p such that the n numbers preceding p and the n numbers following p (i.e., the numbers $p \pm 1, p \pm 2, \ldots, p \pm n$) are all composite.

10.25 Show that if χ is a Dirichlet character modulo m, then

$$|\chi(a)| = 1,$$

for all numbers a satisfying $\gcd(a, m) = 1$.

10.26 Show that the Dirichlet series

$$L(s, \chi) = \sum_{n=1}^{\infty} \frac{\chi(a)}{n^s}$$

associated to a Dirichlet character converges for all $s > 1$.

10.27 Show that a Dirichlet series corresponding to a Dirichlet character χ can be expanded into the product

$$L(s, \chi) = \prod_{p} \frac{1}{1 - \chi(p)p^{-s}},$$

for $s > 1$.

10.28 Show that for any Dirichlet character χ,

$$\log L(s, \chi) = \sum_{p} \frac{\chi(p)}{p^s} + O(1),$$

for $s > 1$.

10.29 There are two Dirichlet characters modulo 3, the trivial character χ_0 and the character defined by

$$\chi_1(n) = \begin{cases} 1 \text{ if } n \equiv 1 \pmod{3} \\ -1 \text{ if } n \equiv 2 \pmod{3} \\ 0 \text{ if } n \equiv 0 \pmod{3}. \end{cases}$$

(a) Show that (10.64) holds for $m = 3$. That is, show that

$$\chi_0(a)\chi_0(b) + \chi_1(a)\chi_1(b) = \begin{cases} 2 \text{ if } a \equiv b \pmod{3} \\ 0 \text{ if } a \not\equiv b \pmod{3} \end{cases}$$

when a and b are not divisible by 3.

(b) Show that $1/2 < L(s, \chi_1) < 1$ for $s > 1$.

(c) Use (10.65) to show that there are an infinite number of primes of the form $a + 3k$ for $a = 1$ and for $a = 2$.

10.8 Notes

Paul Erdős

Paul Erdős (1913–1996) made many important contributions to number theory, particularly in the area of combinatorial applications. Erdős often made reference to "The Book," a collection of the most elegant proofs of theorems that Erdős imagined to be possessed by God. (Though Erdős expressed doubts about the existence of God, he was convinced of the existence of The Book.) In light of this, it is perhaps not surprising that some of his most important work came in providing new proofs for previously accepted results. We have already mentioned his work with Selberg that led to an elementary proof of the Prime Number Theorem. While it is certainly not the case that the original proof was in any way defective, the statement of the Prime Number Theorem is an elementary one, and so the existence of an elementary proof is satisfying. Erdős also found a simpler proof of Bertrand's postulate. He was only 18 at the time of this discovery.

Erdős was one of the most prolific mathematicians of all time, having published over 1500 papers. In fact, since his death in 1996, over 60 papers have appeared in which he was listed as the author or as a coauthor. (Thus, he has had more papers published after his death than most mathematicians publish in their lives.) He had few material possessions and no permanent home, and he spent much of his life traveling the world staying with fellow mathematicians. In his lifetime he collaborated with over 500 colleagues. This staggering number of connections led to the concept of an Erdős number. A mathematician has an Erdős number 1 if he or she coauthored a paper with

Erdős. Mathematicians who have coauthored a paper with someone with Erdős number 1, but not with Erdős himself, have Erdős number 2, and so on. For example, Andrew Wiles (currently) has Erdős number 3, because he wrote a paper with Barry Mazur, who in turn wrote a paper with Andrew Granville, who coauthored a paper with Erdős. For further statistics and information regarding Erdős numbers, see

`http://www.oakland.edu/enp/.`

Those familiar with Kevin Bacon numbers (defined in an analogous way in which two actors are connected if they have appeared in a movie together) will be interested to know that Paul Erdős has Kevin Bacon number 3. We leave the proof as an exercise, but point to the opening scene in the biographical film *N is a Number: A Portrait of Paul Erdős* as a hint.

Chapter 11

Elliptic Curves

> I remember once going to see him when he was lying ill at Putney. I had ridden in taxi cab number 1729 and remarked that the number seemed to me rather a dull one, and that I hoped it was not an unfavorable omen. "No," he replied, "it is a very interesting number; it is the smallest number expressible as the sum of two cubes in two different ways."
>
> [On Ramanujan.]
>
> G. H. HARDY (1877–1947)

In our study of Diophantine equations in Chapter 9, we were able to make precise descriptions of the solution sets of several classes of Diophantine equations. For the most part, however, these tended to be equations of degree one or two. Apart from Fermat's Last Theorem, we have said almost nothing about degree three equations. Such equations do turn out to be more difficult, but their study is now a highly active area of research. Given a degree three Diophantine equation in two variables, we can impose a group structure on the set of solutions. (See the notes in Chapter 3 for a brief discussion of groups.) The group law provides a method of generating new solutions from old ones and often provides insight into what the solution set looks like. More practically, this group structure can be used to provide shortcuts for factoring integers. Recall that the lack of an efficient factoring algorithm is the basis of security for the RSA encryption scheme (see Chapter 4). In fact, one can even use the group structure to obtain a public key encryption scheme that provides a level of security comparable to that of RSA but with smaller keys. The most dramatic application of the study of degree three equations came recently in the completion of the proof of Fermat's Last Theorem. The first few sections of this chapter will serve as a sort of crash course on elliptic curves. For a more complete treatment see [28].

11.1 Cubic curves

A *cubic curve* is the solution set to a degree three polynomial equation in two variables. An *elliptic curve* is a cubic curve that satisfies a "smoothness" condition and is endowed with a group operation that we will describe in detail in later sections. While these curves are not ellipses, they do arise in connection with certain integrals that represent arc lengths of ellipses, and it is from this application that their name is derived. The graph of the set of real solutions in the plane to a degree three equation is in fact a curve.

Example 11.1 A graph of the curve $y^2 = x^3-2x+4$ in the real plane is given in Figure 11.1. One can obtain such a sketch using the usual methods from

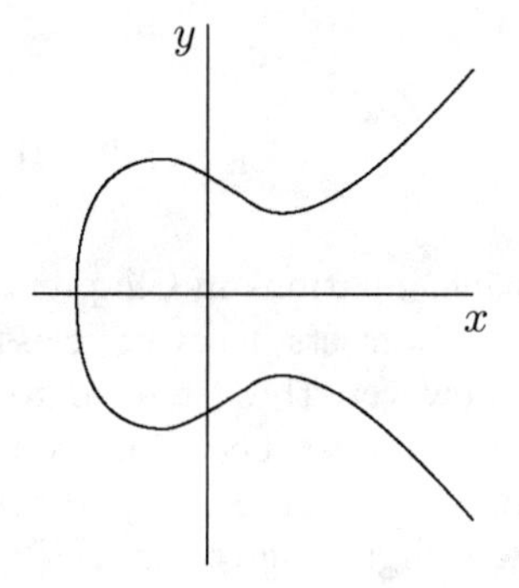

FIGURE 11.1: Graph of $y^2 = x^3 - 2x + 4$.

calculus. The graph can be produced using the plot commands in *Mathematica* or *Maple*. These commands will plot a function over a specified interval. Thus, we will need to solve for y and regard the curve as the graph of two functions. In *Mathematica*, the command

```
Plot[{Sqrt[x^3 - 2x + 4], -Sqrt[x^3 - 2x + 4]}, {x, -2, 3}]
```

will produce a graph resembling the one in Figure 11.1. The *Maple* command

```
plot([sqrt(x^3-2*x+4),-sqrt(x^3-2*x+4)],x=-2..3);
```

will also produce the graph. □

Example 11.2 The set of real solutions to the equation $y^2 = x^3 - 4x$ is pictured in Figure 11.2. Notice that the right side of the equation is only nonnegative when $-2 \le x \le 0$ or $x \ge 2$. This explains why the solution set is broken into two pieces. □

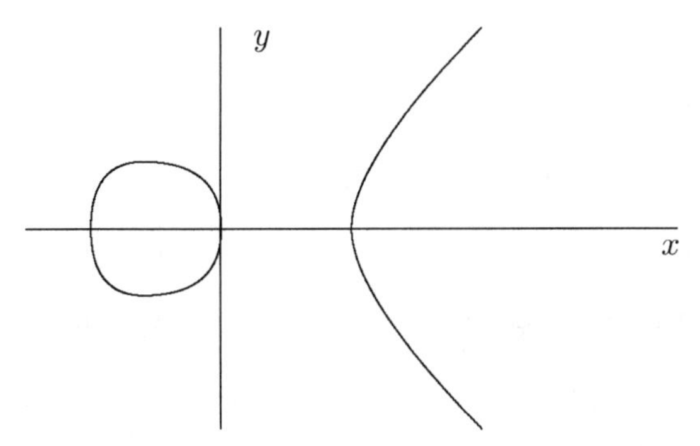

FIGURE 11.2: Graph of $y^2 = x^3 - 4x$.

Notice that we can rewrite the equation giving the curve in Example 11.2 in the form

$$y^2 - x^3 + 4x = 0.$$

In general, any cubic curve is represented by an equation of the form

$$F(x, y) = 0$$

where $F(x, y)$ is a polynomial of total degree three. Both of our examples are curves of the form

$$y^2 - f(x) = 0, \tag{11.1}$$

where $f(x)$ has the form $f(x) = x^3 + b_2x^2 + b_1x + b_0$. We will be primarily focused on curves of this form. While this is not the most general form of an equation for a cubic curve, many questions about cubic curves can be reduced to the class of curves of this form. We note the symmetry of such a curve about the x-axis. In general, if (x_0, y_0) is on the curve $y^2 - f(x) = 0$, then so is $(x_0, -y_0)$.

Exercises

11.1 Sketch graphs of the following curves in the real plane.

(a) $y^2 = x^3$
(b) $y^2 = x^3 + 1$
(c) $y^2 = x^3 - x$
(d) $y^2 = x^3 + x^2$

11.2 Prove that the graph of the set of real solutions to a cubic curve given by an equation $y^2 = f(x)$ intersects a nonvertical line in at most three points.

11.3 Prove that the graph of the set of real solutions to a cubic curve given by an equation $y^2 = f(x)$ intersects a vertical line in at most two points.

†**11.4** The curve $G(x, y) = 0$ is said to be obtained from $F(x, y) = 0$ through a *linear change of variables* if

$$G(x, y) = F(c_1(x, y), c_2(x, y)),$$

where c_1 and c_2 are linear functions in the variables x and y.

(a) Show that by making an appropriate linear change of variables to a curve of the form

$$y^2 + a_4xy + a_3y - x^3 - a_2x^2 - a_1x - a_0 = 0,$$

one can obtain a curve of the form

$$y^2 - x^3 - b_2x^2 - b_1x - b_0 = 0.$$

Hint: Complete the square on the variable y.

(b) Show that by making an appropriate linear change of variables to a curve of the form

$$y^2 - x^3 - b_2x^2 - b_1x - b_0 = 0, \tag{11.2}$$

one can obtain a curve of the form

$$y^2 - x^3 - Ax - B = 0.$$

Hint: Complete the cube on the variable x.

11.2 Intersections of lines and curves

Given a pair of points on a cubic curve, we can often produce a new point on the curve as follows: Construct the line passing through the two points and find a third point of intersection between the line and the curve. Before we investigate this procedure on cubic curves, let's first consider how lines intersect circles. Observe that any line passing though a circle will actually pass though the circle in exactly two points or exactly once at a point of tangency (see Figure 11.3).

We now argue that (almost) any line intersects a cubic curve in (essentially) three points. We first take up the issue of what we mean by "essentially." Consider the line $y = mx + b$ and the cubic curve $y^2 = f(x)$. The points of intersection are determined by the equation

$$(mx + b)^2 = f(x). \tag{11.3}$$

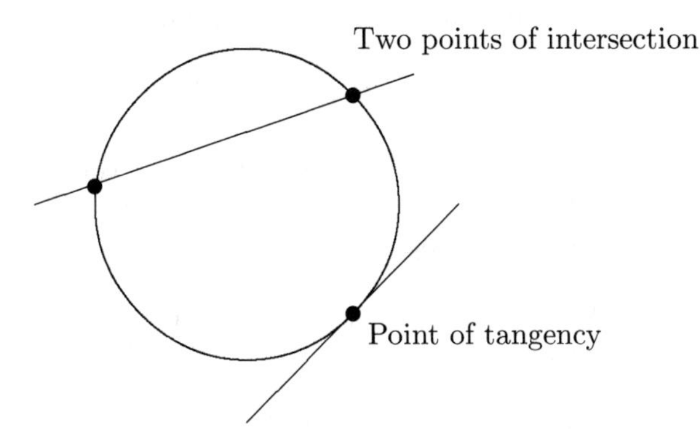

FIGURE 11.3: Intersections of lines and a circle.

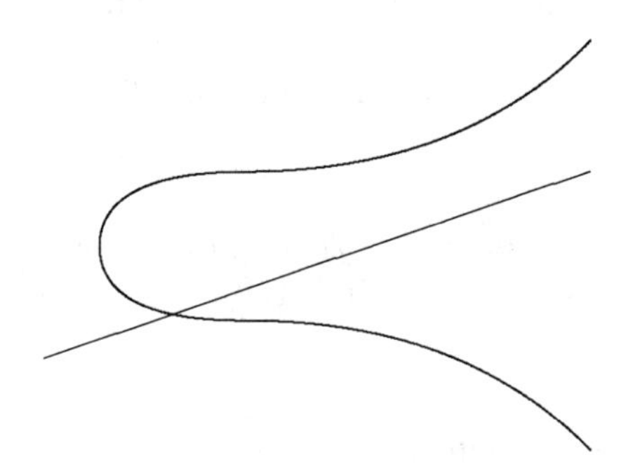

FIGURE 11.4: Intersection of $y^2 = x^3 + 17$ and $y = x - 1$.

This is a polynomial equation of degree three, and so therefore it has at most three solutions. Under what circumstances can a degree three polynomial have fewer than three roots?

Example 11.3 Consider the intersection of the line $y = x - 1$ and the curve $y^2 = x^3 + 17$. (See Figure 11.4.)

To find the points of intersection, we must solve the equation

$$(x - 1)^2 = x^3 + 17.$$

Equivalently, we must find the roots of the polynomial

$$x^3 + 17 - (x - 1)^2 = x^3 - x^2 + 2x + 16.$$

This polynomial has the root $x = -2$. Factoring, we obtain

$$x^3 - x^2 + 2x + 16 = (x + 2)(x^2 - 3x + 8).$$

The quadratic formula provides the remaining two roots,

$$x = \frac{3 \pm \sqrt{-23}}{2}.$$

So the polynomial does indeed have three roots in the set of complex numbers, but only one of the roots is a real number. This explains why the line in Figure 11.4 intersects the curve at only one point in the real plane. □

The Fundamental Theorem of Algebra tells us that if $f(x)$ is any monic cubic polynomial, then it has a factorization

$$f(x) = (x - x_1)(x - x_2)(x - x_3),$$

where the x_i are complex numbers. However, these roots may not be distinct. If $x_1 = x_2 \neq x_3$, then we say that x_1 is a root of $f(x)$ of multiplicity 2. If $x_1 = x_2 = x_3$, then we say that x_1 is a root of multiplicity 3. This motivates the following definition.

DEFINITION 11.4 *The curve $y^2 = f(x)$ and the line $y = mx + b$ have a point $(x_0, mx_0 + b)$ of* intersection multiplicity k *if x_0 is a root of multiplicity k of the polynomial $f(x) - (mx + b)^2$.*

Example 11.5 Consider the intersection of the circle $y^2 = -x^2 + 1$ and the line $y = 1$. (Note that $-x^2 + 1$ is not monic, but the intersection multiplicity definition given above still applies.) The point $(0, 1)$ is a point of intersection multiplicity 2, as $x = 0$ is a double root of $-x^2 + 1 - (1)^2$. In fact, the line $y = 1$ is tangent to the circle at $(0, 1)$ (see Figure 11.5). Later, we will show in general that points of intersection multiplicity greater than 1 correspond to intersections of curves and tangent lines.

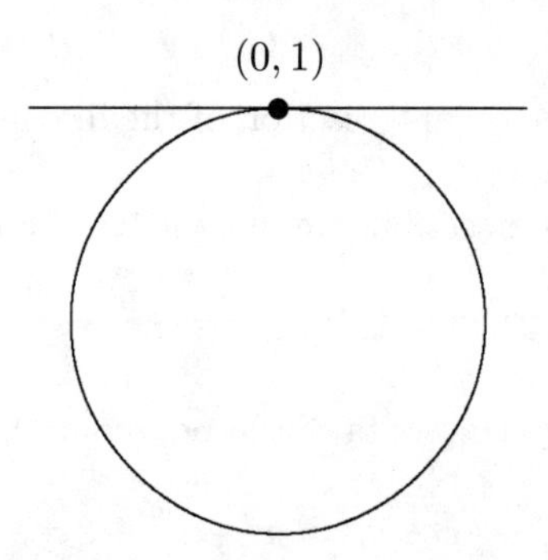

FIGURE 11.5: Point of intersection multiplicity 2.

□

We can extend the definition of intersection multiplicity to include vertical lines as well. Suppose that (x_0, y_0) is a point on the curve $y^2 = f(x)$. We will again say that the line $x = x_0$ intersects the curve at (x_0, y_0) with intersection multiplicity k if the polynomial determining the intersection has a root of multiplicity k. In this case, the intersection is determined by the polynomial $y^2 - f(x_0)$, which is a degree two polynomial in the variable y. Thus, (x_0, y_0) is a point of intersection multiplicity 2 if $y^2 - f(x_0)$ has a double root.

THEOREM 11.6

Let $f(x)$ be a monic degree three polynomial with rational coefficients, and let $y = mx + b$ be a line with rational slope and y-intercept. The cubic curve $y^2 = f(x)$ and the line $y = mx + b$ intersect in exactly three points (which may have complex coordinates), counting intersection multiplicities. If the coordinates of two of the points of intersection are rational, then so are the coordinates of the third point.

PROOF It remains for us to prove the last statement. Let x_1, x_2, and x_3 be the x-coordinates of the points of intersection. Then

$$f(x) - (mx + b)^2 = (x - x_1)(x - x_2)(x - x_3).$$

The coefficient of x^2 on the left side of this equation is a rational number and so the coefficient of x^2 on the right side, $-(x_1 + x_2 + x_3)$, must also be a rational number. If we assume that x_1 and x_2 are rational, then so is x_3. The y-coordinate of the third point of intersection is $mx_3 + b$, and so its rationality follows from the rationality of x_3. ∎

Suppose that $P = (x_1, y_1)$ and $Q = (x_2, y_2)$ are two points with rational coordinates on the curve $y^2 = f(x)$, with $x_1 \neq x_2$. Take $m = (y_2 - y_1)/(x_2 - x_1)$ and $b = y_1 - mx_1$. Then the line $y = mx + b$ intersects the curve $y^2 = f(x)$ at the points P and Q and a third point R whose coordinates are rational. Determination of R will be the key step in defining the "addition" of two points on a cubic curve (see Figure 11.6).

There are two issues that remain to be addressed. First, so far we have considered only intersections of curves with nonvertical lines. It is not hard to see (recall Exercise 11.3) that the curve $y^2 = f(x)$ fails to intersect any vertical line at three points. (This accounts for the 'almost' in our claim above that 'almost' every line intersects a cubic curve in (essentially) three points.) Second, if our goal is to add two points by first constructing the line through the two points, how will we manage this in the case where the two points are the same? We deal with the second issue first. In light of our discussion of intersection multiplicity, the obvious thing to do in this case is to look for a line that passes through the curve at our given point with intersection multiplicity 2. Recall from the example of the circle, $x^2 + y^2 = 1$, we observed

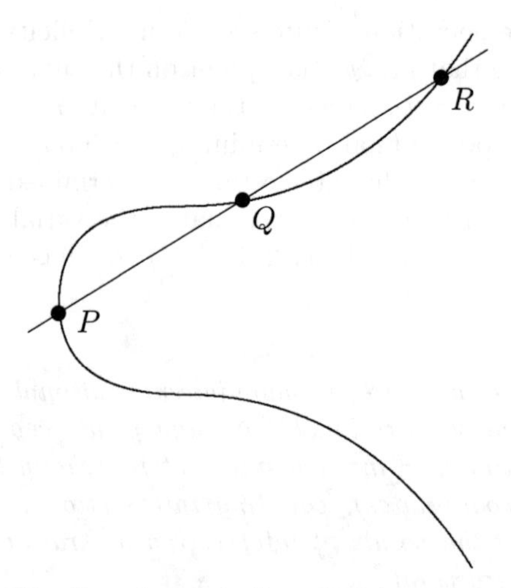

FIGURE 11.6: The line through P and Q intersects the curve at a third point, R.

that a line passes through the circle exactly once at a point of tangency. The same thing will be true in the case of cubic curves.

A tangent line to a curve $F(x, y) = 0$ in the real plane is determined by the direction of the gradient vector. That is, the tangent line is the line passing through the curve at a given point (x_0, y_0) and orthogonal to the gradient vector, $\langle F_x(x_0, y_0), F_y(x_0, y_0)\rangle$ (where F_x and F_y denote the partial derivatives of $F(x, y)$ with respect to x and y, respectively). In order to obtain a unique line orthogonal to the gradient vector $\langle F_x, F_y\rangle$, we need the vector to be nonzero.

DEFINITION 11.7 *A point (x_0, y_0) on a curve $F(x, y) = 0$ is* singular *if*

$$F_x(x_0, y_0) = F_y(x_0, y_0) = 0.$$

A nonsingular *point is a point that is not singular. A curve is* nonsingular *if all points on it are nonsingular.*

Example 11.8 We claim that the curve $F(x, y) = y^2 - x^3 + 4x = 0$ is nonsingular. We first compute $F_x(x, y) = -3x^2 + 4$ and $F_y(x, y) = 2y$. If $F_y(x_0, y_0) = 0$ for some point on the curve, then $y_0 = 0$ and x_0 is a root of $-x^3 + 4x$. The roots of this polynomial are 0, 2, and -2 and are not roots of $-3x^2 + 4$. □

More generally, if $F(x, y) = y^2 - f(x)$, then $F_x(x, y) = -f'(x)$ and $F_y(x, y) = 2y$. Thus, a point (x_0, y_0) on a cubic curve of the form $y^2 = f(x)$ is singular if and only if $y_0 = 0$ and $f(x_0) = f'(x_0) = 0$.

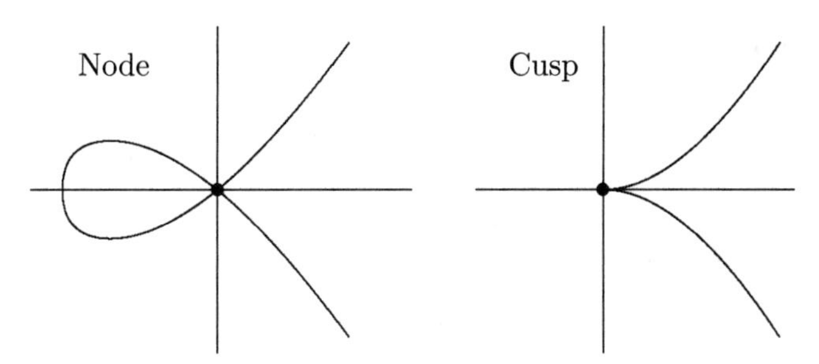

FIGURE 11.7: Singularities on cubic curves.

PROPOSITION 11.9
Let $g(x)$ be a polynomial and let x_0 be a root of $g(x)$. Then x_0 is a root of multiplicity 2 or greater if and only if $g'(x_0) = 0$.

PROOF If x_0 is a root of multiplicity at least 2, then $g(x) = (x - x_0)^2 h(x)$, for some polynomial $h(x)$. By the product rule for derivatives

$$g'(x) = (x - x_0)^2 h'(x) + 2(x - x_0)h(x),$$

from which it follows that $g'(x_0) = 0$.

Conversely, given that x_0 is a root of $g(x)$, we have $g(x) = (x - x_0)k(x)$, for some polynomial $k(x)$. Assuming that $g'(x_0) = 0$, it follows, using the product rule, that $k(x_0) = 0$. Hence, $x - x_0$ is a factor of $k(x)$ and $(x - x_0)^2$ is a factor of $g(x)$. □

As a consequence of Proposition 11.9, singular points on a cubic curve of the form $y^2 = f(x)$ are of the form $(x_0, 0)$, where x_0 is a root $f(x)$ of multiplicity 2 or 3. We immediately see that a cubic curve of this form can have at most one singular point. A singularity of the form $(x_0, 0)$ is called a *node* if x_0 is a double root of $f(x)$ and a *cusp* if x_0 is a triple root (see Figure 11.7).

DEFINITION 11.10 *Let $f(x)$ be a degree n monic polynomial with roots $r_1, r_2, \ldots, r_n$. The* discriminant *of $f(x)$ is the quantity*

$$\triangle_f = \prod_{i \neq j} (r_i - r_j)^2.$$

We note that a root of multiplicity k is included in the list of roots k times. Thus, $\triangle_f = 0$ if and only if $f(x)$ has a multiple root. If $f(x) = x^2 + bx + c$, then the quadratic formula implies that

$$\triangle_f = b^2 - 4c.$$

One can also work out the formula for the discriminant of a cubic polynomial. If the coefficient of x^2 in the cubic is 0, then the formula is quite simple.

THEOREM 11.11
If $f(x) = x^3 + ax + b$, then

$$\triangle_f = -4a^3 - 27b^2. \tag{11.4}$$

PROOF Since $f(x)$ is monic, we have

$$f(x) = (x - r_1)(x - r_2)(x - r_3).$$

Expanding the right side and comparing with the formula $f(x) = x^3 + ax + b$, we have

$$\begin{aligned} 0 &= -r_1 - r_2 - r_3 \\ a &= r_1r_2 + r_1r_3 + r_2r_3 \\ b &= -r_1r_2r_3. \end{aligned}$$

The first equality implies that $r_3 = -(r_1 + r_2)$, and so

$$\triangle_f = (r_1 - r_2)^2(2r_1 + r_2)^2(r_1 + 2r_2)^2. \tag{11.5}$$

We can also rewrite our formulas for a and b as

$$\begin{aligned} a &= -r_1^2 - r_1r_2 - r_2^2 \\ b &= r_1^2r_2 + r_1r_2^2. \end{aligned}$$

Thus, to prove the theorem we need to show that

$$(r_1 - r_2)^2(2r_1 + r_2)^2(r_1 + 2r_2)^2 = -4(-r_1^2 - r_1r_2 - r_2^2)^3 - 27(r_1^2r_2 + r_1r_2^2)^2.$$

Expanding both sides of this equation, we can check that the identity does indeed hold. ∎

COROLLARY 11.12
The curve $y^2 = x^3 + ax + b$ is nonsingular if and only if $-4a^3 - 27b^2 \neq 0$.

Let (x_0, y_0) be a nonsingular point on a curve $F(x, y) = 0$. If $F_y(x_0, y_0) \neq 0$, then a line with slope $-F_x(x_0, y_0)/F_y(x_0, y_0)$ is orthogonal to the gradient, $\langle F_x(x_0, y_0), F_y(x_0, y_0)\rangle$. If $F_y(x_0, y_0) = 0$, then a vertical line is orthogonal to the gradient.

DEFINITION 11.13 *Let (x_0, y_0) be a nonsingular point on a curve $F(x, y) = 0$. The* tangent line *to the curve at the point (x_0, y_0) is the line passing through the point (x_0, y_0) and orthogonal to the vector $\langle F_x(x_0, y_0), F_y(x_0, y_0)\rangle$.*

For a curve of the form $y^2 = f(x)$, the tangent line at a point (x_0, y_0) has the form

$$y - y_0 = \frac{f'(x_0)}{2y_0}(x - x_0) \tag{11.6}$$

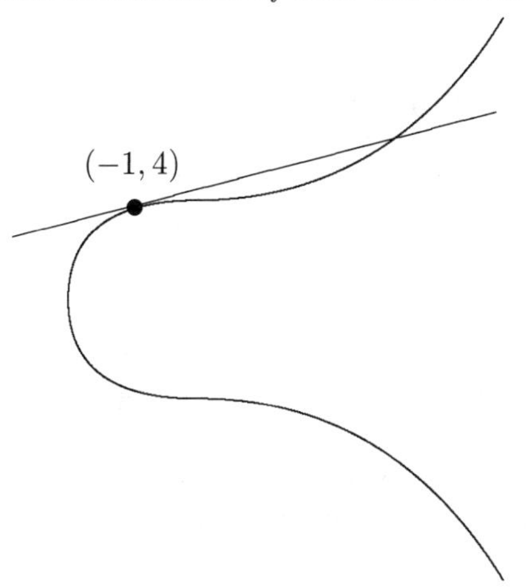

FIGURE 11.8: Tangent line to $y^2 = x^3 + 17$ at $(-1, 4)$.

when $y_0 \neq 0$. The tangent line has the form $x = x_0$ exactly when x_0 is a root of $f(x)$.

Example 11.14 Consider the curve $y^2 = x^3 + 17$. The tangent line to the curve at $(-1, 4)$ is

$$y - 4 = (3/8)(x + 1),$$

or

$$y = (3/8)x + 35/8.$$

Let's compute the intersection multiplicity of the curve and the tangent line at $(-1, 4)$. The equation giving the intersection is

$$x^3 + 17 - ((3/8)x + 35/8)^2 = 0.$$

The derivative of the polynomial on the left hand side is

$$3x^2 - 2((3/8)x + 35/8)(3/8).$$

Evaluating the derivative at -1, we do in fact get 0. Hence, $(-1, 4)$ is a point of intersection multiplicity at least 2. (From Figure 11.8 or by taking another derivative, we see that the intersection multiplicity is exactly 2.) $\square$

We now prove in general that for curves of the form $y^2 = f(x)$, the tangent will intersect the curve at a point of intersection multiplicity at least 2.

THEOREM 11.15

Let (x_0, y_0) be a point on a nonsingular cubic curve $y^2 = f(x)$. The intersection multiplicity of (x_0, y_0) on the curve and the tangent line to the point is 2 or greater. Conversely, if a line intersects the curve at a point (x_0, y_0) of intersection multiplicity 2 or greater, then the line must be the tangent line at (x_0, y_0).

PROOF Let's first dispense with the case of a vertical line. Recall from the discussion following Example 11.5 that a vertical line intersects the curve at a point (x_0, y_0) of intersection multiplicity 2 if and only if $y_0 = 0$. These are exactly the points for which the tangent line is vertical.

Now we turn our attention to nonvertical lines. Let $m = f'(x_0)/(2y_0)$, and so the equation for the tangent line at (x_0, y_0) is $y = m(x - x_0) + y_0$. Set $g(x) = f(x) - (m(x - x_0) + y_0)^2$. We must show that x_0 is a root of multiplicity at least 2 of $g(x)$. Taking a derivative, we have

$$g'(x) = f'(x) - 2m(m(x - x_0) + y_0).$$

Since $f'(x_0) = 2my_0$, we have $g'(x_0) = 0$, and so indeed x_0 is a multiple root of $g(x)$.

Now we consider the converse. Let $y = mx + b$ be a line passing through the point (x_0, y_0) on the curve $y^2 = f(x)$ with intersection multiplicity 2 or greater. We must show that $m = f'(x_0)/(2y_0)$. We have assumed that the polynomial $f(x) - (mx + b)^2$ has a multiple root at x_0, and so the derivative of this polynomial at x_0 is zero. Hence, $f'(x_0) - 2m(mx_0 + b) = 0$, and so $m = f'(x_0)/(2(mx_0 + b)) = f'(x_0)/(2y_0)$. ∎

Exercises

11.5 Determine the intersection multiplicity of the point $(3/5, 4/5)$ on the curve $y^2 = 1 - x^2$ and the line

(a) $y = 3x - 1$;

(b) $y = -3/4x + 5/4$.

11.6 Determine the intersection multiplicity of the point $(2, 1)$ on the intersection of the line $y = 7x - 13$ and the curve $y^2 = x^3 + 2x - 11$.

11.7 Show that the tangent line to the curve $y^2 = x^3 + 1$ at the point $(0, 1)$ transects the curve (passes from one side of the curve to the other) at the point of tangency. What is the intersection multiplicity of this point of intersection?

11.8 Show that the curve $x^2 + y^2 - 1 = 0$ has no singular points.

11.9 Determine whether the curve $x^2 - y^2 = 0$ has any singular points.

11.10 Find all singular points on the curve $y^2 - 2y - x^3 + 12x - 19 = 0$.

11.11 Determine which of the following curves are singular.

(a) $y^2 = x^3 - 12x + 1$

(b) $y^2 = x^3 - 3x + 2$

(c) $y^2 = x^3 - 2x^2 + 3x - 6$

(d) $y^2 = x^3 - 6x^2 + 12x - 8$

11.12 Let (x_0, y_0) be a point on a curve given by the equation $y^2 = f(x)$. Use implicit differentiation to verify the equation for the tangent line given by (11.6) in the case $y_0 \neq 0$.

11.3 The group law and addition formulas

In this section we will describe the group law on a cubic curve and obtain some explicit formulas for computing in this group. Let $y^2 = f(x)$ be a nonsingular cubic curve. Recall from the previous section that almost any line that intersects the curve in two points will necessarily intersect the curve at a third point (counting multiplicities). The exceptions to this rule are vertical lines that intersect the curve in exactly two points. To remedy this "defect" we will add a point, $\mathcal{O}$, to the set of points and imagine this point to be the missing point of intersection between the curve and a vertical line. We refer to this added point as the *point at infinity*. The inclusion of this point at infinity is less mysterious when one considers the curve in projective space. In this setting we also can make sense of a tangent line to the point at infinity, and in fact this tangent line intersects the curve at a point of intersection multiplicity 3 at the point at infinity. (See the notes at the end of the chapter for further details.)

We now describe the procedure for "adding" two points $P = (x_1, y_1)$ and $Q = (x_2, y_2)$ on the curve. First, construct the line through points P and Q (if $P = Q$, this line is the tangent line). Let R be the third point of intersection between the line and the curve. Next, construct the line through R and $\mathcal{O}$. We define the sum of P and Q (and denote it $P + Q$) to be the third point of intersection on the line between R and $\mathcal{O}$ (see Figure 11.9).

Note: A natural question is, why not simply define $P + Q$ as the third point of intersection (rather than the reflection of this point)? The answer is that this definition does not provide a group structure.

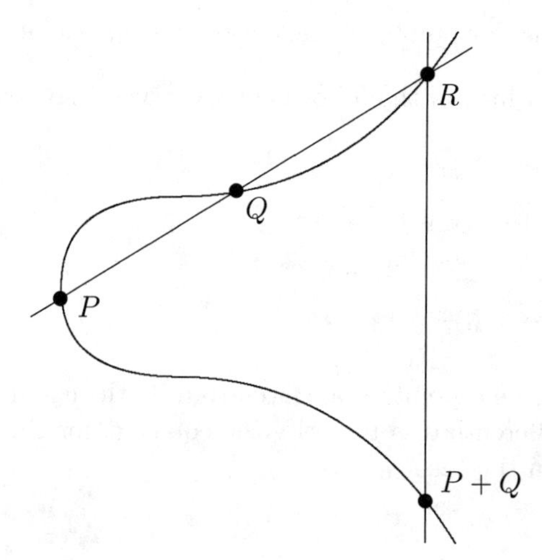

FIGURE 11.9: Addition of points P and Q.

We can simplify this procedure by considering two cases. First, if $R = (x_3, y_3) \neq \mathcal{O}$, then the line through R and $\mathcal{O}$ is a vertical line, and so $P+Q = (x_3, -y_3)$. Second, if $R = \mathcal{O}$, then the line through R and $\mathcal{O}$ is the tangent line at $\mathcal{O}$. As we noted above, $\mathcal{O}$ is a point of intersection multiplicity 3 between the line and curve, and so in this case $P + Q = \mathcal{O}$.

Example 11.16 We will add the points $P = (-1, 4)$ and $Q = (2, 5)$ on the curve $y^2 = x^3 + 17$. The line through P and Q is $y = (1/3)x + (13/3)$. Let (x_3, y_3) be the third point of intersection on the line and the curve. Thus, x_3 together with -1 and 2 are the three roots of the polynomial

$$x^3 + 17 - ((1/3)x + (13/3))^2.$$

Since this polynomial is monic, we have a factorization

$$x^3 + 17 - ((1/3)x + (13/3))^2 = (x + 1)(x - 2)(x - x_3).$$

Equating the coefficient of x^2 on both sides of the above equality, we have

$$-(1/3)^2 = 1 - 2 - x_3,$$

and so $x_3 = -8/9$. Using the equation for the line, we find $y_3 = 109/27$. We conclude that $P + Q = (-8/9, -109/27)$. □

The procedure above includes the case of addition of a point $P = (x_1, y_1)$ and the point $\mathcal{O}$ at infinity. In this case, the line through P and $\mathcal{O}$ is the vertical line through P. The third point of intersection is $(x_1, -y_1)$. (Note that in the event that $y_1 = 0$, the line is tangent at P, and so P is a point

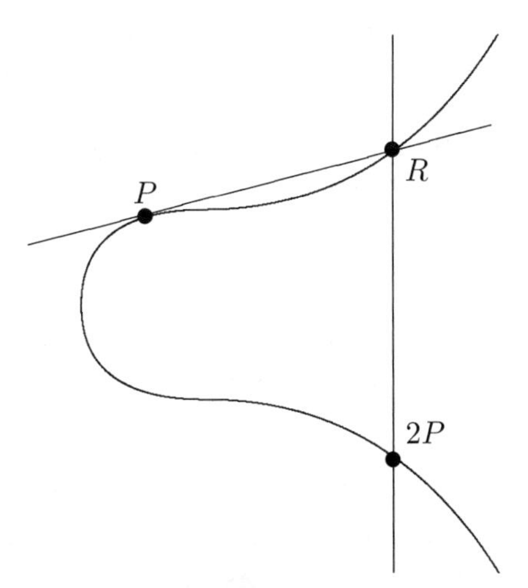

FIGURE 11.10: Addition of point P with itself.

of intersection multiplicity 2.) Thus, we see that $P + \mathcal{O} = P$. To compute $\mathcal{O}+\mathcal{O}$, we use the fact that the tangent line to $\mathcal{O}$ is, as stated at the beginning of the section, a line that intersects the curve with intersection multiplicity 3. Hence, $\mathcal{O}+\mathcal{O} = \mathcal{O}$, and so $\mathcal{O}$ acts as an identity under our addition operation.

Recall from our initial description of the procedure for adding two points that if $\mathcal{O}$ is the third point of intersection on the curve and the line through points P and Q, then $P+Q = \mathcal{O}$. If $P = (x_1, y_1)$ and we take $Q = (x_1, -y_1)$, then the line through P and Q is a vertical line (notice this does hold in the case $y_1 = 0$). Thus, inverses do exist under our operation, and in fact the inverse of a point (x_1, y_1) is just $(x_1, -y_1)$.

Summarizing our discussion so far, we have defined a binary operation for which there is an identity element and for which inverses exist. Further, from the way that it is defined, the operation is clearly commutative. To show that the operation gives the set of points on a nonsingular cubic curve the structure of an abelian group, it remains to prove that the operation is associative. This turns out to be a bit more difficult. We omit the proof of the associativity here, but a sketch of the proof is provided in the notes at the end of this chapter.

We are now ready to give our formal definition of an elliptic curve.

DEFINITION 11.17 *A* (rational) elliptic curve *is the group consisting of the set of solutions to a nonsingular cubic curve of the form $y^2 = f(x)$ together with the point at infinity under the addition operation described above.*

Strictly speaking, we should consider the set of all solutions in which the variables are allowed to take on complex values. However, from Theorem 11.6

or from our addition formulas, we see that the addition operation is closed over the set of rational solutions. Thus, the subset of points with rational coordinates is itself a group (i.e., a subgroup).

Reworking Example 11.16 from a more general perspective, we can obtain formulas for the addition of two points. Let $P = (x_1, y_1)$ and $Q = (x_2, y_2)$ be two points on the elliptic curve

$$y^2 = x^3 + b_2x^2 + b_1x + b_0.$$

Assume that $x_1 \neq x_2$, and let $m = (y_2 - y_1)/(x_2 - x_1)$. Then the equation of the line through P and Q is $y = mx - mx_1 + y_1$. The equation determining the intersection of this line and the curve is

$$x^3 + b_2x^2 + b_1x + b_0 - (mx - mx_1 + y_1)^2 = 0. \tag{11.7}$$

If (x_3, y_3) is the third point of intersection of the line and the curve, then the polynomial on the left side of (11.7) factors as

$$x^3 + b_2x^2 + b_1x + b_0 - (mx - mx_1 + y_1)^2 = (x - x_1)(x - x_2)(x - x_3). \tag{11.8}$$

Equating coefficients of x^2, we obtain

$$b_2 - m^2 = -x_1 - x_2 - x_3.$$

Solving for x_3 we get

$$x_3 = m^2 - b_2 - x_1 - x_2. \tag{11.9}$$

Substituting this expression into the equation for the line, we also arrive at a formula for y_3 in terms of P and Q. We summarize our calculations in the following theorem.

THEOREM 11.18

Let $P = (x_1, y_1)$ and $Q = (x_2, y_2)$ be two points on the elliptic curve

$$y^2 = x^3 + b_2x^2 + b_1x + b_0,$$

with $x_1 \neq x_2$. Then $P + Q = (x_3, -y_3)$, where

$$x_3 = \left(\frac{y_2 - y_1}{x_2 - x_1}\right)^2 - b_2 - x_1 - x_2, \tag{11.10}$$

$$y_3 = \left(\frac{y_2 - y_1}{x_2 - x_1}\right)(x_3 - x_1) + y_1. \tag{11.11}$$

We can also derive a formula for the addition of a point to itself.

THEOREM 11.19

Let $P = (x_1, y_1)$ be a point on the elliptic curve

$$y^2 = x^3 + b_2x^2 + b_1x + b_0.$$

If $y_1 \neq 0$, *then* $2P = (x_3, -y_3)$, *where*

$$x_3 = \left(\frac{3x_1^2 + 2b_2x_1 + b_1}{2y_1}\right)^2 - b_2 - 2x_1, \qquad (11.12)$$

$$y_3 = \left(\frac{3x_1^2 + 2b_2x_1 + b_1}{2y_1}\right)(x_3 - x_1) + y_1. \qquad (11.13)$$

If $y_1 = 0$, *then* $2P = \mathcal{O}$.

The proof of this theorem is similar to that of Theorem 11.18 and is left as an exercise.

With these formulas in hand, it is easy to program *Mathematica* and *Maple* to perform elliptic curve addition computations. Notice from the formulas that the b_0 term is not needed to perform the calculation.

Mathematica

```
ecadd[{x1_, y1_}, {x2_, y2_}, b1_, b2_]:=
  If[x1 == x2,
    If[y1 == -y2,
      "Identity",
      {x3 = ((3x1^2 + 2b2 x1 + b1)/(2y1))^2 - b2 - 2x1,
        y3 = -((3x1^2 + 2b2 x1 + b1)/(2y1)(x3 - x1) + y1)}],
    {x3 = ((y2 - y1)/(x2 - x1))^2 - b2 - x1 - x2,
      y3 = -(((y2 - y1)/(x2 - x1))(x3 - x1) + y1)}]
```

If the x-coordinates of the two points are the same and the y-coordinates are additive inverses, then the command returns the string "Identity." If the x-coordinates of the two points are the same, but the y-coordinates are not additive inverses (i.e., the two points are really the same point), then the command uses the duplication formula from Theorem 11.19. If the x-coordinates of the two points are not the same, then the command uses the formula in Theorem 11.18.

Returning to the curve $y^2 = x^3 + 17$ from Example 11.16 (and noting that $b_1 = b_2 = 0$), we use our `ecadd` to perform the sum of the points $(-1, 4)$ and $(2, 5)$.

Mathematica

```
ecadd[{-1, 4},{2,5},0,0]

{-8/9,-109/27}
```

(This agrees with our calculation in Example 11.16.) Now we try this command to determine $2P$, where P is the point $(2,-3)$, on the curve $y^2 = x^3 - 2x^2 + x + 7$.

Mathematica

```
ecadd[{2,-3},{2,-3},1,-2]

{-47/36,53/216}
```

Finally, we use the command to add $(2,-3)$ to its inverse, $(2,3)$, on the same curve.

Mathematica

```
ecadd[{2,-3},{2,3},1,-2]

Identity
```

Here is the corresponding *Maple* code for the `ecadd` command.

Maple

```
ecadd:=proc(A,B,b1,b2)
  if A[1]=B[1]
    then
      if A[2]=-B[2] then "Identity";
        else [((3*A[1]^2+2*b2*A[1]+b1)/(2*A[2]))^2-b2-2*A[1],
        -(((3*A[1]^2+2*b2*A[1]+b1)/(2*A[2]))
        *(((3*A[1]^2+2*b2*A[1]+b1)/(2*A[2]))^2
        -b2-3*A[1])+A[2])];
      end if;
    else
      [((B[2]-A[2])/(B[1]-A[1]))^2-b2-A[1]-B[1],
      -(((B[2]-A[2])/(B[1]-A[1]))
      *(((B[2]-A[2])/(B[1]-A[1]))^2
      quad-b2-2*A[1]-B[1])+A[2])];
  end if;
end proc:
```

Here are the corresponding *Maple* calculations.

Maple

```
ecadd([-1,4],[2,5],0,0);
[-8/9, -109/27]

ecadd([2,-3],[2,-3],1,-2);
[-47/36, 53/216]

ecadd([2,-3],[2,3],1,-2);
"Identity"
```

Exercises

11.13 Let E be the elliptic curve $y^2 = x^3 + 8x - 8$, and let $P = (1, -1)$ and $Q = (2, 4)$ be points on E.

(a) Compute $P + Q$.

(b) Compute $2P$.

11.14 Let E be the elliptic curve $y^2 = x^3 + 17$, and let $P = (2, 5)$, $Q = (-1, -4)$, and $R = (8, 23)$ be points on E.

(a) Compute $P + Q$.

(b) Compute $2P$.

(c) Compute $P + Q + R$.

†11.15 Prove Theorem 11.19.

11.16 Determine the order of the point $P = (1, 2)$ on the elliptic curve

$$y^2 = x^3 - 2x^2 + 5x.$$

That is, find the smallest positive value of k such that $kP = \mathcal{O}$.

†11.17 Show that an elliptic curve has at most eight points of order 3.

Hint: If P is a point of order 3, then $2P = -P$.

11.4 Sums of two cubes

We turn our attention to a Diophantine equation problem suggested by the quotation at the beginning of the chapter. In the story told by Hardy, Ramanujan explains that in fact the number 1729 is interesting as it is the smallest number that can be expressed as the sum of two (positive) cubes in two different ways. We can write

$$1729 = 1^3 + 12^3 = 9^3 + 10^3.$$

This suggests the following question: Given a positive integer k, does there exist a number n that can be written as the sum of two cubes in k different ways? If so, how can one find such a number n? The theory of elliptic curves allows us to quickly answer the first question in the affirmative and simultaneously provide a procedure for finding such a value of n.

Our goal is to find a positive number n for which the Diophantine equation

$$X^3 + Y^3 = n \tag{11.14}$$

has k different solutions. (We count the solution $X = a$, $Y = b$ to be the same as the solution $X = b$, $Y = a$.) Recasting the problem, we wish to find a value of n for which the cubic curve $F(X,Y) = X^3 + Y^3 - n = 0$ has an appropriate number of points with integers coordinates. This cubic curve does not quite have the form of the curves we have studied in the previous sections. However, we can make a change of variables in equation (11.14) to arrive at an equation of the form $y^2 = f(x)$. We set

$$X = \frac{36n + y}{6x} \tag{11.15}$$

$$Y = \frac{36n - y}{6x}. \tag{11.16}$$

Then (11.14) becomes

$$\left(\frac{36n + y}{6x}\right)^3 + \left(\frac{36n - y}{6x}\right)^3 = n.$$

After expanding and simplifying, the above becomes

$$\frac{432n^3}{x^3} + \frac{ny^2}{x^3} = n.$$

Now multiplying through by x^3/n and rearranging, we arrive at the equation

$$y^2 = x^3 - 432n^2. \tag{11.17}$$

Given a point (x, y) on the curve $y^2 = x^3 - 432n^2$, with $x \neq 0$, the formulas (11.15) and (11.16) provide us with a point on the curve $X^3 + Y^3 = n$. Thus, we have a map from a subset of the points on $y^2 = x^3 - 432n^2$ to the set of points on $X^3 + Y^3 = n$. While this map does not necessarily send points with integer coordinates to points with integer coordinates, it will at least send points with rational coordinates to points with rational coordinates. We claim that this map is one-to-one. To see this, we compute the inverse of the map. Adding the equations (11.15) and (11.16), we obtain $X + Y = 12n/x$, and so

$$x = \frac{12n}{X+Y}. \tag{11.18}$$

Rearranging (11.16) and substituting for x using (11.18), we obtain

$$y = \frac{36n(X-Y)}{X+Y}. \tag{11.19}$$

We now consider how to apply the theory from the earlier sections in the chapter to solve our problem. We wish to find a value of n such that equation (11.17) has k different integer solutions. If we can at least manage to find an n for which the equation has many rational solutions, we can, in some sense, clear the denominators of our list of solutions and obtain a new value of n for which the equation will actually have integer solutions. We will illustrate with an example.

Example 11.20 We will look for a number that can be written as the sum of two cubes in at least three different ways. We begin with consideration of the equation $X^3 + Y^3 = 35$. We note that $(3, 2)$ is a point on this curve. The change of variables above leads us to the elliptic curve

$$E : y^2 = x^3 - 529200.$$

Equations (11.18) and (11.19) tell us that the point $(3, 2)$ corresponds to the point $P = (84, 252)$ on E. We can use the group law to obtain two new points on E. For instance, using Theorem 11.19, we find

$$2P = (1596, -63756).$$

Adding P and $2P$ using Theorem 11.18, we have

$$3P = \left(\frac{1009}{9}, \frac{25327}{27}\right).$$

Using the formulas (11.15) and (11.16), we find that the points $2P$ and $3P$ correspond to the points $(-124/19, 129/19)$ and $(59347/18162, 8693/18162)$ on $X^3 + Y^3 = 35$. Consider the following three equations:

$$\begin{aligned} 3^3 + 2^3 &= 35 \\ (-124/19)^3 + (129/19)^3 &= 35 \\ (59347/18162)^3 + (8693/18162)^3 &= 35. \end{aligned}$$

Multiplying each of the above by $(19 \cdot 18162)^3$, we obtain three integer equations and hence three points with integer equations on the curve $X^3 + Y^3 =$

$35(19 \cdot 18162)^3 = 1438201910159509320$. Thus, we have discovered an integer that can be written as a sum of cubes in three different ways:

$$\begin{aligned} 1438201910159509320 &= 1035234^3 + 690156^3 \\ &= (-2252088)^3 + 2342898^3 \\ &= 1127593^3 + 165167^3. \end{aligned}$$

□

We can generalize the procedure above to obtain an integer that can be written as the sum of two cubes in k different ways. First, choose an integer n and a point (X_1, Y_1) with rational coordinates on $X^3 + Y^3 = n$. Using the formulas (11.18) and (11.19), convert this point to a point P on the curve $y^2 = x^3 - 432n^2$. Now, compute $2P$, $3P$, ..., kP and convert these points to points (X_2, Y_2), (X_3, Y_3), ..., (X_k, Y_k) on $X^3 + Y^3 = n$ using the formulas (11.15) and (11.16). These points have rational coordinates, and so we can let d be the least common multiple of the denominators of these coordinates. The points (dX_1, dY_1), ..., (dX_k, dY_k) all have integer coordinates and lie on the curve $X^3 + Y^3 = nd^3$.

We have produced k ways to write nd^3 as a sum of two cubes. We must note, however, that our procedure does not guarantee that all of the points we obtain are different. If P turns out to be a point of finite order, then our list may have repetition. Fortunately, points of finite order with rational coordinates are rare. In fact, with some additional work, one can prove (the Nagell–Lutz theorem) that rational points of finite order on an elliptic curve must actually have integer coordinates. For instance, in our example above, the point $3P$ is rational but not integral, and so $3P$ (and hence P) is a point of infinite order. We should also point out that, as in our example, the coordinates of the points we obtain are not necessarily all positive. Again, with some additional effort one can prove (assuming that P has infinite order) that the list of points (X_1, Y_1), (X_2, Y_2), ... contains an infinite number of points with positive coordinates. Thus, if we extend our list of points, if necessary, beyond (X_k, Y_k), we are guaranteed to eventually find k points with integer coordinates.

Exercises

◇**11.18** Find the smallest positive integer that can be written as a sum of two cubes of integers (not necessarily positive) in two different ways.

◇**11.19** Find an integer n that can be written as a sum of two cubes in four different ways.

◇**11.20** Find an integer n that can be written as a sum of two *positive* cubes in four different ways.

11.5 Elliptic curves mod p

If $f(x)$ is a cubic polynomial with integer coefficients and p is a prime number, then we may investigate solutions to the congruence

$$y^2 \equiv f(x) \pmod{p}. \tag{11.20}$$

Under suitable hypotheses, this set of solutions will also form a group. In fact, by regarding equations as congruences mod p, all of the theorems from Section 11.2 have a valid interpretation by appropriately replacing $\mathbf{Q}$ with $\mathbf{Z}_p$. As a result, we obtain a group law on the set of solutions to (11.20) when the discriminant of $f(x)$ is not divisible by p (i.e., the discriminant is nonzero when reduced mod p).

DEFINITION 11.21 *Let p be an odd prime number and let $f(x)$ be a cubic polynomial with integer coefficients and discriminant not divisible by p. An* elliptic curve over $\mathbf{Z}_p$ *is the set of solutions to the congruence of the form*

$$y^2 \equiv f(x) \pmod{p}$$

together with a point at infinity under the group law described in Section 11.3.

When the context is clear, we may drop the congruence notation and refer to the curve as being given by an equation of the form $y^2 = f(x)$. Since we have assumed that $p \neq 2$, the formulas for addition given by Theorems 11.18 and 11.19 (again regarded as congruences mod p) apply.

Example 11.22 Consider the curve $y^2 = x^3 + 17$ now over $\mathbf{Z}_{19}$. The x coordinate of the sum of $(-1, 4)$ and $(2, 5)$ is given by

$$x_3 = (5-4)^2((2-(-1))^{-1})^2 - 0 - (-1) - 2 = (3^{-1})^2 - 1.$$

Since $3 \cdot 13 \equiv 39 \equiv 1 \pmod{19}$, the inverse of 3 is 13, and so $x_3 \equiv 13^2 - 1 \equiv 16 \pmod{19}$. We also have

$$-y_3 = -((5-4)(1-(-2))^{-1}(16-(-1))+4) \equiv -((13)(17)+4) \equiv 3 \pmod{19}.$$

Thus, $(-1, 4) + (2, 5) = (16, 3)$. □

Note: One can easily modify the `ecadd` command given in Section 11.3 to work over $\mathbf{Z}_p$. An example of such a function is given in the Notes at the end of the chapter.

One of the advantages of working with elliptic curves over $\mathbf{Z}_p$ is that we now have a finite group to deal with. This proves to be useful for making applications, as we shall see in Sections 11.6 and 11.7.

Example 11.23 Let's determine the set of points on the curve $y^2 = x^3 + 1$ over $\mathbf{Z}_{11}$. First, we have the point at infinity. To determine the points with coordinates in $\mathbf{Z}_{11}$ we will evaluate $x^3 + 1$ for all eleven possible values of x. In *Mathematica*, we can use the command

```
Table[{x, Mod[x^3 + 1, 11]}, {x, 0, 10}] // TableForm
```

In *Maple*, we can use the command

```
for x from 0 to 10 do
  [x,modp(x^3+1,11)]
end do;
```

The results of this calculation are given in the following table.

x	$(x^3+1) \bmod 11$
0	1
1	2
2	9
3	6
4	10
5	5
6	8
7	3
8	7
9	4
10	0

To determine which x values occur as the first coordinate of a point on the curve, we need to know what the squares mod 11 are. We compute these by hand.

$$\begin{aligned} 0^2 &\equiv 0 \pmod{11} \\ 1^2 \equiv 10^2 &\equiv 1 \\ 2^2 \equiv 9^2 &\equiv 4 \\ 3^2 \equiv 8^2 &\equiv 9 \\ 4^2 \equiv 7^2 &\equiv 5 \\ 5^2 \equiv 6^2 &\equiv 3 \end{aligned}$$

Thus, we see that $x^3 + 1$ is a square mod 11 when $x = 0, 2, 5, 7, 9$, or 10, and the corresponding points on the curve are

$$(0,1), (0,10), (2,3), (2,8), (5,4), (5,7), (7,5), (7,6), (9,2), (9,9), (10,0).$$

Adding in the point at infinity, we see that there are twelve points on the curve. □

In general, we can count the points on an elliptic curve $y^2 = f(x)$ over $\mathbf{Z}_p$ by computing $f(x)$ for all x in $\mathbf{Z}_p$. If $f(x_0) \equiv 0 \pmod{p}$, then we get the point $(x_0, 0)$ on the curve. If $f(x_0)$ is a nonzero quadratic residue mod p, then we get two points on the curve, (x_0, y_0) and $(x_0, -y_0)$, where $y_0^2 \equiv f(x_0) \pmod{p}$. If $f(x_0)$ is not a quadratic residue mod p, then there is no point on the curve with x-coordinate x_0. The following *Mathematica* function implements this method to count points on a curve $y^2 = x^3 + b_2x^2 + b_1x + b_0$ over $\mathbf{Z}_p$.

Mathematica

```
countpointsmodp[b0_, b1_, b2_, p_] :=
  Module[{c},
    c = 1;
    Do[c = c + 1 + JacobiSymbol[t^3 + b2 t^2 + b1 t + b0, p],
      {t, 0, p - 1}];
    Return[c]]

countpointsmodp[1, 0, 0, 11]

12
```

Here are the corresponding *Maple* commands.

Maple

```
countpointsmodp:=proc(b0,b1,b2,p)
  local c, t;
  c:=1;
  for t from 0 to p-1 do
    c:=c+1+numtheory[legendre](t^3+b2*t^2+b1*t+b0,p);
    end do;
  c;
end proc:

countpointsmodp(1,0,0,11);

12
```

Notice that when we were calculating values of $x^3 + 1 \mod 11$ for all x in $\mathbf{Z}_{11}$, we got back all eleven residues. Of the ten nonzero residues, exactly half were quadratic residues, and in each of these cases we obtained two points on the curve. In general we have no reason to expect that for an arbitrary cubic polynomial $f(x)$ the complete set values of $f(x) \mod p$ is exactly $\mathbf{Z}_p$. However, we might expect that the values of $f(x)$ are quadratic residues about half the time. In fact, if we estimate that $(p-1)/2$ elements of $\mathbf{Z}_p$ produce nonzero quadratic residues mod p and $f(x) \equiv 0 \pmod{p}$ for one value of x, then the total number of points should be about

$$2\left(\frac{p-1}{2}\right) + 1 + 1 = p + 1.$$

Note that we have included the point at infinity here. If $p \equiv 2 \bmod 3$, then the curve $y^2 = x^3 + 1$ does indeed have $p + 1$ points (Exercise 11.24). However, this is not the case if $p \equiv 1 \pmod{3}$, nor is it the case in general for an arbitrary elliptic curve over $\mathbf{Z}_p$. A theorem of Hasse tells us that our estimate cannot be too far off.

THEOREM (Hasse)

The number of points N on an elliptic curve over $\mathbf{Z}_p$ satisfies

$$|N - (p+1)| < 2\sqrt{p}.$$

Example 11.24 Consider the elliptic curve $y^2 = x^3 + x + 1$ over $\mathbf{Z}_{541}$. If N is the number of points on the curve, then Hasse's theorem tells us that

$$-2\sqrt{541} < N - 542 < 2\sqrt{541}.$$

Since $\sqrt{541} = 23.2\ldots$, we can rewrite this as

$$495 < N < 589.$$

To check, we count the number of points using our *Mathematica* and *Maple* procedures.

Mathematica

```
countpointsmodp[1, 1, 0, 541]

531
```

Maple

```
countpointsmodp(1,1,0,541);

531
```

□

Exercises

11.21 Let E be the elliptic curve $y^2 = x^3 + 1$, defined over $\mathbf{Z}_5$.

(a) List the points on E.

(b) Compute $P + Q$ if $P = (2, 3)$ and $Q = (4, 0)$.

(c) Compute $2P$ if $P = (2, 3)$.

11.22 Let E be the curve $y^2 = x^3 + 4$ over $\mathbf{Z}_{13}$.

(a) List the points on E.

(b) Let $P = (2, 5)$. Determine $2P$.

(c) Let Q be any point on E. Determine $22Q$.
Hint: Use Lagrange's Theorem.

(d) Let $R = (12, 9)$. Determine $11R$.

$\diamond$**11.23** Let E be the curve $y^2 = x^3 + 3x + 7$ defined over $\mathbf{Z}_{229}$.

(a) Verify that E is an elliptic curve by computing the discriminant of $x^3 + 3x + 7 \pmod{229}$.

(b) Use Hasse's theorem to find upper and lower bounds on the number of points on E.

(c) Use a computer to find the exact number of points on E.

11.24 Let p be a prime number satisfying $p \equiv 2 \pmod 3$. Show that the number of points on the elliptic curve $y^2 = x^3 + 1$ over $\mathbf{Z}_p$ is $p + 1$.

11.6 Encryption via elliptic curves

In this section we will give an elliptic curve analog for the Massey–Omura public key exchange presented in Section 4.3. Recall that the basis for the security in the original version of the Massey–Omura exchange is that given a, b, and p, it is not feasible to solve the congruence

$$a^k \equiv b \pmod p$$

for k in a reasonable amount of time. One can hope to extend this idea to any group where it is easy to compute kth powers (or kth multiples in the case of an additive group) of an element, but difficult to solve the corresponding "logarithm" problem. In 1985 Neal Koblitz and Victor S. Miller introduced such a scheme using elliptic curves. Given a point P on an elliptic curve, the formulas from Section 11.3 allow us to easily compute kP for any positive integer k. On the other hand, given points P and Q it is generally very difficult to solve the equation $kP = Q$ for k. The elliptic curve encryption scheme we describe in this section may have a practical advantage over the Massey–Omura and RSA algorithms as it appears one can work with smaller numbers to achieve similar levels of security.

Let E be an elliptic curve over $\mathbf{Z}_p$. In practice, p will be a large prime number, and so by Hasse's theorem, the number of points on the curve will also be large. The first step in the encryption is to convert our "message," m (which we assume to be a number), into a point on E. There are several ways to associate m with a point on E, but most of them are probabilistic. That is, the rule for associating m with a point of E works almost all of the time. In the unlikely event our method fails, we can switch to a different elliptic curve or different prime and try again. We give an example of one such method using an elliptic curve over $\mathbf{Z}_p$, where $p \equiv 3 \pmod 4$. Suppose that E is given by the equation $y^2 = f(x)$. Choose an integer x_0 such that $f(x_0)$ is a quadratic residue mod p, and $1000m < x_0 < 1000(m+1)$. (Such an x_0 might not exist, but the likelihood of 999 consecutive values of x_0 failing to produce such a quadratic residue is extremely small.) Then set $y_0 = f(x_0)^{(p+1)/4}$ and $P = (x_0, y_0)$. We claim that P is a point on E. We compute

$$y_0^2 = (f(x_0))^{(p-1)/2} f(x_0). \tag{11.21}$$

By Euler's criterion (Theorem 5.5),

$$f(x_0)^{(p-1)/2} \equiv \left(\frac{f(x_0)}{p}\right) \pmod p. \tag{11.22}$$

We chose x_0 so that $f(x_0)$ is a square mod p, and so that the right hand side of the above equation is just 1. Hence, $y_0^2 = f(x_0)$, and P is in fact a point on our curve. Notice that we recover m from P by simply deleting the last three digits of x_0.

The curve E and prime p are publicly known as is the procedure for converting messages to points on E. The number of points, n, on the curve is also made publicly known. The sender first chooses a positive number a smaller than n that satisfies $\gcd(a, n) = 1$. The sender then computes and transmits aP to the receiver. The receiver also selects a positive number b smaller than n that satisfies $\gcd(b, n) = 1$ and replies to the first transmission with $b(aP)$. Now the sender solves the congruence

$$a'a \equiv 1 \pmod n$$

and transmits $a'(baP)$. Notice that this last point can be rewritten as $b(a'aP)$.

Since $a'a - 1$ is divisible by the order of E, Lagrange's theorem implies that $a'aP = P$. Thus, in the final transmission the receiver is actually getting bP. To decrypt the message, the receiver solves the congruence

$$b'b \equiv 1 \pmod{n}$$

and computes $b'bP = P$.

Example 11.25 We will implement the encryption procedure using the elliptic curve

$$y^2 = x^3 + 2x + 1.$$

The discriminant of the polynomial $x^3 + 2x + 1$ is -59, and so this curve will be an elliptic curve over $\mathbf{Z}_p$ for all odd primes $p \neq 59$. We will transmit the message "1234." In order to use our procedure for translating 1234 into a point on the curve, we will need a prime larger than 1234999. We will work with $p = 1377359$. Using our `countpointsmodp` command, we find that the number of points on the curve is

$$n = 1375269.$$

There are more sophisticated algorithms to count points on a curve over $\mathbf{Z}_p$ when p is large. In fact, such methods can be used to complete the calculations in a reasonable amount of time for primes p up to 100 digits long. Noting that our prime is congruent to 3 mod 4, we use the procedure described above to convert m to a point on E.

Mathematica

```
converttopoint[m_, b0_, b1_, b2_, p_] :=
  Module[{x0, f},
    x0 = 1000*m;
    f[x_] := x^3 + b2 x^2 + b1 x + b0;
    While[JacobiSymbol[f[x0], p] == -1, x0++];
Print[{x0, PowerMod[f[x0], (p + 1)/4, p]}]]

converttopoint[1234, 1, 2, 0, 1377359]

{1234005, 349433}
```

Here are the corresponding *Maple* commands.

Maple

```
convertopoint:=proc(m,b0,b1,b2,p)
  local x0, f;
  x0:=1000*m;
  f:=x->x^3+b2*x^2+b1*x+b0;
  while not numtheory[legendre](f(x0),p)= 1 do
    x0:=x0+1;
  end do;
  [x0,modp(f(x0)&^((p+1)/4),p)]
end proc:

converttopoint(1234,1,2,0,1377359);

[1234005, 349433]
```

To encrypt this point P for the first transmission, we select $a = 11111$. (Note that 11111 is indeed relatively prime to n.) To make the first transmission, we must compute aP. Using the formulas from Section 11.3, we could compute $P + P$, $2P + P$, $3P + P$, and so on until we reach $11111P$. However, we can actually save ourselves quite a bit of work by regarding 11111 as a sum of powers of 2 and making repeated use of the duplication formula. (Compare with the procedure developed for computing powers mod p in Chapter 3 Notes.) We compute

$$11111 = 2^0 + 2^1 + 2^2 + 2^5 + 2^6 + 2^8 + 2^9 + 2^{11} + 2^{13},$$

and so

$$11111P = P + 2^1P + 2^2P + 2^5P + 2^6P + 2^8P + 2^9P + 2^{11}P + 2^{13}P.$$

Thus, we can actually determine aP with fewer than the 30 calculations. Armed with the duplication and addition formulas, this is an easy task for a computer. Using *Mathematica* or *Maple*, we find $11111P = (1114312, 498654)$. (Further details on how to carry out this computation using *Mathematica* or *Maple* are contained in the Notes section.)

Next we pick a value of b for the receiver to use. Again, we note that 22222 is relatively prime to n. Using the method described above, with some help from *Mathematica* or *Maple*, we obtain $22222(11111P) = (710108, 1324551)$. Now the sender needs to determine the inverse of a mod n. We use the *Mathematica* command `PowerMod` to perform this calculation.

Mathematica

```
PowerMod[11111, -1, 1375269]

283322
```

In *Maple* we can use the `modp` command.

Maple

```
modp(1/11111,1375269);

283322
```

We find $283322(baP) = (1075576, 1307157)$. For the receiver to decrypt the message, we compute the inverse of 22222 mod n.

Mathematica

```
PowerMod[22222, -1, 1375269]

141661
```

Maple

```
modp(1/22222,1375269);

141661
```

The final computation, of $141661(283322baP)$ does in fact return us to $(1234005, 349433)$.

□

Exercises

◇**11.25** This exercise outlines an encrypted exchange using the elliptic curve encryption procedure. The computer procedures in Chapter Notes may be of use for this problem. Let E be the elliptic curve $y^2 = x^3 - x$ defined over $\mathbf{Z}_p$ where $p = 1883639$. Let $P = (809000, 1430400)$, a point on E.

(a) Determine the number of points, n on E.

(b) Let $a = 96383$ and compute aP.

(c) Let $b = 46747$ and compute $b(aP)$.

(d) Find an integer a' such that $aa' - 1$ is divisible by n, and compute $a'(b(aP))$.

(e) Find an integer b' such that $bb' - 1$ is divisible by n and check that $P = b'(a'(b(aP)))$.

◊**11.26** Let $P = (6, 1)$ and $Q = (18, 12)$ be points on the elliptic curve

$$y^2 = x^3 + 2x + 3$$

defined over $\mathbf{Z}_{23}$. For what number k is $Q = kP$?

11.7 Elliptic curve method of factorization

We return now to another example of the applications of elliptic curves that have implications for real world technology. An algorithm developed by Hendrik W. Lenstra makes use of the group structure an elliptic curve possesses to find a factorization of a given integer. Recall that encryption schemes such as RSA rely on the fact that factoring large integers is a difficult problem. The algorithm of Lenstra, while still not providing a rapid method of factorization, is currently among the most efficient factorization procedures known. The elliptic curve method of factorization is an analog of a factorization procedure known as the Pollard $p - 1$ method, which we describe below.

Suppose that we wish to factor a number m. Choose a positive integer k that is the product of small primes. For instance, choose k to be the least common multiple of the numbers 1, 2, ..., n for some integer n, or simply take $k = n!$. We also choose a positive number a less than m. We check that a and m are relatively prime using the Euclidean algorithm. If they are not, the result of the Euclidean algorithm provides a factor of m. Supposing instead that a and m are relatively prime, we compute $\gcd(a^k - 1, m)$. If this greatest common divisor is larger than 1, we have found a factor of m. If not, we choose a larger value of k and try again.

Why should we have any hope that $a^k - 1$ and m will share a common factor? Suppose that m has a factor p such that $p - 1$ is a product of small primes. If we have chosen k large enough, then $p - 1$ will divide k. From Theorem 3.18 (Fermat's little theorem), it follows that

$$a^k \equiv 1 \pmod{p}. \tag{11.23}$$

Hence, p divides $a^k - 1$, from which we conclude that the greatest common divisor of $a^k - 1$ and m is greater than 1.

If a_k is the least positive residue of a^k modulo m, then $\gcd(a^k - 1, m) = \gcd(a_k - 1, m)$. Thus, we do not actually have to compute a^k, but can employ a method such as successive squaring to compute a_k. Further, we know from Proposition 2.7 that the Euclidean algorithm will take less than $\log_\alpha m$ steps where $\alpha = (1 + \sqrt{5})/2$. Thus, the above computation can be completed relatively quickly.

Example 11.26 We implement the procedure in *Mathematica* and *Maple* to factor the number 137703491. We take $a = 2$ and $k = 100!$.

Mathematica

```
GCD[PowerMod[2, 100!, 137703491] - 1, 137703491]

17389
```

Maple

```
gcd(modp(2&^(100!),137703491)-1,137703491);

17389
```

We find that $137703491 = 17389 \cdot 7919$ and, in fact, 17389 and 7919 turn out to be prime. Naturally, we would not ordinarily expect that the procedure would work on the first try. We define a *Mathematica* function to compute $\gcd(a^{n!} - 1, m)$ until a result other than 1 occurs.

Mathematica

```
pollardfactor[m_] :=
  Module[{a, n},
    a = 2;
    n = 1;
    While[GCD[a - 1, m] == 1, a = PowerMod[a, n, m]; n++];
    Print[GCD[a - 1, m]]]
```

Testing the function out with $m = 137703491$, we find the same factor.

Mathematica

```
pollardfactor[137703491]

17389
```

Here are the corresponding commands for *Maple*.

Maple

```
pollardfactor:=proc(m)
  local a, n;
  a:=2;
  n:=1;
  while gcd(a-1,m)=1 do
    a:=modp(a&^n,m);
    n:=n+1;
  end do;
  gcd(a-1,m);
end proc:

pollardfactor(137703491);

17389
```

□

If we vary a, this procedure will likely eventually produce a factor of m whenever m is composite. However, if m does not have a factor p such that $p-1$ is a product of small primes, then the procedure is not any faster than attempting to factor m by trial division. The *Mathematica* command `FactorInteger` does actually make use of the Pollard $p-1$ method. However, the command switches back and forth among the Pollard $p-1$ method, the Pollard rho method, and the continued fraction algorithm after first removing any small prime factors. The `ifactor` command in *Maple* uses the continued fraction factoring algorithm as its base method. The command can also make use of the Pollard rho method or the elliptic curve method if additional parameters are specified.

In the notes at the end of Chapter 3, we observed that Fermat's little theorem can be viewed as a special case of Lagrange's theorem. Since Fermat's little theorem is the key to discovering the factors of m, one might replace the group $\mathbf{Z}_p^*$ with other groups and try the same thing, this time relying on Lagrange's theorem. This idea proves to be fruitful when one uses elliptic curves as the replacement group. We begin with a rational elliptic curve E, given by an equation $y^2 = f(x)$, where $f(x)$ has integer coefficients. As above, we compute a multiple, kP, of our point. Suppose that p is a prime divisor of m. Let $\bar{E}$ be the curve over $\mathbf{Z}_p$ given by $y^2 \equiv f(x) \pmod{p}$. If k is a multiple of the number of points on $\bar{E}$, then p will turn up as a factor of the denominator of the coordinates of kP (see Theorem 11.29 below). An advantage that this method offers over the Pollard $p-1$ method is that we can vary the choice of elliptic curve. Consequently, if a computation with a particular choice of E, P, and k fails to produce a factor of m, we can switch E

and P, and so we are not necessarily forced to take successively larger values of k.

We know from our work in previous sections that if one begins with a point P with integer coefficients, then the coordinates of kP may not be integers. In order to apply the procedure described above, we will need to consider such points as points on the curve $\bar{E}$. We can make sense of this provided that the coordinates of the points have denominators not divisible by p.

DEFINITION 11.27 *Let n be a positive number, and r/s and t/u rational numbers such that n is relatively prime to s and u. Then r/s is* congruent *to t/u* modulo n, *and we write $r/s \equiv t/u \pmod{n}$ if the integers ru and ts are congruent modulo n.*

We note that $r/s \equiv t/u \pmod{n}$ if and only if the numerator of $r/s - t/u$, expressed in lowest terms, is divisible by n. This relation generalizes the usual congruence relation on the integers, and one can easily show that this is an equivalence relation on the set of rational numbers whose denominators are relatively prime to n. Every equivalence class contains integers. For example, if s is relatively prime to n, then there exists an integer s' such that $ss' - 1$ is divisible by n. For such an s', we have $r/s \equiv rs' \pmod{n}$.

Example 11.28 The multiplicative inverse of $5 \pmod{7}$ is 3, and so

$$2/5 \equiv 2 \cdot 3 \equiv 6 \pmod{7}.$$

The *Mathematica* command below generalizes the built-in `Mod` command.

Mathematica

```
genmod[x_, p_] :=
  Mod[x*Denominator[x]*PowerMod[Denominator[x], -1, p], p]

genmod[2/5,7]

6
```

The built-in *Maple* command `modp` is already programmed to deal with fractions.

Maple

```
modp(2/5,7);

6
```

□

One can also show that, as with integers, this generalized congruence relation respects addition and multiplication. That is, if $x_1 \equiv x_2 \pmod{n}$ and $y_1 \equiv y_2 \pmod{n}$, then $x_1 + y_1 \equiv x_2 + y_2 \pmod{n}$ and $x_1y_1 \equiv x_2y_2 \pmod{n}$. We can therefore conclude that if p is a prime number that does not divide the denominators of x_1 and y_1, and $y_1^2 = x_1^3 + ax_1 + b$ for some integers a and b, then $y_1^2 \equiv x_1^3 + ax_1 + b \pmod{p}$. Thus, we can identify a point $P = (x_1, y_1)$ on an elliptic curve E with a point $\bar{P}$ on $\bar{E}$, provided that p does not divide the denominators of x_1 and y_1. We let $\bar{\mathcal{O}}$ be the identity on the curve modulo p.

THEOREM 11.29

Let E be an elliptic curve defined by the equation $y^2 = x^3 + ax + b$, where a and b are integers, and let p be an odd prime number that does not divide $-4a^3 - 27b^2$. Let $P = (x_1, y_1)$ and $Q = (x_2, y_2)$ be points on E with rational coordinates. Suppose that p does not divide the numerator or denominator of x_1, y_1, x_2, and y_2, and suppose that $P + Q \neq \mathcal{O}$. Then the denominator of the x-coordinate of $P + Q$ (expressed in lowest terms) is divisible by p if and only if $\bar{P} + \bar{Q} = \bar{\mathcal{O}}$ on the curve $\bar{E}$.

PROOF We first note that the condition that $-4a^3 - 27b^2$ not be divisible by p ensures that the curve $\bar{E}$ is nonsingular. Since $P + Q \neq \mathcal{O}$, either Theorem 11.18 or Theorem 11.19 provides a formula for the coordinates of $P + Q$, depending on whether $P = Q$.

Suppose that $\bar{P} + \bar{Q} = \bar{\mathcal{O}}$. Then $x_1 \equiv x_2 \pmod{p}$ and $y_1 \equiv -y_2 \pmod{p}$. If $x_1 \neq x_2$, then the formula from Theorem 11.18 applies, and so the x-coordinate of $P + Q$ is

$$\left(\frac{y_2 - y_1}{x_2 - x_1}\right)^2 - x_1 - x_2. \tag{11.24}$$

Now $y_1 - y_2 \equiv 2y_1 \pmod{p}$. Since p does not divide the numerator of y_1 and since $p \neq 2$, we see that p does not divide the numerator of $y_1 - y_2$. On the other hand, p does divide the numerator of $x_2 - x_1$ when expressed in lowest terms. We conclude that when $(y_2 - y_1)/(x_2 - x_1)$ is expressed in lowest terms, p will divide its denominator. It follows that the rational number in (11.24) will also have a denominator divisible by p.

If $x_1 = x_2$, then it follows that $y_1 = \pm y_2$. We assumed that $P + Q \neq \mathcal{O}$, and so $y_1 = y_2$. As above, $y_1 - y_2 \equiv 2y_1 \pmod{p}$. Since $p \neq 2$, we must have $y_1 \equiv 0 \pmod{p}$. However, we ruled out this possibility in the statement of the theorem, and so this case cannot occur.

We now deal with the converse, and so we suppose that p divides the denominator of the x-coordinate of $P + Q$. If $P \neq Q$, then from (11.24) above we see that p must divide the denominator of $(y_2 - y_1)/(x_2 - x_1)$. Since p

does not divide the denominator of y_1 or y_2, it cannot divide the denominator of $y_2 - y_1$. Therefore, p divides the numerator of $x_2 - x_1$, and so $x_1 \equiv x_2 \pmod{p}$. It remains to show that $y_1 \equiv -y_2 \pmod{p}$.

Since (x_1, y_1) and (x_2, y_2) are points on E,

$$y_2^2 - y_1^2 = (x_2^3 + ax_2 + b) - (x_1^3 + ax_1 + b).$$

With some algebra, we can rewrite this as

$$\frac{y_2 - y_1}{x_2 - x_1} = \frac{x_2^2 + x_1x_2 + x_1^2 + a}{y_2 + y_1}. \tag{11.25}$$

We know that the left side of (11.25), expressed in lowest terms, has a denominator divisible by p, and so the same is true of the right side. Since p does not divide the denominators of x_1 and x_2, we see that p does not divide the denominator of $x_2^2 + x_1x_2 + x_1^2 + a$. It follows that the numerator of $y_2 + y_1$ is divisible by p, and so $y_1 \equiv -y_2 \pmod{p}$.

To complete the proof of the converse, we must consider the case $P = Q$. From Theorem 11.19, the x-coordinate of $2P$ is given by

$$\frac{(3x_1^2 + a)^2 - 8x_1y_1^2}{4y_1^2}. \tag{11.26}$$

Since p does not divide the denominators of x_1 and y_1, the denominator of $(3x_1^2 + a)^2 - 8x_1y_1^2$ is also not divisible by p. We also assumed that p does not divide the numerator of y_1 and hence does not divide the numerator of $2y_1$. Thus, p does not divide the denominator of (11.26), and so this case does not occur.

$\square$

We observe that, under the hypotheses of the theorem, when p divides the denominator of the x-coordinate of $P + Q$, it will also divide the denominator of the y-coordinate of $P + Q$.

To implement the algorithm, we first select a family of elliptic curves and a point on each curve. The set of curves of the form

$$y^2 = x^3 + ax - a$$

provides a nice family of curves as $P = (1, 1)$ is a point on each such curve for all a. The discriminant of $y^2 = x^3 + ax - a$ is $-4a^3 - 27a^2$, and so the curve is nonsingular for all integers $a \neq 0$.

Next, we select a value of k, as in the Pollard $p - 1$ method, that is the product of small primes. Before computing kP, we must check that the curve is nonsingular over $\mathbf{Z}_p$ for all p dividing m, the number that we are attempting to factor. This amounts to computing $\gcd(-4a^3 - 27a^2, m)$. If this greatest common divisor is not 1, then we have found a factor of m (which hopefully will be different from m).

Computation of kP involves repeated use of the addition and duplication formulas. For example, to compute $6P$, we might compute $2P$, $2(2P)$, and finally $2P + 2(2P)$. Ultimately, we are not so concerned with the actual coordinates of kP, but rather whether the denominators are divisible by a prime divisor p of m. Thus, we could reduce the coordinates mod p after each step. Of course, we do not know the value of such a p. However, if we reduce the coordinate mod m, then the resulting coordinates will be congruent to the original ones mod p for all prime divisors p of m. Recall that if r/s is a rational number whose denominator is relatively prime to m, then $r/s \equiv r'$ (mod m) for some integer r' satisfying $0 \leq r' < m$. (Note that if at some point we encounter a rational number that cannot be reduced, we will also have found a factor of m.) This reduction is computationally advantageous as the numerators and denominators grow very quickly as one computes increasingly larger multiples of P.

Example 11.30 We use the algorithm to factor the integer 137703491 from Example 11.26. Setting $k = 10!$, we compute kP with $P = (1,1)$ on the curve $y^2 = x^3 + ax - a$ for a variety of choices of a. In the case $a = 1$, we find $\gcd(-31, 137703491) = 1$. One can implement a procedure using *Mathematica* or *Maple* that will compute $2^n P$ for all n such that $2^n \leq k$, and then sum the appropriate subset of these to get kP, reducing each computation modulo m (see Chapter Notes for a procedure to perform this calculation). We can write the procedure in such a way that it will terminate if a factor of m is encountered. For the case $a = 1$, no such factor of m is discovered. We repeat the procedure for additional values of a. Again for $a = 2$, no factor of m is encountered. This continues until we try $a = 17$, at which point the procedure produces the factor 17389. □

Exercises

◇**11.27** Show that the Pollard $p - 1$ method produces a factor of $m = 187$ when $a = 2$ and $k = 4!$.

11.28 Let $m = 604801 \cdot 5570429 = 3369001029629$. Note that 604801 is prime and

$$604800 = 2^7 \cdot 3^3 \cdot 5^2 \cdot 7.$$

Explain why $\gcd(2^k - 1, m) > 1$ with $k = 10!$.

◇**11.29** Implement the elliptic curve method to factor 249853603488473. (You may wish to use the *Mathematica* or *Maple* procedure in the Chapter Notes.)

11.8 Fermat's Last Theorem

Perhaps the most dramatic application of the theory of elliptic curves is the role it played in the proof of Fermat's Last Theorem. In 1985 Gerhard Frey suggested that a nonzero solution to the equation

$$x^p + y^p = z^p, \tag{11.27}$$

where p is an odd prime, would enable one to construct an elliptic curve that is not "modular." A well-known conjecture (the Taniyama–Shimura conjecture) stated that all rational elliptic curves are modular, and so such a solution to (11.27) would lead to a contradiction. Frey's conjecture was proved several years later, and so a proof of the Taniyama–Shimura conjecture would also carry with it a proof of Fermat's Last Theorem as a corollary. This breakthrough inspired Andrew Wiles to begin an attempt to prove the conjecture. Working largely in secrecy for nearly a decade, Wiles achieved success and put an end to a three-and-a-half-century quest to prove Fermat's Last Theorem.

A proper study of the proof of the Taniyama–Shimura conjecture would carry many years of graduate level reading as a prerequisite. In the next few pages, we will provide a brief introduction to some of the ideas that are a part of the conjecture.

A *modular function of weight k and level N* is a complex function $f(z)$ defined for $\mathrm{Im} z > 0$ that satisfies

$$f\left(\frac{az+b}{cz+d}\right) = (cz+d)^k f(z), \tag{11.28}$$

for all integers a, b, c, and d such that $ad - bc = 1$ and $N \mid c$. In particular, taking $a = b = d = 1$ and $c = 0$, all modular functions satisfy

$$f(z+1) = f(z). \tag{11.29}$$

As a result of this identity, a modular function has a Fourier expansion of the form

$$f(z) = \sum_{n=-\infty}^{\infty} c_n e^{2\pi i n z}. \tag{11.30}$$

A modular function is also required to be *meromorphic*, which means that the Fourier expansion above has only a finite number of nonzero c_n for negative values of n. For simplicity, this series expansion of a modular function is often written as

$$f(z) = \sum_{n=-m}^{\infty} c_n q^n, \tag{11.31}$$

where $q = e^{2\pi iz}$. Now the expansion is a Laurent series in the variable q. If $m = 0$ (i.e., the series is a Taylor series), we say that the modular function is a *modular form*.

The Taniyama–Shimura conjecture describes a connection between modular forms and elliptic curves. To investigate this correspondence, we consider an example. The function

$$f(z) = q\prod_{n=1}^{\infty}(1-q^{4n})^2(1-q^{8n})^2 \tag{11.32}$$

is the modular form that corresponds to the elliptic curve $y^2 = x^3 - x$. For primes p, the coefficients c_p of the Fourier series expansion of $f(z)$ are related to the number of points on $y^2 = x^3 - x$ considered as a curve over $\mathbf{Z}_p$.

Let's compute the first few terms of the series expansion of this modular form. Note that $c_1 = 1$. To compute $c_2, c_3, \ldots, c_{10}$, we only need to expand

$$f(z) = q\prod_{n=1}^{2}(1-q^{4n})^2(1-q^{8n})^2,$$

since for $n > 2$, all factors in the product will contribute only to coefficients of degree 13 or higher. We enlist some help from *Mathematica* and *Maple* to perform the algebra.

Mathematica

```
Expand[q*Product[(1 - q^(4n))^2(1 - q^(8n))^2, {n, 1, 2}]]

q - 2 q^5 - 3 q^9 + 8 q^13 - 8 q^21 + 8 q^25 - 8 q^29 -
  6 q^33 + 20 q^37 - 6 q^41 - 8 q^45 + 8 q^49 - 8 q^53 +
  8 q^61 - 3 q^65 - 2 q^69 + q^73
```

Maple

```
sort(expand(q*product((1-q^(4*n))^2*(1-q^(8*n))^2,
  n = 1..2)),q,ascending);

q - 2*q^5 - 3*q^9 + 8*q^13 - 8*q^21 + 8*q^25 - 8*q^29 -
  6*q^33 + 20*q^37 - 6*q^41 - 8*q^45 + 8*q^49 - 8*q^53 +
  8*q^61 - 3*q^65 - 2*q^69 + q^73
```

Since we know that the coefficients of terms of degree 13 or higher are probably not correct, we don't really need to see them. To get only the terms of degree 10 or less, we can use built-in series commands.

Mathematica

```
Series[q*Product[(1 - q^(4n))^2(1 - q^(8n))^2, {n, 1, 2}],
  q, 1, 10]

q - 2q^5 - 3q^9 + O[q]^11
```

The `O[q]^11` at the end of the calculation indicates that the remaining terms in the series are of degree 11 or greater. The `series` command in *Maple* is similar.

Maple

```
series(q*product((1-q^(4*n))^2*(1-q^(8*n))^2,n = 1..2),
  q = 0,10);

q - 2*q^5 - 3*q^9 + O(q^13)
```

In the forthcoming example, we will actually need a few more terms. To get the first 30 terms of the series expansions, we expand the product up to $n = 7$.

Mathematica

```
Series[q*Product[(1 - q^(4n))^2(1 - q^(8n))^2, {n, 1, 7}],
  q, 0, 30]

q - 2q^5 - 3q^9 + 6q^13 + 2q^17 - q^25 - 10q^29 + O[q]^31
```

Maple

```
series(q*product((1-q^(4*n))^2*(1-q^(8*n))^2,n = 1..7),
  q = 0,30);

q - 2*q^5 - 3*q^9 + 6*q^13 + 2*q^17 - q^25 -
  10*q^29 + O(q^33)
```

We now turn to counting the number of points on $y^2 = x^3 - x$ over $\mathbf{Z}_p$ for the first several primes. Let N_p denote the number of points on the curve considered as an elliptic curve over $\mathbf{Z}_p$. For $p = 2$, we have the points $(0, 0)$, $(1, 1)$, and the point at infinity, and so $N_2 = 3$. Using our previously defined `countpointsmodp` function in *Mathematica* or *Maple*, we can easily compute N_p for the first few odd primes. The results are given in Table 11.1. Recall from Section 11.5 that we expect that $N_p \doteq p + 1$. For our curve, $N_p = p + 1$

when $p = 2$ or $p \equiv 3 \pmod 4$). In general, Hasse's theorem tells us that the "error" in this approximation is no more than $2\sqrt{p}$. Table 11.1 shows the values of N_p and $p + 1 - N_p$ for primes up to 29.

p	N_p	$p+1-N_p$
2	3	0
3	4	0
5	8	−2
7	8	0
11	12	0
13	8	6
17	16	2
19	20	0
23	24	0
29	40	−10

TABLE 11.1:
Values of p, N_p, and $p + 1 - N_p$.

Notice that the values of $p + 1 - N_p$ are exactly the values of c_p from our product expansion of f. With a little bit of effort, one can show that this holds true for all primes p. (See Exercise 11.32.)

In predicting that all rational elliptic curves are modular, the Taniyama–Shimura conjecture says that given any elliptic curve, one can find a modular form $f(z) = \sum_{n=0}^{\infty} c_n q^n$ such that

$$c_p = p + 1 - N_p \tag{11.33}$$

for (almost) all primes p. (The relation may not hold for primes p that divide the discriminant; recall that in this case the curve is not an elliptic curve over $\mathbf{Z}_p$.)

Returning to our example, let's see how, given the values of c_p, we can actually derive the values of c_n for all n. It can be shown that the coefficients satisfy the recursive formula

$$c_{p^{r+1}} = c_p c_{p^r} - p c_{p^{r-1}}, \tag{11.34}$$

and so we can compute c_n when n is a power of p. For example,

$$c_9 = c_3 c_3 - 3c_0 = 0 \cdot 0 - 3 \cdot 1 = -3, \text{ and}$$
$$c_{27} = c_3 c_9 - 3c_3 = 0 \cdot 0 - 3 \cdot 0 = 0,$$

which is in agreement with our calculations above. For relatively prime numbers m and n,

$$c_{mn} = c_m c_n. \tag{11.35}$$

This predicts, for our modular form,

$$c_{65} = c_5 c_{13} = (-2)(6) = -12.$$

Let's check this by expanding the product out further. We can ask *Mathematica* or *Maple* for just the coefficient of q^{65}.

Mathematica

```
Coefficient[Expand[q*Product[(1 - q^(4n))^2(1 - q^(8n))^2,
  {n, 1, 16}]], q^65]

-12
```

Maple

```
coeff(expand(q*product((1-q^(4*n))^2*(1-q^(8*n))^2,
  n=1..16)),q^65);

-12
```

Formulas analogous to (11.34) and (11.35) exist for other modular forms as well. (See [18].)

Suppose that, for some prime p, we were able to find nonzero a, b, and c, such that

$$a^p + b^p = c^p.$$

We could then construct an elliptic curve (called a *Frey curve*) of the form

$$y^2 = x(x - a^p)(x + b^p). \tag{11.36}$$

It was this curve that Frey suggested would not be modular. In 1986 work of Jean-Pierre Serre and Kenneth Ribet confirmed that, in fact, if such a curve existed, it would not be modular. In 1995 Wiles completed most of the proof of Taniyama–Shimura conjecture. He was able to show that every "semistable" elliptic curve is modular. As it was also known that the Frey curve, if it existed, would be semistable, Wiles' work implied that no such curve exists and so completed the proof of Fermat's Last Theorem. To learn more about this fascinating story, watch Simon Singh's film *The Proof*.

The history of Fermat's Last Theorem is filled with stories of failed attempts to prove it. Numerous "proofs" of the theorem were presented to the mathematical community over the years only to be exposed as flawed due to some critical error. In fact, during the review process for Wiles' paper, a referee uncovered a gap in the argument. However, with some assistance from former student Richard Taylor, Wiles was able to provide an alternate argument that completed the proof of the Taniyama–Shimura in the semistable case. This corrected version has withstood the scrutiny of experts in the field and has even been extended by others to provide a proof of the full version of the Taniyama–Shimura conjecture.

Exercises

⋆11.30 (a) Let z be a complex number satisfying Im$z > 0$. Show that the series

$$\sum_{(m,n)\neq(0,0)} \frac{1}{(mz+n)^4}$$

converges absolutely. (Note that the sum is over all ordered pairs of integers $(m,n) \neq (0,0)$.)

(b) With Im$z > 0$, let

$$f(z) = \sum_{(m,n)\neq(0,0)} \frac{1}{(mz+n)^4},$$

and show that

$$f\left(\frac{az+b}{cz+d}\right) = (cz+d)^4 f(z),$$

for all integers a, b, c, and d such that $ad - bc = 1$.

11.31 With $q = e^{2\pi iz}$, let

$$f(z) = q\prod_{n=1}^{\infty}(1-q^n)^{24}.$$

Compute the coefficients of q, q^2, q^3, and q^4 by hand. (The function $f(z)$ is a weight 12 modular form. It is conjectured that the coefficient of q^n is nonzero for all $n \geq 1$.)

⋆11.32 Let p be a prime, and let c_p be the coefficient of q^p in the expansion of $q\prod_{n=1}^{\infty}(1-q^{4n})^2(1-q^{8n})^2$. Let N_p be the number of the points on the elliptic curve $y^2 = x^3 - x$ over $\mathbf{Z}_p$ (including the point at infinity).

(a) Show that if $p \equiv 3 \pmod 4$, then $c_p = 0$.

(b) Show that if $p \equiv 3 \pmod 4$, then $N_p = p+1$ (hence, $c_p = p+1-N_p$ for $p \equiv 3 \pmod 4$).

(c) If $p \equiv 1 \pmod 4$, then p has a representation as a sum of two squares. Make a conjecture about the values of c_p and the representation of p as a sum of two squares.

11.9 Notes

Projective space

Let K be a field and n a nonnegative integer. We define *n-dimensional projective space over K*, denoted $\mathbf{P}^n(K)$, to be the set of equivalence classes of ordered $(n+1)$-tuples in K^{n+1} with at least one nonzero coordinate, subject to the following equivalence relation. Two $(n+1)$-tuples $(x_0, x_1, \ldots, x_n)$ and $(x'_0, x'_1, \ldots, x'_n)$ are equivalent if and only if there exists a $\lambda \in K$ such that $x_i = \lambda x'_i$, for $i = 0, 1, \ldots, n$. We write $(x_0 : x_1 : \ldots : x_n)$ for the equivalence class containing $(x_0, x_1, \ldots, x_n)$. Thus, in $\mathbf{P}^2(\mathbf{C})$, for example, $(1 : 2 : 3) = (4 : 8 : 12)$. For the rest of this note, we confine our attention to the complex projective plane (i.e., $\mathbf{P}^2(\mathbf{C})$), although most of what we say holds for any field K.

Notice that in $\mathbf{P}^2(\mathbf{C})$,

$$(x_1 : y_1 : 1) = (x_2 : y_2 : 1)$$

if and only if $x_1 = x_2$ and $y_1 = y_2$. Thus, we can identify the set $\mathbf{C}^2$ of ordered pairs of complex numbers with a subset of the complex projective plane under the map

$$(x, y) \mapsto (x : y : 1).$$

We sometimes refer to $\mathbf{C}^2$ or the subset of $\mathbf{P}^2(\mathbf{C})$ corresponding to $\mathbf{C}^2$ as the *affine* complex plane. If $z_1 \neq 0$, then $(x_1 : y_1 : z_1) = (x_1/z_1 : y_1/z_1 : 1)$. Thus, the points in the complex projective plane in the complement of the affine plane have the form $(x : y : 0)$. Such points are often referred to as *points at infinity*. With the exception of the point $(1 : 0 : 0)$, all points at infinity can be written in the form $(x : 1 : 0)$. Furthermore, $(x_1 : 1 : 0) = (x_2 : 1 : 0)$ if and only if $x_1 = x_2$.

To make sense of solution sets of equations in projective space, we restrict our attention to "homogeneous" polynomials. A polynomial $f(x, y, z)$ is homogeneous of degree k if every monomial in f has degree k. For example,

$$f(x, y, z) = zy^2 - x^3 + xz^2 + z^3$$

is a homogeneous polynomial of degree 3. If $f(x, y, z)$ is a homogeneous polynomial of degree k, then

$$f(\lambda x, \lambda y, \lambda z) = \lambda^k f(x, y, z).$$

Thus, if $f(x_1, y_1, z_1) = 0$, then f also vanishes for all ordered triples in the class $(x_1 : y_1 : z_1)$. We define a projective curve to be the solution set in the projective plane to an equation $f(x, y, z) = 0$ for some homogeneous polynomial $f(x, y, z)$.

Let's consider the points on the curve

$$zy^2 - x^3 + xz^2 + z^3 = 0. \tag{11.37}$$

The point $(x_1 : y_1 : 1)$ is on the curve if and only if

$$y_1^2 = x_1^3 - x_1 - 1.$$

When $z = 0$, equation (11.37) simplifies to $x^3 = 0$, which forces $x = 0$. We conclude that $(0 : 1 : 0)$ is the only point at infinity on the curve. Thus, we can describe the points on the curve as the set consisting of the point $(0 : 1 : 0)$ together with points in the plane that correspond to points on the curve $y^2 = x^3 - x - 1$ in the affine complex plane $\mathbf{C}^2$.

More generally, given an affine curve

$$y^2 = x^3 + b_2x^2 + b_1x + b_0, \tag{11.38}$$

we can "homogenize" the equation to obtain a curve

$$y^2z = x^3 + b_2x^2z + b_1xz^2 + b_0z^3 \tag{11.39}$$

in $\mathbf{P}^2(\mathbf{C})$. The set of points on the projective curve given by (11.39) consists of the point $(0 : 1 : 0)$ together with points $(x : y : 1)$ that correspond to the points on the affine curve given by (11.38).

Given a homogeneous polynomial $f(x, y, z)$, one can define partial derivatives f_x, f_y, and f_z as before. A point $(x_0 : y_0 : z_0)$ on the curve $f(x, y, z) = 0$ is said to be *nonsingular* if not all partial derivatives of f are zero at $(x_0 : y_0 : z_0)$. A curve is said to be *nonsingular* if all points on the curve are nonsingular. We may now more properly define an elliptic curve as a nonsingular projective curve $f(x, y, z) = 0$, where $f(x, y, z)$ is a degree 3 homogeneous polynomial.

A projective line in the complex projective plane is simply the solution set to an equation of the form $g(x, y, z) = 0$, where $g(x, y, z)$ is a homogeneous polynomial of degree 1. That is, a projective line is the solution set in $\mathbf{P}^2(\mathbf{C})$ to an equation of the form

$$ax + by + cz = 0. \tag{11.40}$$

The tangent line to a projective curve $f(x, y, z) = 0$ at a point $(x_0 : y_0 : z_0)$ is the projective line given by the equation

$$f_x(x_0, y_0, z_0)x + f_y(x_0, y_0, z_0)y + f_z(x_0, y_0, z_0)z = 0. \tag{11.41}$$

One can check (Exercise 11.34) that if one "dehomogenizes" the equation for the tangent line at a point $(x_0 : y_0 : 1)$, then the usual tangent line equation for affine curves results. That is, the equation

$$f_x(x_0, y_0, 1)x + f_y(x_0, y_0, 1)y + f_z(x_0, y_0, 1) = 0 \tag{11.42}$$

is the same as the one indicated in Definition 11.13.

We can also generalize the notion of intersection multiplicity of a line and a curve at a point in projective space. Recall that to compute the intersection multiplicity in the affine case we substituted the equation for the line into the equation for the curve. If the x-coordinate (or y-coordinate in the case of a vertical line) of the point is a root of multiplicity k of the resulting equation, then we say that the point has intersection multiplicity k. A similar definition works in the projective case. We illustrate by computing the intersection multiplicity of an elliptic curve

$$y^2 z - (x^3 + b_2 x^2 z + b_1 x z^2 + b_0 z^3) = 0 \tag{11.43}$$

with the tangent line to the curve at the point $(0 : 1 : 0)$. Recall that in Section 11.3 we needed to use the fact that this intersection multiplicity is 3. We first compute the equation for the tangent line. We have

$$\begin{aligned} f_x(x, y, z) &= -3x^2 - 2b_2 xz - b_1 z^2 \\ f_y(x, y, z) &= 2yz \\ f_z(x, y, z) &= y^2 - b_2 x^2 - 2b_1 xz - 3b_0 z^2. \end{aligned}$$

Evaluating at the point $(0 : 1 : 0)$, we find that

$$\begin{aligned} f_x(0, 1, 0) &= 0 \\ f_y(0, 1, 0) &= 0 \\ f_z(0, 1, 0) &= 1. \end{aligned}$$

Thus, the equation for the tangent at $(0 : 1 : 0)$ is $z = 0$. Substituting the equation for this tangent line into equation (11.43), we get

$$-x^3 = 0.$$

As $x = 0$ is a triple root of the polynomial $-x^3$, the curve and the tangent line have a point of intersection multiplicity 3 at $(0 : 1 : 0)$.

Associativity of the group law

In this note we sketch the proof of the associativity of the group law on an elliptic curve. We will make some assumptions about our points, and we will need to quote a few theorems. The interested reader can find a complete proof in [18] that does not require much in the way of algebraic geometry prerequisites. Recall that $\mathcal{O}$ is the additive identity. Suppose that A, B, and C are points on an elliptic curve. We know that there is a unique third point of intersection on the curve and the line connecting A and B. We denote this point $A * B$. Since the x-coordinates of $A * B$ and $A + B$ are the same, the line through these two points has affine equation $x = x_0$, and so the projective equation is $x = x_0 z$. Thus, the third point of intersection between this line and the curve is the point at infinity, which means that

$$A + B = (A * B) * \mathcal{O}.$$

Further observe that

$$(A + B) + C = ((A + B) * C)) * \mathcal{O},$$
$$A + (B + C) = (A * (B + C)) * \mathcal{O}.$$

Thus, the associativity of the group follows from the theorem stated below.

THEOREM
If A, B, and C are points on an elliptic curve, then

$$(A + B) * C = A * (B + C).$$

The proof of the theorem relies greatly on counting the number of intersections of plane curves. Earlier in the chapter we saw that any line and elliptic curve in the projective plane intersect in exactly three points (counting multiplicities). The following result provides a more general formula for counting the number of intersections of projective plane curves.

THEOREM (Bezout's theorem)
Let $F(x, y, z) = 0$ and $G(x, y, z) = 0$ be curves in $\mathbf{P}^2(\mathbf{C})$, where F and G have degrees d_F and d_G, respectively. If $F(x, y, z)$ and $G(x, y, z)$ have no (non-constant) common factors, then the two curves have exactly $d_F \cdot d_G$ points in common.

Bezout's theorem remains true if one replaces $\mathbf{C}$ with any algebraically closed field. A consequence of the theorem is that any two distinct elliptic curves will intersect in $3 \cdot 3 = 9$ points. Additionally, one can prove the following result, a special case of a result known as the Cayley–Bacharach theorem.

THEOREM
Given eight distinct points $A_1, \ldots, A_8$ on an elliptic curve E, there exists a unique point A_9 such that any cubic curve that passes through $A_1, \ldots, A_8$ must also pass through A_9.

The set of all homogeneous polynomials in three variables of degree three forms a 10-dimensional vector space. (The ten dimensions correspond to the ten choices of coefficients of terms in a degree three homogeneous polynomial in three variables). The condition that a curve pass through a given point puts a linear condition on the space of coefficients. One can show that the hypothesis that all eight points lie on an elliptic curve implies that the eight linear conditions given by the eight points are independent. Thus, the space of polynomials passing through the set of points is 2-dimensional. Suppose that F and G are polynomials forming a basis for this space. By Bezout's theorem, the curves $F = 0$ and $G = 0$ intersect in a ninth point. Any cubic curve is given by a linear combination of F and G, that is,

$$\lambda_1 F + \lambda_2 G = 0,$$

and so must also pass through the ninth point of intersection of F and G. (Complete proofs of the above theorems can be found in [12].)

To apply the above special case of the Cayley–Bacharach theorem to our situation, we use the eight points

$$\mathcal{O},\ A,\ B,\ C,\ A * B,\ A + B,\ B * C,\ B + C. \tag{11.44}$$

To make use of the theorem, we must assume that the eight points are distinct. (It is possible to deduce that associativity holds in general if one knows it holds under this assumption on the points.) Let D be the ninth point provided by the theorem. We define a set of three lines as follows:

$L_1 = 0$ is the line passing through A and B,
$L_2 = 0$ is the line passing through $B + C$ and $B * C$,
$L_3 = 0$ is the line passing through $A + B$ and C.

Let $F = L_1 \cdot L_2 \cdot L_3$, and so the curve $F = 0$ is a cubic curve that, as a set, is the union of the three lines $L_1 = 0, L_2 = 0$, and $L_3 = 0$. Notice that $F = 0$ contains all eight points in the list above, as $A * B$ is on $L_1 = 0$ and $\mathcal{O}$ is on $L_2 = 0$. By the theorem, we conclude that D must also be on the curve $F = 0$. Notice that the point $(A + B) * C$ lies on $L_3 = 0$ and hence is also on the intersection of $F = 0$ and our curve E. Since Bezout's theorem implies that the two curves intersect in only nine points, we must have $(A + B) * C = D$.

Now consider a second set of lines:

$L'_1 = 0$ is the line passing through B and C,
$L'_2 = 0$ is the line passing through $A + B$ and $A * B$,
$L'_3 = 0$ is the line passing through A and $B + C$.

If we set $G = L'_1 \cdot L'_2 \cdot L'_3$, then again we obtain a cubic curve $G = 0$ that contains all eight points in (11.44), and so contains D. The point $A * (B + C)$ lies on L'_3 and hence on $G = 0$. Again by Bezout's theorem, this point must coincide with D. So we now have $(A + B) * C = D = A * (B + C)$, as desired.

Elliptic curve calculations

In this section we will define functions to carry out the computations used in Sections 11.6 and 11.7. To make it easier to work with the point at infinity, we first give a projective version of the ecadd command. This amounts to using $(0, 1, 0)$ to represent the point at infinity and adding a third coordinate of "1" for ordinary (affine) points.

Mathematica

```
ecaddprojective[{x1_, y1_, z1_}, {x2_, y2_, z2_},
b1_, b2_] :=
  If[z1 == 0, {x2, y2, z2},
  If[z2 == 0, {x1, y1, z1},
  If[x1 == x2,
  If[y1 == -y2, {0, 1, 0},
    {x3 = ((3x1^2 + 2b2 x1 + b1)/(2y1))^2 - b2 - 2x1,
    y3 = -(((3x1^2 + 2b2 x1 + b1)/(2y1))
      (x3 - x1) + y1), 1}],
    {x3 = ((y2 - y1)/(x2 - x1))^2 - b2 - x1 - x2,
    y3 = -(((y2 - y1)/(x2 - x1))(x3 - x1) + y1), 1}]]]
```

Maple

```
ecaddprojective:=proc(A,B,b1,b2)
  if A[3]=0 then return(B) else
  if B[3]=0 then return(A) else
  if A[1]=B[1] then
  if A[2]=-B[2] then
    return([0,1,0]) else
    return([((3*A[1]^2+2*b2*A[1]+b1)/(2*A[2]))^2-b2-2*A[1],
    -(((3*A[1]^2+2*b2*A[1]+b1)/(2*A[2]))*
    (((3*A[1]^2+2*b2*A[1]+b1)/(2*A[2]))^2-b2-3*A[1])+A[2]),1])
  end if;
  else return([((B[2]-A[2])/(B[1]-A[1]))^2-b2-A[1]-B[1],
    -(((B[2]-A[2])/(B[1]-A[1]))*
    (((B[2]-A[2])/(B[1]-A[1]))^2-b2-2*A[1]-B[1])+A[2]),1])
  end if;
  end if;
  end if;
end proc:
```

We now make use of this procedure to define a function to compute the quantity kP for a point $P = (x : y : z)$ on the curve $y^2 = x^3 + b_2x^2 + b_1x + b_0$.

Mathematica

```
ecpointmultiple[{x_, y_, z_}, b1_, b2_, k_] :=
  Module[{q, r, l},
    l = IntegerDigits[k, 2];
    q = {0, 1, 0};
    r = {x, y, z};
    Do[
      If[l[[i]] == 1, q = ecaddprojective[q, r, b1, b2]];
      r = ecaddprojective[r, r, b1, b2],
      {i, Length[l], 1, -1}];
    Return[q]]
```

We compute $3P$ with $P = (-1 : 4 : 1)$ (i.e., the point $(-1, 4)$) on the curve $y^2 = x^3 + 17$.

Mathematica

```
ecpointmultiple[{-1, 4, 1}, 0, 0, 3]

{298927/40401, 166830380/8120601, 1}
```

Here are the corresponding *Maple* commands.

Maple

```
ecpointmultiple:=proc(A,b1,b2,k)
  local q, r, l, i;
  l:=convert(k,base,2);
  q:=[0,1,0];
  r:=[A[1],A[2],A[3]];
  for i from 1 to nops(l) do
    if l[i]= 1 then
      q:=ecaddprojective(q,r,b1,b2)
    end if;
    r:=ecaddprojective(r,r,b1,b2);
  end do;
  return(q);
end proc:

ecpointmultiple([-1,4,1],0,0,3);

[298927/40401, 166830380/8120601, 1]
```

In practice, the numerators and denominators grow too quickly as k grows for this command to have much practical value. If we compute $13P$ from the above example, the numerator of the x-coordinate has over 100 digits! However, we can modify the command to be of great use when working on elliptic curves over $\mathbf{Z}_p$ (as we need to do for elliptic curve encryption and the elliptic curve method of factorization). We first give a modification of the `ecaddprojective` command for use over $\mathbf{Z}_p$ (for $p \neq 2$).

Mathematica

```
ecaddprojectivemodp[{x1_, y1_, z1_}, {x2_, y2_, z2_},
b1_, b2_, p_] :=
  If[z1 == 0, {x2, y2, z2},
  If[z2 == 0, {x1, y1, z1},
  If[Mod[x1 - x2, p] == 0,
  If[Mod[y1 + y2, p] == 0, {0, 1, 0},
    Mod[{x3 = ((3x1^2 + 2b2 x1 + b1)*PowerMod[2y1, -1, p])^2
      - b2 - 2x1,
    y3 = -(((3x1^2 + 2b2 x1 + b1)*PowerMod[2y1, -1, p])
      (x3 - x1) + y1), 1}, p]],
    Mod[{x3 = ((y2 - y1)*PowerMod[x2 - x1, -1, p])^2 - b2
      - x1 - x2,
    y3 = -(((y2 - y1)*PowerMod[x2 - x1, -1, p])
      (x3 - x1) + y1), 1}, p]]]]
```

Now we give the mod p analog of `ecpointmultiple`.

Mathematica

```
ecpointmultiplemodp[{x1_, y1_, z1_}, b1_, b2_, k_, p_] :=
  Module[{q, r, l},
    l = IntegerDigits[k, 2];
    q = {0, 1, 0};
    r = {x1, y1, z1};
    Do[If[l[[i]] == 1,
      q = ecaddprojectivemodp[q, r, b1, b2, p]];
      r = ecaddprojectivemodp[r, r, b1, b2, p],
      {i, Length[l], 1, -1}];
    Return[q]]
```

Using this function, we can make the computations given at the end of Section 11.6.

Mathematica

```
ecpointmultiplemodp[{1234005,349433,1},2,0,11111,1377359]

{1114312,498654,1}

ecpointmultiplemodp[{1114312,498654,1},2,0,22222,1377359]

{710108,1324551,1}

ecpointmultiplemodp[{710108,1324551,1},2,0,283322,1377359]

{1075576,1307157,1}

ecpointmultiplemodp[{1075576,1307157,1},2,0,141661,1377359]

{1234005,349433,1}
```

Here is the corresponding *Maple* code.

Maple

```
ecaddprojectivemodp:=proc(A,B,b1,b2,p)
  if A[3]=0 then return(B) else
  if B[3]=0 then return(A) else
  if modp(A[1],p)=modp(B[1],p) then
  if modp(A[2],p)=modp(-B[2],p) then
    return([0,1,0]) else
    return(
      [modp(((3*A[1]^2+2*b2*A[1]+b1)/(2*A[2]))^2-b2-2*A[1],p),
      modp(-(((3*A[1]^2+2*b2*A[1]+b1)/(2*A[2]))*
      (((3*A[1]^2+2*b2*A[1]+b1)/(2*A[2]))^2-
        b2-3*A[1])+A[2]),p),1])
  end if;
  else return(
      [modp(((B[2]-A[2])/(B[1]-A[1]))^2-b2-A[1]-B[1],p),
      modp(-(((B[2]-A[2])/(B[1]-A[1]))*
      (((B[2]-A[2])/(B[1]-A[1]))^2-
        b2-2*A[1]-B[1])+A[2]),p),1])
  end if;   end if;   end if;
end proc:
```

Maple

```
ecpointmultiplemodp:=proc(A,b1,b2,k,p)
  local q, r, l, i;
  l:=convert(k,base,2);
  q:=[0,1,0];
  r:=[A[1],A[2],A[3]];
  for i from 1 to nops(l) do
    if l[i]= 1 then
      q:=ecaddprojectivemodp(q,r,b1,b2,p)
    end if;
    r:=ecaddprojectivemodp(r,r,b1,b2,p);
  end do;
  return(q);
end proc:
```

Maple

```
ecpointmultiplemodp([1234005, 349433, 1],2,0,11111,1377359);

[1114312, 498654, 1]

ecpointmultiplemodp([1114312, 498654, 1],2,0,22222,1377359);

[710108, 1324551, 1]

ecpointmultiplemodp([710108, 1324551, 1],2,0,283322,1377359);

[1075576, 1307157, 1]

ecpointmultiplemodp([1075576, 1307157, 1],
2,0,141661,1377359);

[1234005, 349433, 1]
```

Next, we use a modification of `ecpointmultiplemodp` to implement the elliptic curve factoring algorithm. The procedure below will compute kP on the curve $y^2 = x^3 + ax - a$ with $P = (1, 1)$ and $k = (a + 10)!$. The procedure begins with $a = 1$ and may continue as far as $a = 40$. In this example, we attempt to find a factor of $m = 562325707971011$. At each step in the loop that computes kP, we check the coordinates of our points for common factors with m. If we find none, then we reduce the coordinates mod m. If at any point we encounter a factor of m greater than 1, the procedure will terminate and return the factor that was discovered.

Mathematica

```
ecfactor[m_] :=
  Module[{k, b0, b1, b2},
    Do[k = (a + 10)!;
      b2 = 0;
      b1 = a;
      b0 = -a;
      If[GCD[4a^3 - 27a^2, m] > 1,
        Return[Print[GCD[4a^3 - 27a^2, m]]]];
      q = {1, 1, 1};
      r = {0, 1, 0};
      kbinary = IntegerDigits[k, 2];
      For[n = 0, n < Length[kbinary], n++,
      If[kbinary[[Length[kbinary] - n]] == 1,
      r = ecaddprojective[r, q, b1, b2];
      If[GCD[Denominator[r[[1]]], m] > 1,
      Return[Print[GCD[Denominator[r[[1]]], m]]],
      r[[1]] = genmod[r[[1]], m]];
      If[GCD[Denominator[r[[2]]], m] > 1,
      Return[Print[GCD[Denominator[r[[2]]], m]]],
      r[[2]] = genmod[r[[2]], m]];];
      q = ecaddprojective[q, q, b1, b2];
      If[GCD[Denominator[q[[1]]], m] > 1,
      Return[Print[GCD[Denominator[q[[1]]], m]]],
      q[[1]] = genmod[q[[1]], m]];
      If[GCD[Denominator[q[[2]]], m] > 1,
      Return[Print[GCD[Denominator[q[[2]]], m]]],
      q[[2]] = genmod[q[[2]], m]]],
      {a, 1, 40}]]

ecfactor[137703491]

17389
```

Maple

```
ecfactor:=proc(m)
  local k, kbinary, b0, b1, b2, n, a, q, r;
  for a from 1 to 40 do
    k:=(a+10)!;
    b2:=0;
    b1:=a;
    b0:=-a;
    if 1<gcd(-4*a^3-27*a^2,m) then
      return(gcd(-4*a^3-27*a^2,m)); end if;
    q:=[1,1,1];
    r:=[0,1,0];
    kbinary:=convert(k,base,2);
    for n from 1 to nops(kbinary) do
      if kbinary[n]=1 then
      r:=ecaddprojective(r,q,b1,b2);
      if gcd(denom(r[1]),m)>1 then
      return(gcd(denom(r[1]),m)) else
      r[1]:=modp(r[1],m) end if;
      if gcd(denom(r[2]),m)>1 then
      return(gcd(denom(r[2]),m)) else
      r[2]:=modp(r[2],m) end if;
      end if;
      q:=ecaddprojective(q,q,b1,b2);
      if gcd(denom(q[1]),m)>1 then
      return(gcd(denom(q[1]),m)) else
      q[1]:=modp(q[1],m) end if;
      if gcd(denom(q[2]),m)>1 then
      return(gcd(denom(q[2]),m)) else
      q[2]:=modp(q[2],m) end if;
    end do;
  end do;
return([]);
end proc:

ecfactor(137703491);
17389
```

In this case the procedure succeeds in producing a proper factor, 17389, of m (when $a = 17$). In general, the procedure terminates if m itself turns up as a factor of one of the coordinates of a point. With a few extra commands, the procedure could be modified to skip to the next value of a and continue with the procedure in such a case. One could also modify the procedure to continue until a factor is encountered rather than quit after reaching $a = 40$.

Exercises

†11.33 Let $f(x, y, z)$ be a homogeneous polynomial and let (x_0, y_0, z_0) be a nonsingular point on the curve $f(x, y, z) = 0$. Show that (x_0, y_0, z_0) is a point on the projective line

$$f_x(x_0, y_0, z_0)x + f_y(x_0, y_0, z_0)y + f_z(x_0, y_0, z_0)z = 0.$$

11.34 Let $f(x, y, z) = y^2z - (x^3 + b_2x^2z + b_1xz^2 + b_0z^3)$ and $F(x, y) = f(x, y, 1) = y^2 - (x^3 + b_2x^2 + b_1x + b_0)$. Suppose that $(x_0 : y_0 : 1)$ is a point on $f(x, y, z) = 0$ and hence (x_0, y_0) is a point on the curve $F(x, y) = 0$. Suppose further that $F_x(x_0, y_0) \neq 0$. Show that the equation for the tangent line to $F(x, y) = 0$ given by Definition 11.13 as

$$y - y_0 = -\frac{F_x(x_0, y_0)}{F_y(x_0, y_0)}(x - x_0)$$

is the same as the "dehomogenized" equation of the projective version given in the Notes as

$$f_x(x_0, y_0, 1)x + f_y(x_0, y_0, 1)y + f_z(x_0, y_0, 1) = 0.$$

Chapter 12

Logic and Number Theory

What is proved about numbers will be a fact in any universe.

[Said as a child.]

JULIA BOWMAN ROBINSON (1919–1985)

In this chapter, we investigate connections between Diophantine equations and topics in mathematical logic, particularly the theory of recursive sets and recursively enumerable sets and the theory of undecidability. Studying these connections helps us to determine what is possible and what is impossible in solving Diophantine equations. It also illustrates how two great areas of mathematics, number theory and logic, have complemented each other and enriched each other's development.

12.1 Solvable and unsolvable equations

As we mentioned in the Introduction, one of the main focuses of interest in number theory is the study of Diophantine equations and their solutions. Throughout the book, we have investigated a host of important Diophantine equations. In Chapter 2, we examined the linear equation

$$ax + by = \gcd(a, b). \tag{12.1}$$

In Chapter 9, we studied several Diophantine equations, including the general linear equation

$$ax + by = c, \tag{12.2}$$

the Pythagorean formula

$$x^2 + y^2 = z^2, \tag{12.3}$$

the representations of a number as a sum of two or four squares,

$$n = x^2 + y^2 \quad \text{and} \quad n = w^2 + x^2 + y^2 + z^2, \tag{12.4}$$

a special case of Fermat's Last Theorem,

$$x^4 + y^4 = z^4, \tag{12.5}$$

and Pell's equation

$$x^2 - d\,y^2 = 1. \tag{12.6}$$

In Chapter 11, we considered various cubic curves such as

$$y^2 = x^3 - x + 1. \tag{12.7}$$

Clearly, Diophantine equations have played a central role in our explorations in number theory.

Recall that some of the aforementioned Diophantine equations have solutions (in integers) and some do not. Equation (12.1) is always solvable, while equation (12.2) is solvable if and only if $\gcd(a, b) \mid c$. Equation (12.3) has infinitely many solutions (even with $\gcd(x, y, z) = 1$). In (12.4), the first equation is solvable for some positive integers n and unsolvable for others, while the second equation is solvable for every positive integer n. Equation (12.5) is not solvable (in positive integers), while equation (12.6) is always solvable (in positive integers) if d is a positive nonsquare. Equation (12.7) is solvable.

Considering these diverse examples of Diophantine equations, we wonder whether there is a method by which we can tell whether *every* given Diophantine equation is solvable or not solvable. In 1900 David Hilbert (1862–1943), in a lecture to the International Congress of Mathematicians in Paris, discussed 23 important unsolved problems in mathematics; the tenth problem on his list is our question about Diophantine equations. Hilbert phrased the problem, which he called "Entscheidung der Lösbarkeit einer diophantischen Gleichung" ("Determination of the Solvability of a Diophantine Equation"), as follows:

> Given a Diophantine equation with any number of unknown quantities and with rational integral numerical coefficients: *To devise a process according to which it can be determined by a finite number of operations whether the equation is solvable in rational integers.*

In our terminology, a Diophantine equation has integer coefficients, and a solution is in integers. Hilbert's Tenth Problem calls for a universal procedure which, given any Diophantine equation, will say correctly whether it is solvable or unsolvable. For example, the procedure should be able to determine whether an equation such as

$$3x^5 - 7xy + 8y^3 = y^4 - 7x^2 + 100,$$

has a solution in integers x and y. Hilbert's universal procedure, if it existed, would be extremely useful in number theory.

Eventually, in 1973, it was demonstrated that there is no such procedure. This fact was proved by Yuri Matiyasevich, but the accomplishment was the culmination of the work of several mathematicians over a period of many years. Although the main result is a negative one (there does not exist an algorithm to test Diophantine equations for solvability), there are positive results that derive from the analysis. The main positive result is that a "Diophantine set" (a term to be defined later) can be almost any conceivable set, including, for example, the set of prime numbers. This result was thought quite surprising when it was first proved. At any rate, the solution of Hilbert's Tenth Problem is a beautiful piece of mathematics, synthesizing many topics of number theory, including Fibonacci numbers, Lagrange's four squares theorem, and Pell's equation, as well as major ideas of logic. It is truly a *tour de force* of mathematics.

In this chapter we sketch the solution of Hilbert's Tenth Problem, attempt to understand its connections to mathematical logic, and investigate some of the positive results that spring from it. This mathematical inquiry fills us with awe at the reasoning and proofs, as well as excitement at the prospect of a never-ending quest for the solutions of Diophantine equations. A wealth of material regarding the theory introduced in this chapter is available in [24].

Exercises

12.1 (a) Is the Diophantine equation

$$x^2 + y^2 = 2x + 4y + 5$$

solvable in integers?

(b) For what values of k is the Diophantine equation

$$x^2 + y^2 = 2x + 4y + k$$

solvable in integers?

◇**12.2** Use a computer to search for integer solutions to the equation

$$y^2 = x^3 + k,$$

for various values of k.

It is known that there are finitely many solutions for a given k.

12.2 Diophantine equations and Diophantine sets

In this section, we outline the general theory of Diophantine equations and Diophantine sets.

Diophantine equations

We record the definition of a Diophantine equation.

DEFINITION 12.1 *A* Diophantine equation *is a polynomial equation*

$$D(x_1, x_2, \ldots, x_n) = 0 \tag{12.8}$$

whose coefficients are integers. The values of the variables x_1, x_2, $\ldots$, x_n in a solution of a Diophantine equation are integers.

Notice that the stipulation that the right side of the above equation be 0 is unimportant, for if there were nonzero terms on the right side, they could be transfered to the left side.

Often, we are interested in solving a system of Diophantine equations. The system of m Diophantine equations

$$\begin{aligned} D_1(x_1, x_2, \ldots, x_n) &= 0 \\ D_2(x_1, x_2, \ldots, x_n) &= 0 \\ &\vdots \\ D_m(x_1, x_2, \ldots, x_n) &= 0 \end{aligned} \tag{12.9}$$

has a solution if and only if the single Diophantine equation

$$D_1(x_1, x_2, \ldots, x_n)^2 + D_2(x_1, x_2, \ldots, x_n)^2 + \cdots + D_m(x_1, x_2, \ldots, x_n)^2 = 0 \tag{12.10}$$

has a solution. The reason is that the square of any nonzero integer is positive. Thus, the problem of solving arbitrary systems of Diophantine equations is equivalent to the problem of solving arbitrary single Diophantine equations.

We say that the "degree" of a Diophantine equation is the maximum combined degree of a monomial in the equation. For example, the Diophantine equation

$$x^2y^3z^4 + 3x^2z^2 + xyz + y^5 + 8 = 0$$

has degree 9. One might guess that the decision problem for Diophantine equations of high degree is more difficult than the decision problem for Diophantine equations of low degree, but this is not the case. In fact, the general question of solving arbitrary Diophantine equations is equivalent to the problem of solving Diophantine equations of degree 4. We show an example that illustrates this equivalence.

Consider this Diophantine equation of degree 6:

$$x^2y^3z + 3x^2z - 5xyz - z + 5 = 0. \tag{12.11}$$

Using simple equations involving addition, subtraction, and multiplication, we obtain a system of Diophantine equations (with 12 variables) that is equivalent to equation (12.11):

$$\begin{aligned}
&a_1 = x^2, \quad a_2 = a_1 y, \quad a_3 = a_2 y, \quad a_4 = a_3 y, \quad a_5 = a_4 z; \\
&b_1 = 3x^2, \quad b_2 = b_1 z; \\
&c_1 = 5xy, \quad c_2 = c_1 z; \\
&a_5 + b_2 - c_2 - z + 5 = 0. \qquad (12.12)
\end{aligned}$$

In the manner of the transformation of the system of equations (12.9) into equation (12.10), we now turn the above system into an equivalent single equation of degree 4:

$$\begin{aligned}
&(a_1 - x^2)^2 + (a_2 - a_1 y)^2 + (a_3 - a_2 y)^2 + (a_4 - a_3 y)^2 + (a_5 - a_4 z)^2 \\
&\quad + (b_1 - 3x^2)^2 + (b_2 - b_1 z)^2 + (c_1 - 5xy)^2 + (c_2 - c_1 z)^2 \\
&\quad + (a_5 + b_2 - c_2 - z + 5)^2 = 0. \qquad (12.13)
\end{aligned}$$

The problem of determining whether this degree 4 equation has integer solutions is equivalent to the problem of determining whether the original degree 6 equation (12.11) has integer solutions.

Since the decision problem for general Diophantine equations is unsolvable, the decision problem for Diophantine equations of degree 4 is also unsolvable. It is not known whether the decision problem restricted to equations of degree 3 is solvable.

We have required that the solutions to a Diophantine equation be integers, but as far as the decision problem is concerned, we can make the stronger requirement that solutions be nonnegative integers. The two decision problems are equivalent, for equation (12.8) is solvable in integers if and only if equation

$$D(x_1 - x_1', x_2 - x_2', \ldots, x_n - x_n') = 0 \qquad (12.14)$$

is solvable in nonnegative integers. On the other hand, (12.8) is solvable in nonnegative integers if and only if the system

$$\begin{aligned}
D(x_1, x_2, \ldots, x_n) &= 0 \\
x_1 &= y_{1,1}^2 + y_{1,2}^2 + y_{1,3}^2 + y_{1,4}^2 \\
x_2 &= y_{2,1}^2 + y_{2,2}^2 + y_{2,3}^2 + y_{2,4}^2 \\
&\vdots \\
x_n &= y_{n,1}^2 + y_{n,2}^2 + y_{n,3}^2 + y_{n,4}^2 \qquad (12.15)
\end{aligned}$$

is solvable in integers. This observation follows from Lagrange's theorem (Theorem 9.41), which says that every positive integer (and why not include 0, too?) is the sum of four squares of integers. From now on, we assume that "solution" means solution in nonnegative integers.

Diophantine sets and relations

In studying Diophantine equations, it is useful to shift attention from the equations themselves to their solution sets. The definitions of this subsection formulate our inquiry in this context.

DEFINITION 12.2 *A* family *of Diophantine equations with* parameters a_1, a_2, ..., a_m *and* unknowns x_1, x_2, ..., x_n *is a Diophantine equation*

$$D(a_1, a_2, \ldots, a_m, x_1, x_2, \ldots, x_n) = 0.$$

Our point of view in this definition is that if values of the parameters are specified, there results a specific Diophantine equation in the family.

Example 12.3 The Diophantine family

$$D(a, b, c, x, x', y, y') = a(x - x') + b(y - y') - c = 0$$

represents the linear equation recalled in (12.2). □

Example 12.4 The Diophantine family

$$D(n, x, y) = n - x^2 - y^2 = 0$$

represents the sum of two squares equation recalled in (12.4). □

DEFINITION 12.5 *A set S of m-tuples is* Diophantine *if there is a Diophantine family*

$$D(a_1, a_2, \ldots, a_m, x_1, x_2, \ldots, x_n) = 0$$

that has a solution x_1, x_2, ..., x_n if and only if $(a_1, a_2, \ldots, a_m) \in S$.

Note: The case $m = 1$ is especially important: A set S of nonnegative integers is Diophantine if there is a Diophantine family $D(a, x_1, x_2, \ldots, x_n) = 0$ for which there exists a solution x_1, x_2, ..., x_n if and only if $a \in S$.

Example 12.6 Here are some Diophantine sets and their Diophantine family representations:

- Even numbers: $a - 2x = 0$.
- Odd numbers: $a - (2x + 1) = 0$.
- Perfect squares: $a - x^2 = 0$.
- Composite numbers: $a - (x + 2)(y + 2) = 0$.

- Numbers that are not powers of 2: $a - (2x+3)y = 0$.
- Numbers whose prime factors of the form $4k+3$ all appear to even powers: $a - x^2 - y^2 = 0$.

  (This characterization follows from Theorem 9.40.)

In the case of a relation, the Diophantine set in question is the collection of m-tuples which constitute it. In particular, for a Diophantine function f, the set $\{(m,n)\colon\ n = f(m)\}$ is Diophantine. Here are some particularly useful examples (now we freely put terms on both sides of the '=' sign):

- $a < b$: $\quad a + (x+1) = b$,

  $a \le b$: $\quad a + x = b$.
- $a \mid b$: $\quad ax = b$.
- $\gcd(a,b) = 1$: $\quad a(x - x') + b(y - y') = 1$.

  (This characterization follows from Theorem 2.16.)
- $a \equiv b \pmod{m}$: $\quad a - b = m(x - x')$.

$\square$

Many other sets are also Diophantine, but for some sets this is quite difficult to show. For example, the set of powers of 2, i.e., the set

$$\{2, 2^2, 2^3, 2^4, 2^5, 2^6, \ldots\},$$

is Diophantine, but there seems to be no simple way to demonstrate this. We cannot, for instance, use the equation $a = 2^x$ because it is not a polynomial. We saw in Example 12.6 that the set of numbers that are *not* powers of 2 is Diophantine; but the complement of a Diophantine set is not necessarily Diophantine. Indeed, a Diophantine set with a non-Diophantine complement is central to the negative solution of Hilbert's Tenth Problem. The fact that the Diophantine nature of the set of powers of 2 is so difficult, while the same question for the non-powers of 2 is so easy, is yet another illustration of the theme advanced in the Introduction, namely, that simple ideas live among the very complex in the world of number theory.

Although the class of Diophantine sets is not closed under complements, it is closed under intersections and unions. An intersection can be obtained by a sum of squares and a union by a product.

PROPOSITION 12.7

The class of Diophantine sets, for a specific number of parameters, is closed under intersections and unions. That is to say, the logical connectives "and" and "or" are allowed in constructing Diophantine relations.

Note: As we mentioned earlier, the logical operand "not" is not allowed.

Example 12.8 The following sets and relations are Diophantine, as indicated by the given representations:

- Numbers that are perfect squares or perfect cubes:
$$(a - x^2)(a - y^3) = 0.$$
We can also represent this set as
$$a = x^2 \quad \text{or} \quad a = y^3.$$

- Numbers that are perfect squares and not powers of 2:
$$(a - x^2)^2 + [a - (2y + 3)z]^2 = 0.$$
We can also represent this set as
$$a = x^2 \quad \text{and} \quad a = (2y + 3)z.$$

- Greatest common divisor:
$g = \gcd(a, b)$: $\quad a = gx, \quad b = gy, \quad$ and $\quad \gcd(x, y) = 1.$

- Quotient and remainder:
q, r: $\quad a = bq + r$ and $0 \le r < b$.
Note that $q = \lfloor a/b \rfloor$, where $\lfloor x \rfloor$ denotes the "floor function" of x, i.e., the greatest integer less than or equal to x. Similarly, we can represent the "ceiling function" $\lceil a/b \rceil$ (where $\lceil x \rceil$ denotes the least integer greater than or equal to x) as $a = bq - r$ and $0 \le r < b$.

□

Once we know that a function is Diophantine, we can use it in other Diophantine representations. For example, the set of pairs (a, b) such that
$$\gcd(a, b)^3 + \gcd(a, b) = 222$$
is Diophantine. In constructing more complex Diophantine representations, composition of Diophantine functions is allowed. For example, given that the function $f(a, b)$ is Diophantine, then the set of pairs (a, b) for which
$$f(f(a, b^2), a) + f(a, f(b, a^2)) = 100$$
is also Diophantine; for this relation can be written as the system
$$\begin{aligned} f(a, b^2) &= c \\ f(b, a^2) &= d \\ f(c, a) + f(a, d) &= 100. \end{aligned}$$

Exercises

12.3 Given the Diophantine equation

$$x^5 + xyz + z^5 = 100,$$

find an equivalent Diophantine equation of degree 4.

12.4 Formulate the Diophantine family that represents the Pell equation recalled in (12.6).

12.5 (a) Prove that the set of multiples of 3 is Diophantine.

(b) Prove that the set of non-multiples of 3 is Diophantine.

12.6 Prove that the set of pairs of numbers where the first number divides the second number and the first number is composite is Diophantine.

12.7 Prove that the set of non-powers of 3 is Diophantine.

12.8 Given that the set of powers of 2 is Diophantine, prove that the set of powers of 4 is Diophantine.

12.9 Prove that the set of nonsquares is Diophantine.

Hint: Use the fact that a is a nonsquare if and only if there exists z for which $z^2 < a < (z+1)^2$.

12.10 Prove that the relation $l = \text{lcm}(a, b)$ is Diophantine.

Hint: Use the fact that $\gcd(a, b) \cdot \text{lcm}(a, b) = ab$.

12.11 Prove that the functions $\min\{a, b\}$ and $\max\{a, b\}$ are Diophantine.

12.3 Positive values of polynomials

The next result, combined with the later classification of Diophantine sets, has some surprising consequences.

PROPOSITION 12.9

A set of positive integers is Diophantine if and only if it is the set of positive values of a polynomial with nonnegative integer values of the variables.

PROOF Suppose that S is Diophantine with representation $D(a, x_1, \ldots, x_n) = 0$. Then $a \in S$ if and only if there is a solution $x_1, \ldots, x_n$ to this equation. With $x_0 = a$, this solution yields a solution to the equation

$$x_0(1 - D(x_0, x_1, \ldots, x_n)^2) = a.$$

Conversely, since $1 - D(x_0, x_1, \ldots, x_n)^2$ is at most equal to 1, any solution to this latter equation implies that $D(x_0, x_1, \ldots, x_n) = 0$, and hence, $x_0 = a$ and $D(a, x_1, \ldots, x_n) = 0$. Therefore, the polynomial

$$f(x_0, x_1, \ldots, x_n) = x_0(1 - D(x_0, x_1, \ldots, x_n)^2)$$

has the required property.

To complete the proof, it must be shown that the set of positive values of a polynomial with integer values of the variables is Diophantine. This is left as an exercise. $\square$

Example 12.10 There exists a polynomial whose positive values are precisely the composite numbers. This follows from one of the observations in Example 12.6, together with Proposition 12.9. In fact, from the proof of Proposition 12.9, we can write such a polynomial explicitly:

$$x_0\{1 - [x_0 - (x_1 + 2)(x_2 + 2)]^2\}.$$

$\square$

As we shall see, the set of prime numbers is Diophantine, so that there exists a polynomial whose set of positive values is precisely the set of primes. However, there is no polynomial in a single variable representing only primes (see Exercises). In proving that the set of prime numbers is Diophantine, we will use Wilson's theorem, that p is prime if and only if $(p-1)! \equiv -1 \pmod{p}$, but we must first show that the factorial function is Diophantine. We do this in Section 12.6.

Example 12.11 (James P. Jones) The set of positive values of the polynomial

$$2y - x^4y - 2x^3y^2 + x^2y^3 + 2xy^4 - y^5, \tag{12.16}$$

where x and y are positive integers, is precisely the set of (positive) Fibonacci numbers. Furthermore, the polynomial represents each Fibonacci number exactly once. (Recall from Chapter 1 that the Fibonacci numbers f_n are defined by $f_0 = 0$, $f_1 = 1$, and $f_n = f_{n-1} + f_{n-2}$, for $n \geq 2$.)

Let's test this claim using a computer.

Mathematica

```
f[x_, y_] := 2y - x^4y - 2x^3y^2 + x^2y^3 + 2xy^4 - y^5
Do[
  If[f[x, y] > 0, Print[{f[x, y], {x, y}}]],
  {x, 1, 100}, {y, 1, 100}
]

{1, {1, 1}}
{2, {1, 2}}
{3, {2, 3}}
{5, {3, 5}}
{8, {5, 8}}
{13, {8, 13}}
{21, {13, 21}}
{34, {21, 34}}
{55, {34, 55}}
{89, {55, 89}}
```

Maple

```
f:=(x,y)->2*y-x^4*y-2*x^3*y^2+x^2*y^3+2*x*y^4-y^5

for x from 1 to 100 do
  for y from 1 to 100 do
    if f(x,y)>0 then
    print([f(x,y),[x,y]])
    end if;
  end do;
end do;

[1, [1, 1]]
[2, [1, 2]]
[3, [2, 3]]
[5, [3, 5]]
[8, [5, 8]]
[13, [8, 13]]
[21, [13, 21]]
[34, [21, 34]]
[55, [34, 55]]
[89, [55, 89]]
```

We observe that (over the given range) the Fibonacci numbers are produced as the values of y when the input is (x, y), with x and y consecutive Fibonacci numbers. We proceed to prove this.

The polynomial (12.16) is enigmatic, but let's decipher its meaning. We notice that y is a factor of each term, so we write the polynomial as

$$y(2 - x^4 - 2x^3y + x^2y^2 + 2xy^3 - y^4).$$

Next, we put the polynomial in a form similar to that indicated in Proposition 12.9:

$$y[2 - (y^2 - xy - x^2)^2].$$

We now see that the polynomial can have a positive value if and only if $y^2 - xy - x^2 = 0$ or ± 1. But the first case is impossible, for $y^2 - xy - x^2 = 0$ implies that $4y^2 - 4xy - 4x^2 = 0$, and hence, $(2y - x)^2 = 5x^2$, but 5 is not a square. Therefore, the only possibility is

$$y^2 - xy - x^2 = \pm 1. \tag{12.17}$$

This equation is essentially a Diophantine representation of the Fibonacci numbers; that is, y is a Fibonacci number if and only if there exists x that satisfies the relation. A small (unimportant) detail is that in this problem we have stipulated that the variables be positive integers rather than nonnegative integers. (See Exercises.)

We rewrite (12.17) as

$$y^2 - x(y + x) = \pm 1. \tag{12.18}$$

This last equation is reminiscent of the Fibonacci numbers identity (called Cassini's identity), which we saw in Chapter 1:

$$f_n^2 - f_{n-1}f_{n+1} = \begin{cases} 1 & n \text{ odd} \\ -1 & n \text{ even}, \end{cases} \tag{12.19}$$

and this is the key to the problem. By (12.19), the ordered pairs $(x, y) = (f_{n-1}, f_n)$, for $n \geq 1$, satisfy (12.17). We must show that in fact these are the only solutions.

Let's check the case $x = y$. From (12.17), we see that this is possible only for $x = y = 1$.

Now suppose that (x, y) satisfies (12.17). Then $(y - x, x)$ satisfies the same equation:

$$x^2 - (y - x)x - (y - x)^2 = x^2 - (y - x)y = x^2 + xy - y^2 = \mp 1.$$

For $x > 1$, we know that $x \neq y$. Clearly, $y > x$; that is, $x + 1 \leq y$. Also, $y - x < x$, for

$$(x + 1)(y - x) \leq y(y - x) = x^2 \pm 1,$$

and hence, if $y - x \geq x$, then $x^2 \pm 1 \geq x^2 + x$, which is only possible for $x = 1$.

Working "backwards," we see that the descent ends with $(x, y) = (1, 2)$. Going "forwards," the transformation $(x, y) \to (y, x + y)$ yields the sequence (f_{n-1}, f_n), with n even if the $+1$ holds and n odd if the -1 holds. □

Note: The Diophantine representation (12.17) was used in Matiyasevich's proof of the unsolvability of Hilbert's Tenth Problem.

Although we have found a Diophantine representation of the set of Fibonacci numbers (say, $D(a, x) = a^2 - ax - x^2 \pm 1 = 0$), we have not found a representation of the "Fibonacci function" $n = f_m$. This seems to be considerably more difficult to do, and is indeed the crux of Matiyasevich's proof. It is often more useful to have the set of pairs $(x, y = f(x))$ as a Diophantine set, rather than merely the set range(f), for in the case of the pairs one can use both function (y) and argument (x) in other Diophantine representations (as noted in the remark after Example 12.8).

Exercises

12.12 Find a polynomial whose positive values are the non-powers of 2.

12.13 Complete the proof of Proposition 12.9 by showing that the set of positive values of a given polynomial with integer values of the variables is Diophantine.

Hint: Let f be a polynomial whose positive values are the set S. Consider the polynomial

$$D(a, x_1, x_2, \ldots, x_n) = f(x_1, x_2, \ldots, x_n) - a.$$

◇**12.14** Write a computer program to verify (over a suitable range) that the polynomial of Example 12.10 produces precisely the set of composite numbers as its positive values.

12.15 Find a polynomial in x and y whose positive values (for positive x and y) are precisely the non-Fibonacci numbers.

Hint: A positive integer is a non-Fibonacci number if it lies between two consecutive Fibonacci numbers.

12.16 (J. P. Jones) The Lucas numbers L_n are defined by $L_1 = 1$, $L_2 = 3$, and $L_n = L_{n-1} + L_{n-2}$, for $n \geq 2$. Prove that the Lucas numbers are the positive values of the polynomial

$$y\{1 - [(y^2 - xy - x^2)^2 - 25]^2\}.$$

◇**12.17** The polynomial of Example 12.11 takes on the value 0 multiple times, if x and y are allowed to equal 0. Show that the polynomial

$$7y^4x^2 - 7y^2x^4 - 5yx^5 + y^3x^3 + y^5x - 2y^6 + 3yx + 2y^2 + 2y - x^6 + x^2 + x$$

(also found by Jones) takes the value of each Fibonacci number (including 0) exactly once, where x and y are nonnegative integers. Such a polynomial is called a "singlefold" Diophantine representation.

Hint: Use a computer to find an appropriate factorization of the polynomial similar to the one in the example.

12.4 Logic background

In this section we give background on several results in logic that are applicable to number theory and in particular to the solution of Hilbert's Tenth Problem.

The liar's paradox

Many exciting developments in mathematical logic of the twentieth century share a lineage that can be traced back to an ancient conundrum known as the liar's paradox. This paradox is credited to Epimenides, a sixth century B.C.E. philosopher from Crete. Epimenides claimed that "All Cretans are liars.... One of their own poets has said so." According to this statement, the Cretan poet in question made a self-referential assertion that all Cretan's (including himself) are liars. The truth of this assertion would seem to invalidate the assertion itself. However, the assertion is not a true paradox because it is possible that some Cretans tell the truth and the poet lied. There are additional complications, too: since Epimenides was a Cretan, we would need to evaluate the truth or falsity of his assertions relative to the truth or falsity of the statements that all Cretans are liars and that a Cretan poet said so; and we must determine what "liar" means when multiple assertions are made.

In the fourth century B.C.E., the philosopher Eubulides offered a simpler, more direct version of the liar's paradox: "This statement is false." This statement is a true paradox in that its truth or falsity cannot be evaluated. If it were true then it would be false, and if it were false then it would be true.

This basic paradox is at the heart of modern results in logic due to Georg Cantor (1845–1918), Alan Turing (1912–1954), and Kurt Gödel (1906–1978). These results "mathematicize" the paradox in various settings.

The diagonal method

A significant theorem of Cantor concerns the cardinality (size) of sets (which are allowed to be infinite). In Cantor's set theory, two sets have the same cardinality if and only if there is a bijection between them. Furthermore, if there is no onto function from one set to another set, then the second set is larger than the first set. The "power set" $\mathcal{P}(S)$ of a set S is the collection of all subsets of S.

***THEOREM* (Cantor)**
Given any set S, the cardinality of $\mathcal{P}(S)$ is greater than the cardinality of S.

Note: If S is finite, i.e., $|S| = n \geq 0$, then the claim is simply that $2^n > n$, which can be proved easily by induction (Exercise 1.11). The import of the theorem is in the case where S is infinite.

PROOF We must demonstrate that there is no onto function from S to $\mathcal{P}(S)$. We give a proof by contradiction. Assume that g is such a function. Let

$$T = \{s \in S \colon s \notin g(s)\}.$$

Since g is onto, the set T is the image under g of some $t \in S$, i.e., $g(t) = T$. We ask, is $t \in T$? As in the liar's paradox, there is a dilemma: if $t \in T$, then, by definition, $t \notin T$; while if $t \notin T$, then, by definition, $t \in T$. We have contradicted the assumption that there is an onto function from S to $\mathcal{P}(S)$. Therefore, there is no such function. ∎

An intriguing consequence of Cantor's theorem is that there is no largest infinite set.

The method of proof used above is called the "diagonal method," since, in the finite or countably infinite case, the ordered pairs $(s, g(s))$ constitute an array displaying the definition of g, and T is the complement of the main diagonal of this array. An application of the diagonal method (and indeed a special case of the above theorem) is the proof that the set of real numbers ($\mathbf{R}$) is uncountable. The diagonal method is employed often in set theory and logic. We will see how the idea comes to fruition in the study of algorithms, and then ultimately in the classification of Diophantine sets.

Algorithms

Hilbert's Tenth Problem asks if there exists an algorithm to decide whether a given Diophantine equation is solvable. The resolution of Hilbert's problem depends on an analysis of what an algorithm is. The seminal work in this area was done by Turing in the 1930s. We now turn our attention to this analysis.

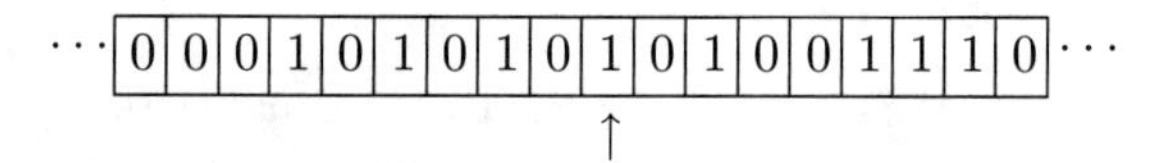

FIGURE 12.1: Turing machine.

Turing envisioned an abstract, general-purpose computer that could compute all possible algorithms. It should be able to perform calculations, store and produce data, and make decisions about what steps to perform next. The computer, now called a "Turing machine," consists of a tape with cells containing the symbols 0 or 1 and a pointer (↑) that points to one cell at a time. (See Figure 12.1.) For our purposes, we do not need to give the details of a Turing machine. Readers who wish to study these details are encouraged to read the informal account in [8] or the more rigorous treatment in [29]. It is enough to think of an algorithm as some definite procedure capable of generating positive numbers.

For example, the following algorithm, called `Short and Sweet`, produces the number 23 and stops:

```
Algorithm: Short and Sweet
  Step 1: Print 23.
  Step 2: Stop.
```

This algorithm does not accomplish much (other than printing a prime number).

Here is a more sophisticated algorithm that produces squares of numbers in increasing order:

```
Algorithm: Squares
  Step 1: Input n.
  Step 2: Let k=1.
  Step 3: Output k^2.
  Step 4: Let k=k+1.
  Step 5: If k<=n, go to Step 3.
  Step 6: Stop.
```

This algorithm inputs a number n (supplied by the user), outputs the squares from 1 to n^2, and then stops.

Note: We allow only a single `Input` statement in our algorithms. This rule prevents an input from occuring in a loop in which the user supplies values that are immediately output, thus obviating the need for the algorithm to perform computations.

This algorithm, like the one before it, uses fairly simple commands. Indeed, the "Turing language" consists of a small number of different commands.

These commands usually allow for moving the pointer, for printing 0 and 1, for controlling the flow of the algorithm, and for stopping the computation. Complex commands, such as "`Let k=k+1`," can be built up from the simpler Turing language commands. Again, the details are unimportant for our purposes. What is important is that there exists a language in which we can write our algorithms. (When written in a specific language, algorithms are usually called programs.)

For a general discussion of algorithms, it is of particular interest whether an algorithm halts given its input. Let's look again at our two examples. The algorithm `Short and Sweet` prints the number 23 and stops. The algorithm `Squares` also always stops (no matter what positive number is input). As another example, consider the following algorithm, called `Short Loop`:

```
Algorithm: Short Loop
  Step 1: Go to Step 1.
  Step 2: Stop.
```

In this algorithm, the `Stop` command is never encountered; instead, the algorithm loops endlessly through Step 1.

From our examples, we see that (given their input) some algorithms stop and some algorithms do not stop.

While the preceding three algorithms are relatively simple, complex programs can be built in the Turing language. In fact, every deterministic algorithm that has been so far imagined by computer scientists can be performed on a Turing machine. The hypothesis that this is always true is known as Church's Thesis, after Alonzo Church (1903–1995).

Church's Thesis: Any algorithm that is computable is computable by a Turing machine.

An important step in Turing's analysis of algorithms is the realization that programs can be encoded as data (numbers) and operated upon by other programs. Each type of command (`Print`, `Stop`, etc.) is given a numerical code, and the entire algorithm consists of all the codes put together. The end result is a number that defines a unique algorithm. It is convenient to indicate the code number of a program P as $code\,(P)$.

We represent a code P together with its data d as the pair (P, d). Now we can envision a program P together with its input data d encoded as a single number. We represent this number as $code\,(P, d)$. (We are intentionally letting *code* be a function of just a program or of a program together with data.) Going in the other direction, given a coded string we can determine the program and data that it represents. Therefore, some numbers can be decoded as algorithms together with data upon which they operate. All other numbers are gibberish in terms of representing algorithms because they do not obey the rules of the code.

With our encoding concept, it is easy to imagine a number that translates to an algorithm and data such that the algorithm does not halt given the data. For example, the above algorithm `Short Loop`, with no data at all, is of this type.

It is possible to write a program in the Turing language to emulate the action of any Turing program on data. Such a program is called a "universal Turing program." Basically, the program works by following the steps of a given program on its data. (The input must first be parsed into the commands comprising the program plus the input data.) The details of a universal Turing program are complicated, but it should be clear that such a program could, at least in theory, be created. An important point to note is that the universal program would run in an infinite loop if a `Stop` command isn't encountered. Another key point is that the universal Turing program could (according to Church's Thesis) be written in the Turing language itself. We record this observation as a theorem.

THEOREM 12.12
There exists a (universal) program U with the property that, for all P and d, the result of P acting on d is the same as the result of U acting on $code(P, d)$.

The program U is special, and we put it to good use later.

A major revelation in Turing's analysis of algorithms is the observation that there exist certain algorithms that one would like to create but cannot in fact exist. A specific example of such an algorithm is known as the "Halting program."

THEOREM 12.13
There does not exist an algorithm H (a Halting program) that accepts as input P and d and determines whether (P, d) halts.

PROOF This is a proof by contradiction (using the diagonal method). Suppose that such an algorithm H exists. Consider the following algorithm X.

```
Algorithm X:
  Step 1: Input code(P).
  Step 2: Use H to investigate P acting on code(P).
          If it halts, go to Step 2.
  Step 3: Stop.
```

Let $P = X$ to get a contradiction. How does X act on X itself? Since H is an algorithm that can tell whether another algorithm halts (given its data), H can decide whether X halts given itself as input. Suppose that X does halt. Then, following the directions in X, we go into an infinite loop. Hence, X

does not halt. This is a contradiction. Suppose that X does not halt. Then X halts, again a contradiction. Therefore, the assumption that H exists is false. ∎

We have in fact proved the stronger statement that there does not exist an algorithm that accepts P and determines whether P acting on $code(P)$ halts.

In order to derive the most powerful consequences of Turing's analysis, we must strengthen the previous theorem so that it applies to a specific input program, namely, the universal program U.

THEOREM 12.14
There does not exist an algorithm H that, given any v, tells whether (U, v) halts.

PROOF Suppose that such an H exists, and consider the following algorithm X:

```
Algorithm: X
Step 1: Input v.
Step 2: If possible, find a program P such that code(P)=v.
        If this is impossible, go to Step 2.
Step 3: Use H to investigate (P,v)=(v,v).
        If (U,vv) halts, go to Step 3.
Step 4: Stop.
```

Now (X, v) halts if and only if v is the code for P and (U, Pv) does not halt. Hence, (X, v) halts if and only if (P, v) does not halt. Let v be the code for X to obtain a contradiction. ∎

We reiterate that the above argument shows that there exists a program, namely, the universal algorithm U, for which there is no halting algorithm (an algorithm that would tell whether that program halts given its data).

Recursive and recursively enumerable sets

The collection of sets of natural numbers that can be produced using Turing machines is very broad. It allows for virtually all imaginable operations that can be performed on sets. We call a set of natural numbers that can be produced via a Turing machine a "recursively enumerable set." Recursively enumerable sets are also called "listable," "computable," or "effectively enumerable." A different category of sets is also of interest. Given a set A, we might want to know whether a particular natural number is a member of A. If a Turing machine can answer this question, for all natural numbers, then we call A a "recursive set." Clearly, a recursive set is recursively enumerable

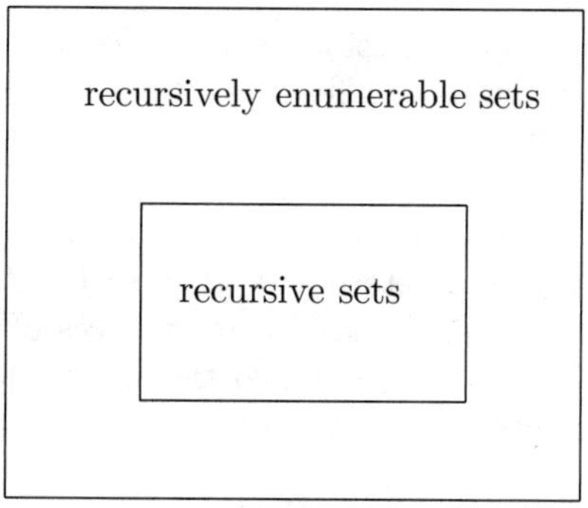

recursively enumerable sets

recursive sets

FIGURE 12.2: Recursive and recursively enumerable sets.

(simply construct a Turing machine that outputs the number a if and only if $a \in A$). But there are recursively enumerable sets that are not recursive. (Due to the nonexistence of the Halting program, the set of numbers corresponding to programs that halt is an example of such a set. To show the details of this, one could introduce a coding scheme that assigns a unique number to each program.) Figure 12.2 shows the relationship between the two types of sets.

The collection of recursively enumerable sets is closed under unions and intersections, but not complements. Of course, these facts are consistent with the conclusion that the collection of recursively enumerable sets is identical to the collection of Diophantine sets.

The collection of recursive sets is closed under unions, intersections, and complements. The fact that it is closed with respect to unions and intersections can be seen from imagining a recursive algorithm combining recursive algorithms for each of the two given sets; the outputs of the two algorithms are combined with logical "and" or "or." The fact that the complement of a recursive set is recursive follows from the definition of recursive set: If we have a decision procedure for telling whether an element is a member of our set, then we can take the logical "not" of the output of this procedure and this will tell whether an element is in the complement of our set.

Some sets are not even recursively enumerable. An example is the problem of determining whether a given program will run forever. This is different from the Halting problem, which asks whether a given program will halt.

The reader is asked in an exercise to prove the following classification of recursive sets.

THEOREM
A set A of natural numbers is recursive if and only if both A and the complement of A are recursively enumerable.

For our purposes, it is important to realize that every Diophantine set is recursively enumerable. Given a Diophantine family, one can simply check all n-tuples in an orderly way to see if they are solutions. We offer the following algorithm that produces values of a parameter for which there is a solution to an arbitrary Diophantine equation.

```
Algorithm: Solution Generator
  Step 1: Input (mypolynomial,numberofunknowns).
  Step 2: Let sumofvariables=0.
  Step 3: For all (a,x_1,x_2,...,x_numberofunknowns) such that
          a+x_1+x_2+...+x_numberofunknowns=sumofvariables Do:
            If mypolynomial(a,x_1,x_2,...,x_unknowns)=0,
            Output a.
  Step 4: Let sumofvariables=sumofvariables+1.
  Step 5: Go to Step 3.
```

If a set of nonnegative integers is Diophantine, then it is represented by the first coordinate in the solutions to such an equation. The critical point is that although all solutions will be reported eventually (perhaps multiple times), we do not know how long we have to wait for a particular solution. Therefore, the algorithm does not tell us whether a given number is a member of the Diophantine set. (If the algorithm were capable of answering that question, then every Diophantine set would be recursive.)

We give practical versions of the above algorithm in the following *Mathematica* and *Maple* programs, which find any and all nonnegative solutions (within a given bound) to a given Diophantine equation in two variables.

Mathematica

```
solutionsearch[mypolynomial_, bound_] := Module[{i, j},
  Do[
  If[mypolynomial[i, j] == 0, Print[{i, j}]],
  {i, 0, bound},
  {j, 0, bound}
  ]
]
```

Maple

```
solutionsearch:=proc(mypolynomial,bound)
  local i, j;
  for i from 0 to bound do
      for j from 0 to bound do
      if mypolynomial(i,j)=0 then
      print([i,j])
      end if;
    end do;
  end do;
end proc:
```

Let's define a polynomial to test the procedure. We choose the one from equation (12.7).

```
f(x, y) = y^2 - x^3 + x - 1

solutionsearch(f,100)

{0,1}
{1,1}
{3,5}
{5,11}
```

The program has found all ordered pairs (x, y) with $f(x, y) = 0$ and $0 \leq x, y \leq 100$. There are four such ordered pairs. Will an ordered pair with $x > 100$ ever appear in a solution to $f(x, y) = 0$? We cannot tell from our procedure (other than by running it and maybe receiving a definite answer); that would require a mathematical investigation of the particular polynomial function that we have chosen.

Exercises

12.18 Explain why there does not exist a Halting program for the class of algorithms with no data.

12.19 Prove the following classification of recursive sets: A set A of natural numbers is recursive if and only if both A and the complement of A are recursively enumerable.

†12.20 Use the diagonal method to show that there exists a recursively enumerable set that is not recursive.

Hint: Consider all possible algorithms that define recursive sets of integers. Arrange them in a list (there are countably many). Suppose that the nth machine decides, at step n, whether the number n is or is not a member of its defined set. Now create a new set S based on the diagonal of this list. Explain why S is recursively enumerable but not recursive.

12.5 The negative solution of Hilbert's Tenth Problem

The combination of work by Martin Davis, Hilary Putnam, Robinson, Matiyasevich, and others shows that every recursively enumerable set is Diophantine. Since all Diophantine sets are recursively enumerable (as indicated in the previous section), the class of Diophantine sets is the same as the class of recursively enumerable sets. It follows that Hilbert's Tenth Problem is not solvable. For let S be a recursively enumerable set that is not recursive. Then, since S is Diophantine, it is represented by a Diophantine family $D(a, x_1, x_2, \ldots, x_n) = 0$; that is to say, $a \in S$ if and only if there is a solution x_1, x_2, $\ldots$, x_n. But if Hilbert's Tenth Problem were solvable, then the decision problem would be solved for this family (a contradiction). Therefore, Hilbert's Tenth Problem is not solvable.

In this section, we give an overview of the chain of arguments that lead to the negative solution of Hilbert's problem and provide the details for the proofs that the special sequence (to be defined) and the exponential function are Diophantine.

Although the answer to Hilbert's Tenth Problem is a negative one (there exists no algorithm that decides the solvability of arbitrary polynomials in integers), the main result (all recursively enumerable sets are Diophantine) is a positive one that allows for interesting applications. Perhaps the most surprising result is the existence of a Diophantine representation of the set of prime numbers, as shown in the next section.

Overview

Historically, the proof that all recursively enumerable sets are Diophantine occurred in several steps. In 1961 Davis, Putnam, and Robinson proved that all recursively enumerable sets are "exponential Diophantine." An exponential Diophantine set is defined similarly to a Diophantine set except that the defining function can be built up from an exponential function, such as $a = \lambda^b$, as well as the usual polynomial operations. One can have towers of exponents in such a function. It is easy to demonstrate (see Exercises) that if the exponential function $a = \lambda^b$ is Diophantine, then all exponential Diophantine sets (including those defined by towers of exponents) are Diophantine. It remained to be proved that the exponential function $a = \lambda^b$ is Diophantine. This was done indirectly. It had already been observed that if some exponentially growing function is Diophantine, then the function $a = \lambda^b$

is Diophantine. The crucial final step, therefore, was to prove that some particular exponentially growing function is Diophantine. In 1970 Matiyasevich proved that the exponentially growing function $n = f_{2m}$, where f_{2m} is the $2m$th Fibonacci number, is Diophantine. (Surprisingly, it seems to be easier to work with an exponentially growing sequence such as the Fibonacci numbers than with an exponential function itself.) Thus, Matiyasevich completed the argument that the function $a = \lambda^b$ is Diophantine, thereby establishing that the class of Diophantine sets is the same as the class of recursively enumerable sets and proving the unsolvability of Hilbert's Tenth Problem.

The special sequence

After Matiyasevich's initial proof that $n = f_{2m}$ is Diophantine, simpler arguments were found based on solutions to the "special Pell equation"

$$x^2 - (a^2 - 1)y^2 = 1$$

(see Example 9.51 of Section 9.6). Later, a further streamlined argument focused on solutions to equation

$$x^2 - \lambda xy + y^2 = 1. \tag{12.20}$$

This equation is a slight alteration of the special Pell equation, as one can see by making the change of variables $(x, y, a) \leftarrow (x - \lambda y/2, y, \lambda/2)$. The fact that the equation is symmetric in x and y is convenient in the proof. We will show the details of the proof that the sequence of solutions to (12.20) is Diophantine.

The solutions to this equation, like those of the special Pell equation, are described by a simple second-order recurrence relation.

PROPOSITION 12.15
Let $\lambda \geq 2$. Then x and y satisfy the equation

$$x^2 - \lambda xy + y^2 = 1$$

if and only if $(x, y) = (a_n, a_{n+1})$ or $(x, y) = (a_{n+1}, a_n)$, for some $n \geq 0$, where the sequence $\{a_n\}$ is defined by the recurrence relation

$$\begin{aligned} a_n &= \lambda a_{n-1} - a_{n-2}, \quad n \geq 2, \\ a_0 &= 0, \ a_1 = 1. \end{aligned}$$

PROOF We could deduce this result from our knowledge of the solutions to the special Pell equation, but we prefer to give a self-contained proof of the kind employed in Example 12.11.

Let's consider the possibility that $x = y$. The equation becomes $(2-\lambda)x^2 = 1$, but this is a contradiction since $\lambda \geq 2$. So we cannot have $x = y$. Assume that $x < y$ (the case $y < x$ is symmetric).

The ordered pair $(x, y) = (0, 1)$ satisfies the equation. Also, (x, y) is a solution if and only if $(y, \lambda y - x)$ is a solution:

$$\begin{aligned} y^2 - \lambda xy + x^2 = 1 &\Longleftrightarrow \\ y^2 + (\lambda y - x)(\lambda y - x - \lambda y) = 1 &\Longleftrightarrow \\ y^2 - \lambda y(\lambda y - x) + (\lambda y - x)^2 = 1. & \end{aligned}$$

Hence, all the pairs (a_n, a_{n+1}), with $n \geq 0$, satisfy the equation.

If $x = 0$, then $y = 1$. Assume that (x, y) satisfies the equation and $1 \leq x < y$. The inverse change of variables to $(x, y) \to (y, \lambda y - x)$ is $(x, y) \to (\lambda x - y, x)$. We find that $0 \leq \lambda x - y < x$. This can be seen when the equation is written in the form

$$x^2 = y(\lambda x - y) + 1.$$

Hence, working backwards, we end at the beginning of the sequence with $(x, y) = (0, 1)$. ∎

DEFINITION 12.16 *For $\lambda \geq 2$, the* special sequence $\{a_n(\lambda)\}$ *satisfies the recurrence relation*

$$\begin{aligned} &a_n(\lambda) = \lambda a_{n-1}(\lambda) - a_{n-2}(\lambda), \quad n \geq 2, \\ &a_0(\lambda) = 0, \ a_1(\lambda) = 1. \end{aligned} \tag{12.21}$$

(The notation emphasizes the role of λ in the recurrence relation; we may omit this argument if there is no confusion.)

Our goal is to prove that the function $a = a_n(\lambda)$ is Diophantine; that is, the set of ordered pairs (a, n) (for each fixed λ) constituting the function is a Diophantine set. According to Proposition 12.15, given our equation

$$x^2 - \lambda xy + y^2 = 1,$$

there exists $n \geq 0$ with $x = a_n(\lambda)$. However, the parameter n is "hidden," i.e., it does not appear explicitly in the relation. We must find a way to bring the n to light.

The recurrence relation (12.21) defines a sequence "backwards" as well as "forwards." It follows that the sequence $\{a_n\}$ is "purely periodic" (i.e., it is periodic with no initial sequence before the period) modulo any number v. Moreover, the backwards recurrence is the same as the forwards one:

$$a_{n-2} = \lambda a_{n-1} - a_n, \quad n \geq 2. \tag{12.22}$$

We give counting proofs of some of the forthcoming results about the special sequence $\{a_n\}$. All the results can be proved by induction. In the counting proofs, a_n is viewed as the number of objects of a certain type.

DEFINITION 12.17 *Consider finite strings whose elements are chosen from the symbols* 0, 1, ..., $\lambda - 1$. *A* good string *is one that begins with a* 1 *and does not contain the substring* 01. *A* bad string *is one that is not good.*

LEMMA 12.18
The number of good strings of length n is a_n.

PROOF The number of good strings of lengths 0 and 1 are 0 and 1, respectively, so the claim is true for $n = 0$ and $n = 1$. (The reason for decreeing that good strings begin with a 1 is to make the initial conditions match those of $\{a_n\}$.) Now, assuming that the result is true for all $k < n$, we will show that a_n is the number of good strings of length n. Every good string of length $n - 1$ can be extended to a string of length n in λ ways (by adding a symbol at the end). However, not all these strings are good. The bad strings end in 01, and there are a_{n-2} of these. Hence, the number of good strings of length n is $\lambda a_{n-1} - a_{n-2}$, which is a_n. So, by induction, the formula is correct for $n \geq 0$. ▯

The sequence $\{a_n\}$ satisfies an addition law similar to that of the Pell equation solutions (see (9.33)).

LEMMA 12.19
For $i \geq 1$ and $j \geq 0$,

$$a_{i+j} = a_i a_{j+1} - a_{i-1} a_j.$$

PROOF The left side is the number of good strings of length $i+j$. A string of length $i+j$ consists of a substring of length i concatenated with a substring of length j. There are a_i good strings of length i and a_{j+1} strings of length j that do not contain 01. However, not all concatenations of these strings are good. The bad concatenation strings are those in which the first substring ends with 0 and the second substring begins with 1. There are $a_{i-1}a_j$ of these strings, and they are subtracted in the formula. ▯

LEMMA 12.20
For $n \geq 0$,

$$a_n(2) = n.$$

One can easily prove this result by means of string counting or by induction.

LEMMA 12.21
Consecutive terms of the special sequence are relatively prime.

This result is obtained instantly from Proposition 12.15 or by induction.

The next observation follows directly from the recurrence relation (12.21).

LEMMA 12.22
For any $v > 1$, if $\alpha \equiv \beta \pmod{v}$, then, for all $n \geq 0$,

$$a_n(\alpha) \equiv a_n(\beta) \pmod{v}.$$

In particular, if $\alpha \equiv 2 \pmod{v}$, then

$$a_n(\alpha) \equiv a_n(2) = n \pmod{v}.$$

PROOF The recurrence relations that define $\{a_n(\alpha)\}$ and $\{a_n(\beta)\}$ are the same modulo v. The second statement follows from Lemma 12.20. ∎

Researchers guessed early on that analysis of the periods of the special sequence (in the context of Pell's equation) modulo various numbers would be helpful in solving Hilbert's Tenth Problem. As an example of this type of analysis, let's consider the period modulo $v = a_n$, for $n \geq 2$. This is a warm-up computation not directly related to the proof of the main theorem.

The first terms are

$$0,\ 1,\ a_2,\ a_3,\ \ldots,\ a_{n-1},\ 0.$$

We use the "backwards recurrence" (12.22) to find the next terms:

$$-a_{n-1},\ -a_{n-2},\ \ldots,\ -a_3,\ -a_2,\ -1.$$

And the cycle repeats with a period of length $2n$. Hence, if $i \equiv j \pmod{2n}$, then $a_i \equiv a_j \pmod{v}$.

In a similar manner, we obtain the period for the modulus used in the proof of the main theorem.

LEMMA 12.23
Let $v = a_{n+1} - a_{n-1}$, for $n \geq 1$. Then the special sequence $\{a_n\} \pmod{v}$ has a period of length $4n$:

$$0,\ 1,\ a_2,\ a_3,\ \ldots,\ a_{n-1},\ a_n,\ a_{n-1},\ \ldots,\ a_3,\ a_2,\ 1,$$
$$0,\ -1,\ -a_2,\ -a_3,\ \ldots,\ -a_{n-1},\ -a_n,\ -a_{n-1},\ \ldots,\ -a_3,\ -a_2,\ -1.$$

Hence, if $i \equiv j \pmod{4n}$*, then* $a_i \equiv a_j \pmod{v}$.

Next, we need a result about divisibility of the elements of $\{a_n\}$.

LEMMA 12.24
For all $m, n \geq 0$, $m \mid n$ *if and only if* $a_m \mid a_n$.

The forward direction can be proved by counting (see Exercises), but we give an induction proof (for both parts) here.

PROOF Suppose that $m \mid n$. We show that $a_m \mid a_{km}$, for all $k \geq 1$. We proceed by induction on k. The result is obvious if $k = 1$. Assume that $a_m \mid a_{km}$. By Lemma 12.19,

$$a_{(k+1)m} = a_{km+m} = a_{km}a_{m+1} - a_{km-1}a_m,$$

and it follows that $a_m \mid a_{(k+1)m}$. Hence, $a_m \mid a_n$.

Now suppose that $a_m \mid a_n$. Let $n = km + j$, with $0 \leq j < m$. By Lemma 12.19,

$$a_n = a_{km+j} = a_{km}a_{j+1} - a_{km-1}a_j,$$

and, by Lemma 12.21 and since $a_m \mid a_{km}$, this implies that $j = 0$. Hence, $m \mid n$. ∎

Our goal now is to establish a key lemma that allows us to "synchronize" the period of the special sequence values with the period of the indices, so that the former is a multiple of the latter. In his original argument, Matiyasevich proved the relation for Fibonacci numbers:

$$f_m^2 \mid f_n \Longrightarrow f_m \mid n.$$

First we need a technical lemma.

LEMMA 12.25

$$a_{km} \equiv k a_m a_{m+1}^{k-1} \pmod{a_m^2} \tag{12.23}$$

PROOF Consider good strings of length km. These strings are composed of k substrings of length m, which we call "blocks." Let $A_{i,j}$ be the number of such strings in which the jth block is the rightmost one that begins with 1 and the ith block is the rightmost one such that all blocks from the 1st to the ith begin with 1. Note that $A_{i,j} = 0$ if $j = i + 1$. It is easy to see that $a_{km} = \sum_{1 \leq i \leq j \leq k} A_{i,j}$. For $i < j$, each term $A_{i,j}$ is 0 modulo a_m^2. The reason is that there are two contributed factors of a_m to $A_{i,j}$, one from the ith block and one from the jth block. For $i = j$, we obtain

$$A_{i,i} = (a_m - a_{m-1})^{i-1} a_m (a_{m+1} - a_m)^{k-i} \equiv a_m a_{m+1}^{k-1} \pmod{a_m^2}.$$

(We have used the fact that $a_{m+1} \equiv -a_{m-1} \pmod{a_m}$.) Our result now follows as there are k such terms. ▯

Now we prove our key lemma.

LEMMA 12.26
If $a_m^2 \mid a_n$, then $a_m \mid n$.

PROOF Assume that $a_m^2 \mid a_n$. By Lemma 12.24, we see that $m \mid n$, and hence $n = mk$, for some $k \geq 1$. From Lemma 12.25 and the fact that $\gcd(a_m, a_{m+1}) = 1$, it follows that $a_m \mid k \mid n$. ▯

The main result

We are now ready to prove that the special sequence is Diophantine.

THEOREM 12.27
The function $a = a_b(\lambda)$, with $\lambda \geq 4$, is Diophantine. Specifically, the relation holds if and only if there is a solution to the system of relations
(1) $x^2 - wxy + y^2 = 1$; $w \geq 2$
(2) $w \equiv \lambda \pmod{v}$
(3) $w \equiv 2 \pmod{u}$
(4) $a \equiv x \pmod{v}$
(5) $b \equiv x \pmod{u}$
(6) $s^2 - \lambda rs + r^2 = 1$; $r < s$
(7) $u^2 - \lambda ut + t^2 = 1$
(8) $v = \lambda s - 2r$
(9) $u > 2a$; $v > 2a$
(10) $u > 2b$
(11) $u^2 \mid s$.

PROOF First we prove that if there is a solution to the given system, then the relation $a = a_b(\lambda)$ holds.

From (1), there exists $n \geq 0$ for which

$$x = a_n(w). \tag{12.24}$$

From (6), there exists $m \geq 1$ with

$$s = a_m(\lambda),\ r = a_{m-1}(\lambda). \tag{12.25}$$

By (8),

$$\begin{aligned} v &= \lambda s - 2r \\ &= \lambda a_m(\lambda) - 2a_{m-1}(\lambda) \\ &= a_{m+1}(\lambda) - a_{m-1}(\lambda). \end{aligned} \tag{12.26}$$

Let

$$n = 2mk \pm j, \quad j \leq m. \tag{12.27}$$

From (12.26) and the fact that the special sequence is increasing, we obtain

$$\begin{aligned} v &= \lambda a_m(\lambda) - 2a_{m-1}(\lambda) \\ &\geq 4a_m(\lambda) - 2a_m(\lambda) \\ &= 2a_m(\lambda) \\ &\geq 2a_j(\lambda). \end{aligned} \tag{12.28}$$

Now, by (4), (12.24), (2), Lemma 12.22, (12.27), (12.26), and Lemma 12.23,

$$a \equiv x \equiv a_n(w) \equiv a_n(\lambda) \equiv a_{2mk\pm j}(\lambda) \equiv \pm a_j(\lambda) \pmod{v}. \tag{12.29}$$

By (9) and (12.28),

$$a = a_j(\lambda). \tag{12.30}$$

We make a similar argument modulo u. By (7), (11), and Lemma 12.26,

$$u \mid m. \tag{12.31}$$

By (9) and (12.30),

$$u > 2a = 2a_j(\lambda) \geq 2j. \tag{12.32}$$

Now, by (5), (12.24), (3), Lemma 12.22, (12.27), (12.31), and Lemma 12.23,

$$b \equiv x \equiv a_n(w) \equiv a_n(2) \equiv n \equiv 2mk \pm j \equiv \pm j \pmod{u}. \tag{12.33}$$

Hence, by (10) and (12.32),

$$b = j. \tag{12.34}$$

Therefore, putting (12.30) and (12.34) together,

$$a = a_b(\lambda). \tag{12.35}$$

Now assume that the relation $a = a_b(\lambda)$, with $\lambda \geq 4$, holds. We show that there is a solution to the system of relations (1)–(11).

Choose $u = a_k(\lambda) > 2a$, with u odd, so that the first inequality in (9) holds. (To see that we can take u odd, look at the special sequence modulo 2.) Choose $t = a_{k+1}(\lambda)$, so that (7) holds. Choose $m = uk$. And choose $r = a_{m-1}(\lambda)$ and $s = a_m(\lambda)$, so that (6) holds.

We see from Lemma 12.25 that

$$s = a_{uk}(\lambda) \equiv u a_k(\lambda) a_{k+1}(\lambda)^{u-1} \equiv 0 \pmod{u^2}. \tag{12.36}$$

So $u^2 \mid s$ ((11) holds). Now let $v = \lambda s - 2r$, so that (8) holds.

We show that u and v are coprime. If $d \mid u$ and $d \mid v$, then $d \mid s$, so that $d \mid 2r$ and $d \mid r$ (since u is odd); hence, $d \mid 1$ (since s and r are coprime, as they are consecutive terms in the special sequence).

Now from the Chinese remainder theorem (Theorem 3.27), there exists $w \geq 2$ with $w \equiv \lambda \pmod{v}$ and $w \equiv 2 \pmod{u}$, so that (2) and (3) hold.

Choose $x = a_b(w)$ and $y = a_{b+1}(w)$, so that (1) holds.

By (12.28),

$$v \geq 2a_m > m > u > 2a, \tag{12.37}$$

so that the second inequality in (9) holds, and

$$u > 2a \geq 2b \tag{12.38}$$

(so that (10) holds).

Now, by Lemma 12.22,

$$x \equiv a_b(w) \equiv a_b(\lambda) \equiv a \pmod{v} \tag{12.39}$$

and

$$x \equiv a_b(w) \equiv a_b(2) \equiv b \pmod{u}, \tag{12.40}$$

so that (4) and (5) hold. □

The exponential function

Now that we have proved that the function $a = a_b(\lambda)$ is Diophantine, let's prove that the exponential function $a = \lambda^b$ is Diophantine. Assume that $\lambda > 1$.

We begin by noting the inequalities

$$(\lambda - 1)^{n-1} \leq a_n(\lambda) \leq \lambda^{n-1}, \quad n \geq 1. \tag{12.41}$$

These inequalities are immediately apparent from the string counting concept. (For the lower bound, omit the '0' symbol.)

We claim that for all sufficiently large x,

$$\lambda^b - 1 < \frac{a_{b+1}(\lambda x)}{a_{b+1}(x+1)} \leq \lambda^b. \tag{12.42}$$

The upper bound on the quotient follows from (12.41):

$$\frac{a_{b+1}(\lambda x)}{a_{b+1}(x+1)} \leq \frac{(\lambda x)^b}{x^b} = \lambda^b. \tag{12.43}$$

To establish the lower bound, we use Bernoulli's inequality (Exercise 1.12): for all real numbers $x > -1$ and all positive integers n,

$$(1+x)^n \geq 1 + nx. \tag{12.44}$$

We find that

$$\begin{aligned}
\frac{a_{b+1}(\lambda x)}{a_{b+1}(x+1)} &\geq \frac{(\lambda x - 1)^b}{(x+1)^b} \quad \text{(by (12.41))} \\
&> \lambda^b \left(\frac{x-1}{x+1}\right)^b \quad \text{(since } \lambda > 1\text{)} \\
&= \lambda^b \left(1 - \frac{2}{x+1}\right)^b \\
&\geq \lambda^b \left(1 - \frac{2b}{x+1}\right) \quad \text{(by Bernoulli's inequality)} \\
&> \lambda^b - 1.
\end{aligned} \tag{12.45}$$

The last inequality holds for $x + 1 > 2b\lambda^b$. Using (12.41), we ensure the inequality in a Diophantine sense by taking $x > 2ba_{b+1}(\lambda + 1) - 1$.

Finally, from (12.42), we write the exponential function in terms of the ceiling function of the quotient:

$$\lambda^b = \lceil a_{b+1}(\lambda x)/a_{b+1}(x+1) \rceil, \quad x > 2ba_{b+1}(\lambda+1) - 1. \tag{12.46}$$

We have already proved that the ceiling function (of a quotient) is Diophantine (see Example 12.8).

THEOREM 12.28
The function $a = \lambda^b$ is Diophantine.

Thus, we have completed the last step in the negative solution of Hilbert's Tenth Problem.

Exercises

12.21 Assuming that the function $f(a, b) = a^b$ is Diophantine, prove that the solution set to the equation

$$x^{(y+1)^x} + (y^2 + yz + z^2)^x + y^{z^x} = 0$$

is Diophantine.

Hint: Rewrite the equation as

$$f(x, f(y+z, x)) + f(y^2 + yz + z^2, x) + f(y, f(z, x)) = 0.$$

Then, write the above equation as a system of Diophantine equations and finally as a single Diophantine equation.

12.22 Let $\{a_n\}$ be the special sequence. Prove the following two identities:

(a) $a_{2n+1}^2 = a_n^2 + a_{n+1}^2$
(b) $a_n^2 = a_{n-1}a_{n+1} + 1$.

12.23 Prove Lemma 12.19 by induction.

12.24 Prove Lemma 12.21 by induction.

$\star$**12.25** Prove the special sequence identity

$$a_{k+m}a_{k+n} = a_{k+m+n}a_k + a_m a_n, \quad k, m, n \geq 0.$$

12.26 In the case $\lambda = 3$, prove that $v_n = a_{n+1} - a_{n-1}$ is equal to the Lucas number L_{2n}, for $n \geq 1$.

12.27 (a) Define the sequence $\{a_n\}$ by the recurrence relation $a_n = 3a_{n-1} - a_{n-2}$, for $n \geq 2$, and $a_0 = 0$, $a_1 = 1$. Prove that $a_n = f_{2n}$ (the $2n$th Fibonacci number).

(b) Define the sequence $\{a_n\}$ by the recurrence relation $a_n = 3a_{n-1} - a_{n-2}$, for $n \geq 2$, and $a_0 = 1$, $a_1 = 1$. Prove that $a_n = f_{2n+1}$ (the $(2n+1)$st Fibonacci number).

12.28 (a) Prove the Fibonacci numbers identity

$$f_{m+n} = f_{m+1}f_n + f_m f_{n-1}, \quad m \geq 0, n \geq 1.$$

(b) Prove that $f_m \mid f_n$ if and only if $m \mid n$.

(c) Prove that there are arbitrarily long sequences of consecutive Fibonacci numbers that are composite.

$\star$**12.29** Prove the Fibonacci numbers identity

$$\gcd(f_a, f_b) = f_{\gcd(a,b)}.$$

Does a similar property hold for the special sequence $\{a_n\}$?

$\star$**12.30** Give counting arguments for the following facts regarding the special sequence $\{a_m\}$:

(a) $a_{km+1} \equiv a_{m+1}^k \pmod{a_m^2}$
(b) $m \mid n$ implies that $a_m \mid a_n$.

$\star$**12.31** Establish a bijection to show that the equation $x^2 - \lambda xy + y^2 = 1$ is satisfied when x and y are consecutive terms of the sequence $\{a_n\}$ of good strings of length n.

$\star$**12.32** Prove that the period of the special sequence $\{a_n\} \pmod{a_m^2}$ has length ma_m or $2ma_m$.

$\star$**12.33** (Perrin sequence) Define $f(0) = 3$, $f(1) = 0$, $f(2) = 2$, and $f(n) = f(n-2) + f(n-3)$, for $n \geq 3$. Prove that $p \mid f(p)$ for p prime. Use a computer to find a composite value of n such that $n \mid f(n)$.

12.6 Diophantine representation of the set of primes

According to Matiyasevich's analysis, every set of positive integers that can be generated by an algorithm (i.e., recursively enumerable set) is Diophantine. The set of prime numbers, as mysterious and complicated as it may be, can certainly be generated by a simple algorithm: just check each positive integer in turn for nontrivial factors; if it has none, include it. Therefore, the set of prime numbers is Diophantine. Already that is a nice surprise. But a further surprise follows by Proposition 12.9: there exists a polynomial whose set of positive values (for nonnegative integer values of the variables) is precisely the set of prime numbers. This was considered counter-intuitive when it was first proved (via Matiyasevich's analysis).

How do we find such a polynomial? The construction of an exponential Diophantine representation of an arbitrary recursively enumerable set is complicated, and although we could in principle use it to formulate a Diophantine representation of the set of primes, the resulting polynomial would be monstrously large. A more deft approach uses special properties of the set of primes. We follow this approach, first proving that binomial coefficients are Diophantine, then showing that the factorial function is Diophantine, and finally deducing (as indicated in Section 12.2), that the set of primes is Diophantine. The steps in this process are constructive, yielding a polynomial (still with a fairly large number of variables) representation of the set of primes.

We have already proved (in the previous section) that the exponential function is Diophantine. The proof that binomial coefficients are Diophantine derives from the observation that since

$$(b+1)^n = \sum_{i=0}^{n} \binom{n}{i} b^i,$$

the binomial coefficients $\binom{n}{i}$ are the base-b digits of $(b+1)^n$, provided that b is sufficiently large. In order to select the binomial coefficient $\binom{n}{m}$, we rewrite the above equation as follows:

$$(b+1)^n = \sum_{i=0}^{m-1} \binom{n}{i} b^i + \binom{n}{m} b^m + \sum_{i=m+1}^{n} \binom{n}{i} b^i.$$

Let b be large enough so that there are no "carries" (i.e., we need $\binom{n}{m} < b$); it suffices to take $b = 2^n$. Based on this analysis, we obtain a Diophantine representation of $\binom{n}{m}$:

$$l = \binom{n}{m}: \qquad (b+1)^n = x + lb^m + yb^{m+1}, \quad x < b^m,\ l < b,\ b = 2^n.$$

Now we proceed to show that the factorial function is Diophantine. We establish the inequalities

$$m! \leq \frac{n^m}{\binom{n}{m}} < m! + 1, \tag{12.47}$$

for n sufficiently large (a bound n_0 is given below) and thereby obtain the following Diophantine representation:

$$m! = \left\lfloor n^m / \binom{n}{m} \right\rfloor, \quad n \geq n_0. \tag{12.48}$$

Recall that we proved that the floor function (of a quotient) is Diophantine (see Example 12.8).

To prove (12.47), note that

$$\frac{n^m}{m!\binom{n}{m}} = \frac{n^m}{n(n-1)\dots(n-m+1)}. \tag{12.49}$$

The lower bound in (12.47) follows immediately. We now establish the upper bound:

$$\begin{aligned}\frac{n^m}{n(n-1)\dots(n-m+1)} &\leq \frac{n^m}{(n-m+1)^m} \\ &= \left(1 + \frac{m-1}{n-m+1}\right)^m \\ &= 1 + \sum_{k=1}^{m} \binom{m}{k} \left(\frac{m-1}{n-m+1}\right)^k.\end{aligned}$$

We will take n much larger than m so that $\frac{m-1}{n-m+1} < 1$. Now

$$\begin{aligned}\frac{n^m}{n(n-1)\dots(n-m+1)} &< 1 + \sum_{k=1}^{m} \binom{m}{k} \left(\frac{m-1}{n-m+1}\right) \\ &= 1 + (2^m - 1)\frac{m-1}{n-m+1}.\end{aligned}$$

We obtain the upper bound in (12.47) by requiring that

$$1 + (2^m - 1)\frac{m-1}{n-m+1} < 1 + \frac{1}{m!}$$

or

$$m!(2^m - 1)(m-1) + m - 1 < n.$$

We ensure this inequality in a Diophantine sense by taking

$$n_0 = m^m(2^m - 1)(m-1) + m - 1. \tag{12.50}$$

The last piece of the constructive proof now follows from Wilson's theorem (Theorem 3.35): n is prime if and only if $n > 1$ and $n \mid (n-1)! + 1$. We have completed the proof of our final theorem.

THEOREM 12.29
The set of prime numbers is Diophantine.

Following the methods described in the previous section and this one, one can explicitly produce a polynomial whose positive values are precisely the set of prime numbers. However, the number of variables, while not astronomical, is fairly large (see Exercise 12.38). In 1976 Jones, Daihachiro Sato, Hideo Wada, and Douglas Wiens found this more compact polynomial:

$$\begin{aligned}(k+2)\{1-([wz+h+j-q]^2 &+ [(gk+2g+k+1)(h+j)+h-z]^2\\ &+ [16(k+1)^3(k+2)(n+1)^2+1-f^2]^2 + [2n+p+q+z-e]^2\\ &+ [e^3(e+2)(a+1)^2+1-o^2]^2 + [(a^2-1)y^2+1-x^2]^2\\ &+ [16r^2y^4(a^2-1)+1-u^2]^2\\ &+ [((a+u^2(u^2-a))^2-1)(n+4dy)^2+1-(x+cu)^2]^2\\ &+ [(a^2-1)l^2+1-m^2]^2 + [ai+k+1-l-i]^2 + [n+l+v-y]^2\\ &+ [p+l(a-n-1)+b(2an+2a-n^2-2n-2)-m]^2\\ &+ [q+y(a-p-1)+s(2ap+2a-p^2-2p-2)-x]^2\\ &+ [z+pl(a-p)+t(2ap-p^2-1)-pm]^2)\}. \qquad (12.51)\end{aligned}$$

Since the creation of this polynomial, various researchers have found new prime representing polynomials with fewer variables and/or lower degree.

Notice that the prime representing polynomial (12.51) gives a bound on the number of operations necessary in a proof that a number is prime. If p is a prime number, then there is a proof of that fact using 87 additions and multiplications.

Hilbert's Tenth Problem is also unsolvable in the ring of Gaussian integers. (See [24].) However, the corresponding Hilbert's Tenth Problem for rational numbers is a major open problem.

Exercises

12.34 (a) Show that the polynomial (due to Euler) $x^2 - x + 41$ takes on prime values for all positive integers x less than 41.

(b) Prove that no nonconstant polynomial $p(x)$ takes on prime values for all positive integers x.

12.35 Show that n is prime if and only if $n > 1$ and $\gcd(n, (n-1)!) = 1$. Which method is better to represent prime numbers, this one or Wilson's theorem?

12.36 Prove that the set of Mersenne primes and the set of Fermat primes are Diophantine.

$\star$**12.37** Find a Diophantine representation of the set of Catalan numbers $\{c(n)\}$, defined by $c(0) = 0$, $c(1) = 1$, and $c(n+1) = \sum_{i=0}^{n} c(i)c(n-i)$, for $n \geq 1$.

12.38 The methods of Sections 12.5 and 12.6 yield a polynomial whose positive values are precisely the set of prime numbers. Determine the number of unknowns in this polynomial.

12.39 Compare the type of proof of primality indicated in Section 12.6 with the prime certificates of Chapter 7. Which is better?

12.7 Notes

Julia Bowman Robinson

Julia Bowman Robinson (1919–1985) was a mathematician who specialized in the interplay between mathematical logic and number theory. As the quote at the beginning of this chapter indicates, Robinson was intrigued by both numbers and logic at an early age.

Robinson was married to the mathematician Raphael M. Robinson (1911–1995), a professor at the University of California–Berkeley, and there was a policy that forbade faculty spouses from holding professorships (at least in the same department). Consequently, Julia Robinson worked in statistics and then in some corporate positions before eventually attaining a faculty position at U.C.–Berkeley in 1976.

Although Robinson was the first female mathematician to achieve several distinctions, including being the first woman to be president of the American Mathematical Society, she preferred being known as a mathematician and professional. She summed up her views as follows:

> What I really am is a mathematician. Rather than being remembered as the first woman this or that, I would prefer to be remembered, as a mathematician should, simply for the theorems I have proved and the problems I have solved.

Filmmaker George Csicsery (who directed *N is a Number: A Portrait of Paul Erdős*) is making a film about Julia Robinson. It is to be called *Julia Robinson and Hilbert's Tenth Problem.*

12.26 [illegible] Hoffman [illegible] of the [illegible] Dodecahedron. [illegible]

12.27 [illegible] graph [illegible] to find the [illegible]

12.28 [illegible] and 12.[illegible] precisely the [illegible] of [illegible]. Determine the [illegible] of [illegible] in this polyhedron.

12.29 Compare the [illegible] of [illegible] the prime [illegible] of Theorem 7.[illegible]

12.7 Notes

Julia Bowman Robinson

Julia Bowman Robinson (1919–1985) was [illegible] in [illegible] between [illegible] and [illegible] the beginning of [illegible] indicates [illegible] numbers [illegible] each [illegible]

Robinson [illegible] the [illegible] 1950s [illegible] the same [illegible] Robinson [illegible] [illegible] L. C. [illegible] 1970.

Although Robinson was the first woman [illegible] the first woman to be president of the American Mathematical Society [illegible] known as a mathematician and [illegible]

> What I really want is to be remembered [illegible] as a mathematician [illegible] proved and the problems I have solved.

[illegible] Robinson [illegible] Julia Robinson and Hilbert's Tenth Problem.

Appendix A

Mathematica Basics

> I wish to God these calculations had been executed by steam!
>
> [Finding errors in a manually produced set of mathematical tables.]
>
> CHARLES BABBAGE (1792–1871)

Throughout the text, we demonstrate a variety of ways to make computations using the software system *Mathematica*. This appendix is intended to serve as a brief introduction to *Mathematica*. The primary source for information on *Mathematica* is [34].

Numerical calculations

Upon launching *Mathematica*, we can begin to perform calculations right away. To execute an arithmetic computation, simply type the expression to be evaluated. When you are ready to evaluate, press Shift-Enter (i.e., press the Enter button while holding down the Shift key). For example, if we wish to evaluate the expression $3+5-9/2$, we would type following expression and press Shift-Return (or Enter on an extended keypad).

```
3+5-9/2
```

Mathematica will perform computations obeying the usual order of operations. After we have asked *Mathematica* to evaluate our expression, it will replace what we have typed with the following two lines.

```
In[n]:= 3+5-9/2

Out[n]= 7/2
```

The n indicates the nth command *Mathematica* has executed since the program was launched. Thus, if this was our first command, we would see the following.

```
In[1]:=3+5-9/2

Out[1]= 7/2
```

In further examples, as in the rest of the text, we will simply display the input and output, suppressing the `In[n]` and `Out[n]`.

As expected, we use the symbols +, -, *, / for addition, subtraction, multiplication, and division, respectively. We may also use a space between two numbers or expressions to indicate multiplication.

```
2 3
```

```
6
```

To indicate an exponent, we use the ^ symbol.

```
2^5
```

```
32
```

We can evaluate more involved expressions with the use of parentheses.

```
(1+2^(5-3))/3
```

```
5/3
```

If we ask *Mathematica* to perform a computation involving rational numbers, it will always attempt to return an answer using rational numbers, even if the numbers are very large or small.

```
4^30 + 124/1800982780
```

```
519097944122155518467768351/450245695
```

In the event the results of our calculations produce an irrational number, *Mathematica* will still attempt to return the exact value.

```
2^(3-7/3)
```

```
2^(2/3)
```

If we wish to obtain a numerical approximation, we can add `//N` to the end of our expression.

```
2^(3-7/3)//N
```

```
1.5874
```

In addition to performing these kinds of arithmetic operations, *Mathematica* can evaluate functions, both built-in and user defined. All built-in functions begin with a capital letter and use square brackets to surround the input value. For example, to compute $\cos 0$, we type the following.

```
Cos[0]
```

```
1
```

Mathematica is particular about case and brackets. If we violate the rules, it will produce an error message or not evaluate an expression in the way we expect.

Algebraic calculations

Mathematica can perform a wide range of algebraic manipulations on variables that have not been assigned values. The `Expand` command will "multiply out" an expression.

```
Expand[(x+2)(3-y)]
```

```
6 + 3 x - 2 y - x y
```

The `Factor` command attempts to factor an algebraic expression.

```
Factor[x^2-4]
```

```
(-2 + x) (2 + x)
```

We are free to use any string not representing a built-in *Mathematica* function as a variable or function, and we may define these however we wish. For instance, we can define $x = 3$.

```
x = 3
```

```
3
```

Now the value 3 will be substituted for all further occurences of x.

```
x^2
```

```
9
```

If at some point we no longer want x to be defined as 3, we may restore x as an unassigned variable using the `Clear` command.

```
Clear[x]
```

Now we can again work with `x` as an undetermined quantity.

```
2x+1
```

```
1 + 2 x
```

We may also use a string consisting of several letters as a variable name.

```
num = 5
```

```
5
```

As before we can calculate with our defined variable.

```
3(num+1)^2-1
```

```
107
```

The expressions `x y` and `xy` are recognized as distinct by *Mathematica.* The space between the `x` and the `y` in `x y` indicates that this is the product of the variables `x` and `y`, while the expression `xy` represents a single variable called "xy."

Functions

To define a function in *Mathematica*, we use square brackets around the input variable with the input variable followed by a "_". We use ":=" instead of "=."

```
f[x_] := 2x+1
```

Once the function has been defined, we may use it in the same way we use a built-in function.

```
f[3]
```

```
7
```

We may also insert expressions in place of the input variable.

```
f[t-3]
```

```
1 + 2 (-3 + t)
```

We will frequently need to generate a list of data by performing some sort of procedure several times. The `Table` command allows us to perform such a sequence of computations with just one instruction. The command has (up to) five pieces of input and has the following structure.

```
Table[(expression),{(variable),(start),(final),(step)}]
```

The ⟨expression⟩ indicates the *Mathematica* expression that we wish to evaluate. The ⟨variable⟩ denotes the quantity in our expression that we will be changing in our sequence of computations. The ⟨start⟩ is the starting value of the variable, and the ⟨final⟩ the final value of the variable. The ⟨step⟩ represents the step size of the variable. We may omit the step size, and a default value of 1 will be used. For example, to list the squares of the first five nonnegative even integers, we give the following command.

```
Table[n^2,{n,0,8,2}]
```

```
{0,4,16,36,64}
```

If we add `//TableForm` to the end of the command, *Mathematica* arranges the results in a single column.

```
Table[n^2,{n,0,8,2}]//TableForm
```

```
0
4
16
36
64
```

We may even evaluate multiple expressions in one command. For example, the command below will list the first five nonnegative even integers along with their squares and cubes.

```
Table[{n,n^2,n^3},{n,0,8,2}]//TableForm
```

```
0 0 0
2 4 8
4 16 64
6 36 216
8 64 512
```

It is sometimes convenient to create functions that are used only once and have no name. Such functions are called "pure functions."

The *Mathematica* syntax for applying a pure function `f` to a set `A` is `f& /@ A`. For example, to compute the squares of the first five positive integers, we apply the function `#^2` to the set $\{1, 2, 3, 4, 5\}$, using the construction `#^2& /@ {1,2,3,4,5}`. The symbol `&` says that the function is a pure function. The marker `#` takes the value of each element of $\{1, 2, 3, 4, 5\}$.

```
#^2& /@ {1, 2, 3, 4, 5}
```

```
{1, 4, 9, 16, 25}
```

As a deeper example, suppose that we have a list S containing some duplicated elements, say,

$$S = \{1, 1, 1, 1, 3, 4, 4, 4, 5, 5, 6, 7, 7, 7\}.$$

We would like to produce a list S' that displays the different elements of S along with their multiplicities:

$$S' = \{\{1, 4\}, \{3, 1\}, \{4, 3\}, \{5, 2\}, \{6, 1\}, \{7, 3\}\}.$$

(Such a construction is used in the prime factorization algorithms of Chapter 7.) We accomplish this task with a pure function applied to the set of distinct elements of S. The set of distinct elements of a list is given by the `Union` command.

```
s = {1, 1, 1, 1, 3, 4, 4, 4, 5, 5, 6, 7, 7, 7};
```

```
Union[s]
```

```
{1, 3, 4, 5, 6, 7}
```

We now apply the pure function `{#,Count[s,#]}&` to `Union[s]`, producing the set of pairs $\{x, \mu(x)\}$, where $\mu(x)$ is the multiplicity of x in S. (The command `Count[s,x]` gives the number of occurrences of x in S.)

```
sprime = {#,Count[s,#]}& /@ Union[s]
```

```
{{1, 4}, {3, 1}, {4, 3}, {5, 2}, {6, 1}, {7, 3}}
```

Besides being possibly more elegant than regular functions, pure functions are also normally more concise. Another example of the use of pure functions is in selecting some elements of a list to be reported. Say, for instance, that we would like to report the numbers less than 100 that divide 1000 and do not end in 5. We can use a pure function over the indicated range in conjunction with the `Select` command.

```
Select[Range[1000], (IntegerQ[1000/#] && Mod[#,10]!=5)&]
```

```
{1, 2, 4, 8, 10, 20, 40, 50, 100, 200, 250, 500, 1000}
```

We could also have selected directly over the set of divisors of 1000 by using the command `Divisors[1000]`.

Programs

As we begin to develop some of the programming capabilites of *Mathematica* in the text, we will need to evaluate multi-line expressions. That is, we will write *Mathematica* programs with several different commands and ask *Mathematica* to evaluate the set of commands all at once. To do this, we simply press Return after each line (instead of Shift–Return). After we have typed our last line, we press Shift–Return as usual. A simple example is given below.

```
a=2
b=a+1
a*b
```

```
2
3
6
```

Often we are only interested in the final result of a procedure. A semicolon at the end of a line in a procedure instructs *Mathematica* to suppress the output for that command.

```
a=2;
b=a+1;
a*b
```

```
6
```

Programs may also include `Do`, `For`, or `While` loops, which operate as they do in other programming languages.

For example, the following code produces the number of the first n odd primes of the forms $4k+1$ and $4k-1$, for the first few powers of n (see Introduction).

```
primes4kplus1 = 0;
Do[
  If[Mod[Prime[n + 1], 4] == 1, primes4kplus1++];
  If[MemberQ[{10, 100, 1000, 10000, 100000}, n],
    Print[{n, primes4kplus1, n - primes4kplus1}]],
  {n, 1, 10^5}
  ]
```

```
{10, 4, 6}
{100, 47, 53}
{1000, 495, 505}
{10000, 4984, 5016}
{100000, 49950, 50050}
```

We can also compute, say, the last line of this result very quickly by combining *Mathematica* commands as follows.

```
i = 5;
{10^i,
primes4kplus1 = Count[Mod[Prime[Range[2, 10^i + 1]], 4], 1],
10^i - primes4kplus1}

{100000, 49950, 50050}
```

As an exercise, lest you think that primes of the form $4k + 3$ are always more numerous than primes of the form $4k + 1$, try the above calculations looking at the first 2945 odd primes.

Another important consideration in writing procedures is the use of "modules." The `Module` command introduces a procedure in which the values of variables are "local," i.e., they are not defined outside of the module.

```
a=100;

Module[{a,b}, a=17; b=a^2]

289

a

100
```

The commands and examples in this appendix barely scratch the surface of the capabilities of *Mathematica*. Further applications are developed in the text. We hope that after you work through these, you will be able to venture off to explore number theory using *Mathematica*. Most implementations of *Mathematica* have the entire manual (over 1000 pages) available online. In addition to explaining all the built-in features, the manual also contains a large number of examples.

Appendix B

Maple Basics

> [A good calculating machine could] weave algebraic patterns just as the Jacquard loom weaves flowers and leaves.
>
> AUGUSTA ADA BYRON (1815–1852)

This appendix is intended to serve as a brief introduction to *Maple*.

Numerical calculations

Much of the syntax of *Maple* is different from that of *Mathematica*. For example, to show the result of a command we end the command with a semicolon (;), while to suppress the result we use a colon (:). Also, to form the product of two expressions, we must use an asterisk (*).

```
3*4;
```

```
12
```

As in *Mathematica*, we use the ^ symbol to indicate an exponent.

```
3^4;
```

```
81
```

We can evaluate more involved expressions with the use of parentheses.

```
(11+2^(5-3))/3;
```

```
5
```

Maple contains built-in functions such as the sine function.

```
sin(Pi/2);
```

```
1
```

To evaluate an expression numerically, we use the `evalf` command.

```
evalf(Pi);
```

```
3.141592654
```

We can generate any number of digits by including the desired number of digits in the `evalf` command.

```
evalf(Pi,30);
```

```
3.14159265358979323846264338328
```

Algebraic calculations

In order to assign a value to a variable, we use the := command.

```
a: = 5: a^2;
```

```
25
```

To clear the value of a variable, we use the `unassign` command.

```
unassign('a'):
```

```
a + a;
```

```
2a
```

We can expand and factor polynomials.

```
expand((x+1)^3);
```

```
x^3+3*x^2+3*x+1
```

```
factor(%);
```

```
(x+1)^3
```

And we can simplify complicated expressions with the `simplify` command.

```
simplify((x+1)^3+(x-1)^3);
```

```
2*x^3+6*x
```

Functions

To define a function, we use the := command.

```
f := x-> x^3+x+1:
f(3);
```

```
31
```

We can evaluate functions at expressions. Here we combine this type of operation with the `simplify` command.

```
simplify(f(a^2+a+1));
```

```
a^6+3*a^5+6*a^4+7*a^3+7*a^2+4*a+3
```

We can also compute sums and products.

```
sum(n^2, n=1..5);
```

```
55
```

```
product(i, i=1..4);
```

```
24
```

We can create a sequence of terms by using the `seq` command.

```
a := seq(ithprime(n), n=1..10):
a; 3*a;
```

```
2, 3, 5, 7, 11, 13, 17, 19, 23, 29
6, 9, 15, 21, 33, 39, 51, 57, 69, 87
```

We can convert a sequence into a list or a set by using the constructions `[]` or `{}`, respectively.

```
b:=[a]: b;
```

```
[2, 3, 5, 7, 11, 13, 17, 19, 23, 29]
```

```
c:={a}: c;
```

```
{2, 3, 5, 7, 11, 13, 17, 19, 23, 29}
```

We can add sequences of the same length by simply using `+` and multiply a sequence by a scalar by using `*` as in the example above.

In order to apply an operation to each term of a set or list, we use the `map` command.

```
map(x->x^2,b);
```

```
[4, 9, 25, 49, 121, 169, 289, 361, 529, 841]
```

A useful command, called **nops** (number of operands), gives the number of elements in a set or list.

```
nops(%);
```

```
10
```

Programs

Programming in *Maple* is similar to programming in *Mathematica.* Let's take the ourgcd program of Chapter 2 as an illustrative example.

```
ourgcd := proc(a,b)
  local atemp, btemp;
  (atemp, btemp) := (a, b);
  while btemp > 0 do
    (atemp, btemp) := (btemp,modp(atemp,btemp));
  end do;
  atemp;
end proc:
```

In *Maple*, programs are called procedures. We use the construction `proc()` / `end proc` to begin and end a procedure. We also indicate the inputs in the procedure (in our case, `a` and `b`). On the second line, we declare the variables to be used in the procedure (`atemp` and `btemp`).

Our example contains a declaration of a pair of variables on the third line. The fourth line begins with a `while` loop which has a test condition (`btemp > 0`) and a command to be performed (another declaration of a pair of variables). The command is bookended by `do` and `end do`. Finally, an output (the value of `atemp`) is given.

You can get more information about a *Maple* command by typing `?` before the command. For instance, typing `? nops` will give you information about the `nops` command.

Appendix C

Web Resources

> It is not once nor twice but times without number that the same ideas make their appearance in the world.
>
> ARISTOTLE (384–322 B.C.E.)

We present links to *Mathematica* notebooks and *Maple* worksheets that contain the programs included in this book and hyperlinks to web pages that contain useful information about number theory and mathematics in general. You can find the web pages for this book at

`www2.truman.edu/~erickson/introduction_to_number_theory/`.

Mathematica notebooks and Maple worksheets

Mathematica notebooks and *Maple* worksheets, containing all *Mathematica* and *Maple* calculations in the text, are available at the web pages for this book (see address above).

Websites

Internet addresses are listed here for web pages that contain useful information about mathematics in general and number theory specifically. For your convenience, hyperlinks are furnished at the web pages for this book (see address above).

General Mathematics:

- `www-groups.dcs.st-and.ac.uk/~history/index.html`

  "The MacTutor History of Mathematics archive," a site containing biographies of mathematicians (including dates and full names).

- `mathworld.wolfram.com`

  Eric Weisstein's "MathWorld," a site containing mathematical information and resources in an encyclopedia format.

- www.maa.org/

  The Mathematical Association of America, a site containing articles on mathematics and mathematics education for a general audience.

- www.ams.org/

  The American Mathematical Society, a site containing articles of interest to mathematicians, mathematics educators, and students.

- math.furman.edu/~mwoodard/mquot.html

  "The Mathematical Quotations Server," a collection of quotes from mathematicians and other mathematics-minded people.

Number Theory (general):

- www.numbertheory.org/ntw/web.html

  "The Number Theory Web," a site devoted to the various aspects of number theory.

- www.wikipedia.org/wiki/Number_theory

  "Number theory—Wikipedia," a collection of pages on number theory in a free encyclopedia.

Prime Numbers:

- www.utm.edu/research/primes/

  "The Prime Pages," a site containing facts and resources about prime numbers, including special kinds of primes (e.g., Mersenne numbers and Fermat numbers) and prime number records.

- users.forthnet.gr/ath/kimon/PNT/Prime%20Number%20Theorem.htm

  "Prime Number Theorem," contains an article on the PNT from the Encyclopædia Britannica.

- www.mersenne.org/prime.htm

  "GIMPS (The Great Internet Mersenne Prime Search)," a site containing historical and computational information on the search for Mersenne primes, including instructions on how to participate in the search.

Factoring:

- www.fermatsearch.org/

  "Distributed Search for Fermat Number Divisors," a site that organizes volunteer searches for divisors of Fermat numbers.

- www.prothsearch.net/fermat.html

  "Prime Factors $k.2n + 1$ of Fermat numbers F_m and Complete Factoring Status," a site containing current knowledge about Fermat number factorizations.

Hilbert's Tenth Problem:

- logic.pdmi.ras.ru/Hilbert10/

  "Hilbert's Tenth Problem page," a site devoted to research on connections between Diophantine equations and logic.

***Mathematica*:**

- www.wolfram.com/

  Wolfram Research, the definitive resource for *Mathematica* mathematical software.

- documents.wolfram.com/v4/index3.html

  "The Mathematica Book," an on-line version of the principal book about *Mathematica*.

***Maple*:**

- www.maplesoft.com/

  MapleSoft, the definitive resource for *Maple* mathematical software.

Appendix D

Notation

> A good notation has a subtlety and suggestiveness which at times make it almost seem like a live teacher.
>
> BERTRAND RUSSELL (1872–1970)

Sets

$\mathbf{N}$	natural numbers, 6
$\mathbf{Z}$	integers, 8
$\mathbf{Q}$	rational numbers, 99
$\mathbf{R}$	real numbers, 87
$\mathbf{C}$	complex numbers, 310
$[k]$	residue class, 59
$\mathbf{Z}_n$	equivalence classes modulo n, 59
R	quadratic residues, 131
N	quadratic nonresidues, 131
RR, NN, RN, NR	consecutive quadratic residues, quadratic nonresidues, etc., 152
$\mathbf{Z}[i]$	Gaussian integers, 310

Relations

$\mid$	divides, 15
$\nmid$	does not divide, 15
$\equiv$	congruence, 57

Functions

$n!$	factorial, 10
$\binom{n}{k}$	binomial coefficient, 13
$\lfloor x \rfloor$	floor, 13
$\lceil x \rceil$	ceiling, 13
gcd	greatest common divisor, 21
lcm	least common multiple, 24
$\pi(x)$	prime counting function, 46
$\phi(n)$	Euler's ϕ function, 72
$\left(\frac{x}{p}\right)$	Legendre symbol, 132
$\left(\frac{x}{n}\right)$	Jacobi symbol, 143
$P_a(x)$	number of pseudoprimes, 72
$\sigma(n)$	sum of divisors, 162
$\tau(n)$	number of divisors, 170
θ	constant 1 function, 170
$\sigma_k(n)$	sum of kth powers of divisors, 170
$\theta_k(n)$	n^k, 170
$\mu(n)$	Möbius function, 175
$\omega(n)$	number of distinct prime divisors, 177
$\Omega(n)$	number of prime divisors (counting multiplicity), 177
$p(n)$	partition number, 186
$p(n,k)$	partition number, 186
$\zeta(n)$	Riemann zeta function, 181
$N(w)$	norm of complex number, 311
$\vartheta(x)$	Chebyshev ϑ function, 364
$\psi(x)$	Chebyshev ψ function, 364
$\Lambda(n)$	von Mangoldt function, 379

Sequences

f_n	Fibonacci numbers, 11
p_n	prime numbers, 49
B_n	Bernoulli numbers, 385
L_n	Lucas numbers, 465
$a_n(\lambda)$	special sequence, 477

Groups

$\mathbf{Z}_n$	cyclic group of order n, 59
$\mathbf{Z}_n^*$	multiplicative group of order $\phi(n)$, 73

Miscellaneous

$\doteq$	approximately equal, 11
min	minimum, 44
max	maximum, 44
l-AP	l-term arithmetic progression, 47
$\sim$	asymptotic, 203
$O(f(x))$	big-O notation, 362

References

[1] W. R. Alford, A. Granville, and C. Pomerance. There are infinitely many Carmichael numbers. *Annals of Mathematics Second Series*, 139(3):703–722, 1994.

[2] T. Apostol. *Introduction to Analytic Number Theory.* Springer-Verlag, New York, first edition, 1976.

[3] G. Bachman. Flat cyclotomic polynomials of order three. *Bulletin of the London Mathematical Society*, 38:53–60, 2006.

[4] C. B. Boyer and U. C. Merzbach. *A History of Mathematics.* Wiley, New York, second edition, 1991.

[5] S. A. Burr, editor. *The Unreasonable Effectiveness of Number Theory*, volume 76 of *Proceedings of Symposia in Applied Mathematics.* American Mathematical Society, 1992.

[6] C. K. Caldwell. The primes pages: Prime number research, records, and resources, 2003. `http://www.utm.edu/research/primes/`.

[7] J. R. Chen. On the representation of a large even integer as the sum of a prime and the product of at most two primes. *Scientia Sinica*, 16:157–176, 1973.

[8] M. Davis. What is a computation? In L. A. Steen, editor, *Mathematics Today: Twelve Informal Essays.* Vintage Books, New York, 1980.

[9] H. M. Edwards. *Fermat's Last Theorem: A Genetic Introduction to Algebraic Number Theory.* Springer-Verlag, New York, second edition, 1996.

[10] R. L. Graham, B. L. Rothschild, and J. H. Spencer. *Ramsey Theory.* Wiley, New York, second edition, 1990.

[11] H. Halberstam and H. E. Richert. *Sieve Methods.* Academic Press, New York, 1974.

[12] R. Hartshorne. *Algebraic Geometry.* Springer-Verlag, New York, 1977.

[13] T. L. Heath, editor. *Euclid, The Thirteen Books of The Elements*, volume 1: Books I and II. Dover, New York, second edition, 1956.

[14] T. L. Heath, editor. *Euclid, The Thirteen Books of The Elements*, volume 2: Books III–IX. Dover, New York, second edition, 1956.

[15] T. L. Heath, editor. *Euclid, The Thirteen Books of The Elements*, volume 3: Books X–XIII. Dover, New York, second edition, 1956.

[16] T. L. Heath, editor. *A Manual of Greek Mathematics*, volume I: From Thales to Euclid. Dover, New York, first edition, 1963.

[17] I. N. Herstein. *Abstract Algebra.* Prentice–Hall, Upper Saddle River, NJ, third edition, 1996.

[18] A. Knapp. *Elliptic Curves.* Springer-Verlag, New York, 1992.

[19] J. M. Kubina and M. C. Wunderlich. Extending Waring's Conjecture to 471,600,000. *Mathematics of Computation*, 55(192):815–820, 1990.

[20] M. Laczkovich. On Lambert's proof of the irrationality of π. *American Mathematical Monthly*, 104(5):439–443, 1997.

[21] T. Y. Lam and K. H. Leung. On the cyclotomic polynomial $\phi_{pq}(x)$. *American Mathematical Monthly*, 103(7):565–567, 1996.

[22] E. Lehmer. On the magnitude of coefficients of the cyclotomic polynomials. *Bulletin of the American Mathematical Society*, 42:389–392, 1936.

[23] N. Levinson. A motivated account of an elementary proof of the prime number theorem. *The American Mathematical Monthly*, 76(3):225–245, 1969.

[24] Y. V. Matiyasevich. *Hilbert's Tenth Problem.* MIT Press, Cambridge, MA, first edition, 1993.

[25] M. A Morrison and J. Brillhart. A method of factoring and the factorization of F_7. *Mathematics of Computation*, 29(129):183–205, 1975.

[26] P. Ribenboim. *The New Book of Prime Number Records.* Springer, New York, first edition, 1996.

[27] K. H. Rosen. *Elementary Number Theory and Its Applications.* Addison–Wesley, Boston, fourth edition, 2000.

[28] J. H. Silverman and J. Tate. *Rational Points on Elliptic Curves.* Springer-Verlag, New York, 1992.

[29] M. Sipser. *Introduction to the Theory of Computation.* PWS, Boston, first edition, 1997.

[30] R. P. Stanley. *Enumerative Combinatorics*, volume 1. Cambridge University Press, New York, 1999.

[31] I. Stewart. *Galois Theory.* Chapman and Hall, London, second edition, 1989.

[32] A. Weil. *Number Theory: An Approach Through History from Hammurapi to Legendre.* Birkhäuser, Boston, 1984.

[33] E. Weisstein. *MathWorld.* Wolfram Research, 2007. `http://mathworld.wolfram.com/`.

[34] S. Wolfram. *The Mathematica Book.* Cambridge University Press, New York, fourth edition, 1999.

Index